Planetary Geoscience

For many years, planetary science has been taught as part of the astronomy curriculum, from a very physics-based perspective, and from the framework of a tour of the Solar System – body by body. Over the past decades, however, spacecraft exploration and related laboratory research on extraterrestrial materials have given us a new understanding of planets and how they are shaped by geologic processes.

Based on a course taught at the University of Tennessee, Knoxville, this is the first textbook to focus on geologic processes, adopting a comparative approach that demonstrates the similarities and differences between planets, and the reasons for these. Profusely illustrated, and with a wealth of pedagogical features, this book provides an ideal capstone course for geoscience majors – bringing together aspects of mineralogy, petrology, geochemistry, volcanology, sedimentology, geomorphology, tectonics, geophysics, and remote sensing.

Harry Y. McSween, Jr. is Chancellor's Professor Emeritus of Planetary Geoscience at the University of Tennessee, Knoxville. He holds degrees from The Citadel (BS), the University of Georgia (MS), and Harvard University (PhD). His research focuses on meteorites and has resulted in the publication of hundreds of scientific papers on the subject. He has also authored three popular books on planetary science, as well as textbooks in geochemistry and cosmochemistry. He has served as co-investigator for many NASA spacecraft missions, including Mars Pathfinder, Mars Exploration Rovers, Mars Odyssey orbiter, and Dawn asteroid orbiter. McSween has been elected President of the Meteoritical Society and of the Geological Society of America, and is a fellow of the American Academy of Arts and Sciences. He is the recipient of the Leonard Medal (Meteoritical Society), the J. Lawrence Smith Medal (US National Academy of Sciences), and the Whipple Award (American Geophysical Union), and is the namesake for asteroid 5223 McSween.

Jeffrey E. Moersch is Professor of Planetary Science at the University of Tennessee, Knoxville. He holds degrees from Cornell University (BA, MS, and PhD) and Arizona State University (MS). His research focuses on remote sensing, planetary surface geology, instrument development, and terrestrial analog field work. He has served on the science teams for many NASA spacecraft missions, including the Mars Exploration Rovers, Mars Science Laboratory, and Mars Odyssey. Along with Professor McSween, he originally developed the planetary geology course from which this book is derived. Professor Moersch has authored and co-authored more than 80 peer-reviewed scientific articles and book chapters, and served for five years as the Mars editor for the scientific journal *Icarus*.

Devon M. Burr is Associate Professor of Planetary Science at the University of Tennessee, Knoxville. She holds degrees from the United States Naval Academy (BS), St. John's College, Santa Fe (MA), University of Iowa (MS), and University of Arizona (PhD). Her research has focused on planetary geomorphology and she is currently conducting research on fluvial, aeolian (wind-driven), and tectonic landscapes and processes on planetary bodies, using image analysis and wind tunnel experiments (featured on the cover of *National Geographic* magazine for kids, entitled Dr. E's Super Stellar Solar System!). Burr is the lead editor of the book *Megaflooding on Earth and Mars* (Cambridge University Press, 2009). She is also a member of the Geological Society of America (and former member of the GSA Planetary Geology Division management board), the AAS Division of Planetary Sciences (serving as Science Organizing Committee Chair for DPS 2018), the American Geophysical Union, and the namesake of asteroid DevonBurr.

William M. Dunne is Professor of Geology at the University of Tennessee, Knoxville. He received his BS and PhD in geology from the University of Bristol, England. His research concerns the development of mountain belts and characterization of fracture networks in rocks. The latter interest has practical importance with regard to hydrocarbon exploration and groundwater remediation efforts, leading to collaborations with civil engineers. Dunne serves as an editor of the *Journal of Structural Geology*, is a fellow of the Geological Society of America, and has served as chair of the GSA Structural Geology and Tectonics Division and the ASEE Engineering Research Council. He has received teaching awards from the department, college and university.

Joshua P. Emery is the Lawrence A. Taylor Associate Professor of Planetary Science at the University of Tennessee, Knoxville. He received his BS from Boston University and PhD from the University of Arizona. His research focuses on investigating the formation and evolution of the Solar System and the distribution of organic material. As an observational planetary astronomer, he applies the techniques of reflection and emission spectroscopy of primitive and icy bodies in the near- (0.8–5.0 micron) and mid-infrared (5–50 micron). Current projects include the Jupiter Trojan asteroids, Kuiper Belt objects, icy satellites, and other asteroids. He is leader of the Thermal Analysis Working Group on the OSIRIS-REx asteroid sample return mission and the Surface Composition Working Group on the Lucy Trojan asteroid flyby mission.

Linda C. Kah is Professor of Carbonate Sedimentology and Geochemistry at the University of Tennessee, Knoxville. She received concurrent BS and MS degrees in geology from MIT, followed by a PhD from Harvard. Her research integrates sedimentology, stratigraphy, geochemistry, and paleobiology to understand the evolution of the Earth's biosphere. Current projects include reconstructing the ocean–atmospheric oxygenation and the redox structure of Mesoproterozoic shallow marine systems, exploring the effects of changing ocean circulation on the Great Ordovician Biodiversification Event (GOBE), and characterizing microbe–mineral interactions in the mineralization of Holocene lacustrine microbialites. In addition to Earth-based research projects, she also investigates potential habitable environments as co-investigator on the Mars Science Laboratory Mission.

Molly C. McCanta is Associate Professor of Mineralogy and Petrology at the University of Tennessee, Knoxville. She received her BS from the University of Oregon and concurrent MSc and PhD from Brown University. Her research focuses on the record of igneous processes retained in mineral grains as a means of better understanding geologic processes in planetary interiors. Current projects include experimentally investigating mineral and melt redox conditions as a function of planetary evolution, constraining the timing of eruptive hazards at several Costa Rican volcanoes, determining weathering geochemistry on the surface of Venus, and identifying cryptotephra layers in deep sea drill cores.

Planetary Geoscience

Harry Y. McSween, Jr.
UNIVERSITY OF TENNESSEE, KNOXVILLE

Jeffrey E. Moersch
UNIVERSITY OF TENNESSEE, KNOXVILLE

Devon M. Burr
UNIVERSITY OF TENNESSEE, KNOXVILLE

William M. Dunne
UNIVERSITY OF TENNESSEE, KNOXVILLE

Joshua P. Emery
UNIVERSITY OF TENNESSEE, KNOXVILLE

Linda C. Kah
UNIVERSITY OF TENNESSEE, KNOXVILLE

Molly C. McCanta
UNIVERSITY OF TENNESSEE, KNOXVILLE

CAMBRIDGE
UNIVERSITY PRESS

Shaftesbury Road, Cambridge CB2 8EA, United Kingdom

One Liberty Plaza, 20th Floor, New York, NY 10006, USA

477 Williamstown Road, Port Melbourne, VIC 3207, Australia

314–321, 3rd Floor, Plot 3, Splendor Forum, Jasola District Centre, New Delhi – 110025, India

103 Penang Road, #05–06/07, Visioncrest Commercial, Singapore 238467

Cambridge University Press is part of the University of Cambridge.

It furthers the University's mission by disseminating knowledge in the pursuit of education, learning, and research at the highest international levels of excellence.

www.cambridge.org
Information on this title: www.cambridge.org/9781107145382
DOI: 10.1017/9781316535769

First published 2019 (version 2, October 2023)

Printed in Great Britain by CPI Group (UK) Ltd, Croydon CR0 4YY

A catalogue record for this publication is available from the British Library.

Library of Congress Cataloging-in-Publication Data
Names: McSween, Harry Y., author.
Title: Planetary geoscience / Harry Y. McSween, Jr. (University of Tennessee, Knoxville) [and six others].
Description: Cambridge ; New York, NY : Cambridge University Press, 2019. | Includes end of chapter
    summaries and review questions. | Includes bibliographical references and index.
Identifiers: LCCN 2018047455 | ISBN 9781107145382 (hardback : alk. paper)
Subjects: LCSH: Extrasolar planets–Remote sensing. | Extrasolar planets–Research. | Astronomy–Research. |
    Geology–Research. | Planetary geographic information systems.
Classification: LCC QB820 .P5265 2019 | DDC 523.2/4–dc23
LC record available at https://lccn.loc.gov/2018047455

ISBN 978-1-107-14538-2 Hardback

Additional resources for this publication at www.cambridge.org/Mcsween.

Brief Contents

Contents

Preface: Geologic Processes in the Solar System

When we explore other worlds, what once seemed the only way a planet could be turns out to be somewhere in the middle range of a vast spectrum of possibilities...
Carl Sagan, 1995, Pale Blue Dot: A Vision of the Human Future in Space

When one world is not enough.
Motto of the Geological Society of America's Planetary Geology Division

Planet Earth has always been the geoscientist's laboratory. Now, though, after decades of planetary exploration, geoscientists recognize that our own world is but a singular grand experiment, and that other planets afford opportunities to examine nature's geologic experiments run with different starting compositions and under varying conditions. What used to be considered a part of astronomy now rightly belongs to geology. This substantial shift of scientific real estate has greatly expanded geology's reach, as distant points of light have been transformed into worlds shaped by more-or-less familiar geologic processes.

This book focuses on geologic processes on the planets, moons, and smaller bodies (asteroids, comets) of the Solar System. These processes are revealed and understood through exploration by orbiting or flyby spacecraft, landers and rovers, and, in the case of the Moon, by astronauts. Much of planetary geoscience involves remote sensing – visible observations, as well as measurements that use other parts of the electromagnetic spectrum ranging from gamma rays to radio waves. The application of geochemical and geophysical methods and numerical models also play significant roles. And, as in terrestrial geology, extraterrestrial samples analyzed in the laboratory constrain and quantify planetary geologic processes.

The book begins with a Grand Tour, an overview of the geologic bodies in the Solar System. Rather than taking the traditional approach of stepping outward planet by planet, we consider the temporal history of Solar System exploration in the spacecraft era and the critical role that has been played by geoscience.

Of necessity, planetary geoscientists often employ different tools than those used by geologists on Earth. Some

of these will be familiar to geology students, but with different twists.

- Imaging and spectroscopy: The principles of visible/ ultraviolet, near-infrared/ thermal infrared, X-ray, gamma ray, neutron, and radio spectroscopy are introduced, focusing on what kinds of data are provided by remote-sensing measurements at various wavelengths and how they are interpreted.
- Geochronology: The application of stratigraphic principles and crater density measurements provide the means of determining the relative ages of planetary surfaces, supplemented by radiometric dating of samples, where available, for absolute age determinations.
- Geologic mapping: We will see how spacecraft images are obtained and used to identify and map stratigraphic, structural, and geomorphic features, and how geologic mapping techniques at different spatial scales are complemented by remote sensing data.
- Geophysical methods: Topography, gravity, magnetics, and seismic data, as well as density and moment of inertia, are used to probe the unseen interiors of planets, along with numerical simulations and thermal evolution models.
- Laboratory analysis of planetary materials: Various kinds of directly sampled extraterrestrial materials are available for petrologic and geochemical analyses in the laboratory, and the laboratory has now been carried into the field by landed spacecraft.

Following chapters on the toolkits used by planetary geoscientists, the bulk of the book focuses on processes, which are introduced wherever possible through first principles. Each chapter begins with an overview and ends with a set of questions intended to provide the student with an opportunity to review important concepts and integrate what they have learned into a broad understanding of planetary geoscience. The subject matter follows this structure:

- Cosmochemical processes explain the origin of elements and isotopes as the building blocks of planets and how they were processed and fractionated in the early solar nebula.

- Observations and models help us understand the process of planetary accretion, as well as early collisional processes and orbital evolution.
- To understand planetary thermal evolution, we consider available heat sources and the resulting processes of melting and global differentiation.
- Modeling of planetary interiors requires an understanding of minerals stable at high pressures and temperatures, and how these phases are distributed.
- Planetary geodynamic processes are driven by gravity, heat flow, fluid mechanics, and rheology.
- Planetary tectonic processes differ on bodies with active versus stagnant lids.
- Igneous processes everywhere include magma generation, emplacement, and crystallization, but compositional and tectonic differences among planets yield varying results.
- Impact cratering is not a familiar process, so we consider crater mechanics, explore the geology of craters, and describe shock metamorphism.
- Some planets and many smaller bodies have volatiles sequestered in atmospheres, oceans (often subsurface), and ices.
- Wind-driven processes on bodies with atmospheres produce distinctive erosional and depositional landforms.
- The surfaces of some planets have been shaped by flowing and ponded water and other liquids.
- Physical and chemical weathering produce surficial regoliths and sedimentary deposits; aqueous alteration and thermal metamorphism affect rocks in the interiors of planets and asteroids; and mass wasting modifies planetary surfaces.
- For the emerging field of astrobiology, we review biological requirements and planetary habitability, and consider how life might be detected and how it influences a planet's evolution.

- A case study illustrates how all these kinds of information can be integrated into an understanding of the geologic processes that have shaped the evolution of a planet, in this case Mars.
- We end with an epilogue briefly exploring the diversity of exoplanets recently discovered around other stars, speculating about geologic processes on them.

The intended audience for this book is the undergraduate geology major, who already has had some exposure to mineralogy and petrology, sedimentology and stratigraphy, structure and tectonics, geochemistry, and geophysics. We provide limited background in these subjects as we go along, but some prior geologic coursework is assumed. For a decade, we have taught this course to geology majors at the University of Tennessee, and this experience has shaped our views of how to present this subject. However, we have been frustrated that a textbook focused on geoscience has not been available. Our goal is to help promote the integration of planetary geoscience into the undergraduate geology curriculum, by exploring the generality of geologic processes on other bodies. Although only a modest number of professional geologists actually work in planetary geoscience, ongoing planetary exploration has proved to be a powerful means of motivating the next generation of scientists and engineers.

We hope you will enjoy and learn from this foray into planetary geoscience, and come to appreciate that, although some of the tools may differ, the tried-and-true methods of geology work on other worlds too.

H. McSween, J. Moersch, D. Burr, W. Dunne, J. Emery, L. Kah, and M. McCanta
Department of Earth and Planetary Sciences
University of Tennessee, Knoxville

1

Exploring the Solar System

We present a brief overview of the planets, moons, dwarf planets, asteroids, and comets – intended as a primer for those with limited or no familiarity with planetary science. The terrestrial planets (Earth, Mars, Venus, and Mercury) are rocky bodies having mean densities that indicate metal cores; the giant planets are composed mostly of hydrogen and helium and can be divided into gas giants (Jupiter and Saturn) and ice giants (Uranus and Neptune), based on their physical states. Small bodies, composed of rock and ices, are either differentiated or not, depending on their thermal histories. Each section of this chapter is generally organized in the historical order in which the objects have been explored by spacecraft. We will return to these bodies repeatedly in the book, focusing on understanding their geologic characteristics and materials, and the processes that produced them.

1.1 Planetary Exploration and Explorers

Planetary books traditionally begin with the Grand Tour – an obligatory traverse of the planets, in lockstep from innermost to outermost. However, that route is not how the planets have been explored. So, instead of introducing the planets in order of distance from the Sun, we will discuss planets (and moons and small bodies) in the order in which they have been meaningfully investigated by spacecraft missions. This book is all about planetary geologic processes, which are revealed through exploration.

The explorers have been national space agencies, the only institutions with the financial wherewithal to undertake these challenges. These agencies are best known by their acronyms: NASA (USA's National Aeronautics and Space Administration), ESA (European Space Agency), ROSCOSMOS (Russian State Corporation for Space Activities), JAXA (Japan Aerospace Exploration Agency), CNSA (China National Space Administration), ISRO

(Indian Space Research Organization), and a few others. It is conceivable that private industry may conduct some future planetary exploration efforts, but that is not yet reality.

Spacecraft have flown rapidly by planets and small bodies, orbited them, and landed (or crashed; the euphemistic term is "lithobraking") on them, and in a few cases deposited astronauts or unmanned rovers to explore their surfaces. An assortment of flyby and orbital spacecraft is shown in Figure 1.1, and a family portrait of Mars rovers is shown in Figure 1.2. Exploration by spacecraft is complex, and large multidisciplinary (often international) teams of scientists and engineers have to work together seamlessly. Mission operations can last for decades, sometimes requiring several generations of investigators. This can be heady stuff for geoscientists used to working in isolation and on projects of limited duration.

But sometimes it is disheartening; the history of planetary exploration is littered with as many spacecraft failures as successes. Moreover, many early missions counted as successes returned little or no scientific data or were technology demonstrations. In this chapter, we focus on missions that provided the most useful data for geoscience.

1.2 Poking Around the Neighborhood: The Terrestrial Planets

The so-called **terrestrial planets** are Mercury, Venus, Earth, and Mars (Figure 1.3). Although the Moon is not really a planet, we will include it in this list. These bodies are composed of silicate rock and metal, and have solid surfaces.

All the terrestrial planets (as well as other planets) orbit within a common plane, called the **ecliptic**. This planar orientation is likely inherited from the **protoplanetary**

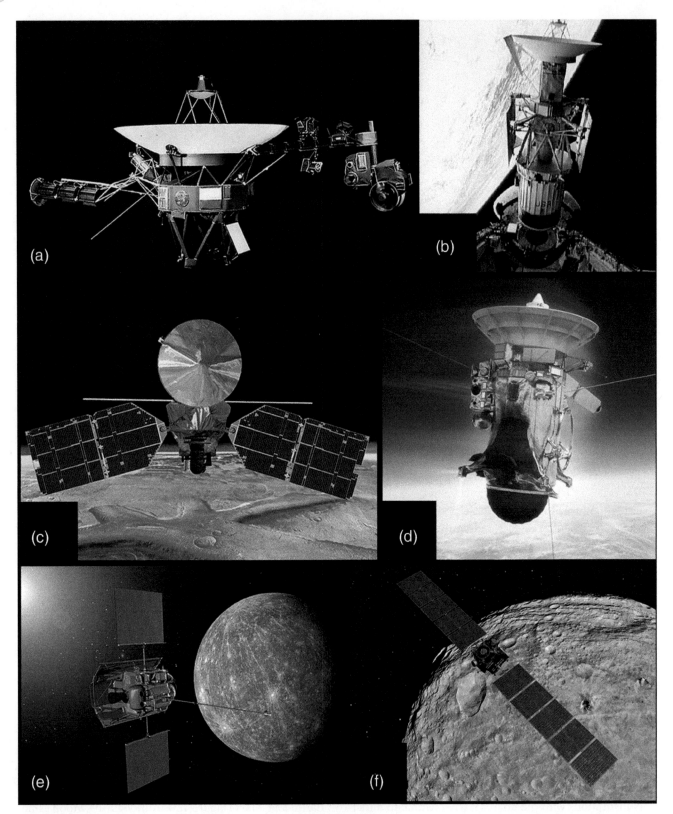

Figure 1.1 A few examples of spacecraft that have flown by or orbited planets and small bodies. (a) *Voyagers*, which conducted the first exploration of planets of the outer Solar System. (b) *Magellan*, being launched from the bay of a space shuttle, toward Venus. (c) *Mars Reconnaissance Orbiter*, mapping the red planet. (d) *Cassini*, a nuclear-powered probe that explored the Saturn system. (e) *MESSENGER*, which recently completed its exploration of Mercury. (f) *Dawn*, with solar panels needed to explore asteroids at great solar distance. NASA images.

BOX 1.1 HOW DO WE GET THERE?

One of the great challenges of exploration by spacecraft is accommodating the combined needs of engineering and science (both can be expressed in terms of mass) versus available power (which is required for both propulsion and operations). The greatest contribution to mass is usually the propellants used to launch the spacecraft and allow it to escape the gravitational grasp of the Earth. Launch vehicles and their fuel are also commonly the most costly parts of a spacecraft mission.

The escape velocity from the Earth's surface is ~11.2 km/s, normally achieved using several rocket stages. Once the spacecraft has left our planet, we stop using the conventional velocity notation. The reason, of course, is that spacecraft do not travel to their targets in straight lines, but instead are placed into elliptical orbits around the Sun so that they spiral outward over time. The spacecraft's velocity is constantly changing, depending on where it is in its orbit, so it is more convenient to speak of ΔV (literally "change in velocity," pronounced "delta V"), a measure of the impulse needed to perform a maneuver, such as launch or insertion into a planetary orbit. ΔV is proportional to the thrust per unit mass and the burn time. Because the relative orbital positions of the planets change over time, different launch dates from Earth have different ΔV requirements; this leads to optimum "launch windows" for each target body.

Rocket engines combust stored propellant, mixed with a source of oxygen, to produce hot gas exhaust, whose expulsion through a nozzle produces thrust. The propellants are usually liquid hydrocarbons. Some spacecraft have utilized electric propulsion, expelling reaction mass (such as heavy ionized atoms) at high speeds. Spacecraft may employ several kinds of engines for use during launch, interplanetary travel, and maneuvering. Many spacecraft now utilize gravity assist – making use of the relative motion and gravity of a planet to alter the path and speed of the spacecraft as it swings by. This saves propellant and reduces mission cost.

The least costly planetary exploration missions are flybys, because they require little or no fuel for operations at the targets. Orbiters require propellant and/or tricky maneuvers that utilize atmospheric drag to slow the spacecraft for orbital insertion, as well as station keeping. Stationary landers must successfully navigate to the ground, using retrorockets, parachutes, and other means. Rovers require additional technology for traversing. The energy needed for all of these spacecraft is provided by photovoltaic solar panels that convert sunlight into electricity or by radioisotope thermoelectric generators (RTGs) that convert heat generated by decay of suitable radioactive materials into electricity. Sending spacecraft data back to Earth occurs by direct-to-Earth radio transmissions or, for landed spacecraft, relay through orbiters that can store information and send it to Earth at a later time. Radio communications are received by the Deep Space Network (DSN), a collection of large antennae located around the planet. Future interplanetary communications may be based on optical lasers.

Planetary geoscience benefits greatly from geologic samples that can be studied in the laboratory. **Sample return missions** require not only the capability to land, rove, or operate in close proximity to the target body, but also the means of acquiring and storing samples and returning them to Earth with minimal damage. Samples from bodies that could potentially harbor life must address rigorous planetary protection protocols.

disk from which the planets formed. Distances in the Solar System are measured in astronomical units (AU), equal to the distance between the Sun and the Earth. Planetary orbits are ellipses, so we describe a planet's distance as the semi-major axis of its orbital ellipse.

Some of the orbital and global characteristics of the terrestrial planets are compared in Table 1.1. Additional data are tabulated in Lodders and Fegley (1998). The tabulated **uncompressed densities** are corrected for self-compression (which is greater for larger planets with higher gravity), and they give a better indication of the relative proportions of dense metal and less dense silicate.

1.2.1 Earth's Moon

The Moon is the largest satellite, relative to the size of the planet it orbits. It rotates synchronously with the Earth, always showing the same face. The lunar surface is heavily cratered, and ejecta from the largest impact basins form the basis for its chronostratigraphy. Gravity anomalies are associated with large basins. The Moon has an ancient feldspar-rich crust ("highlands") that floated in a magma ocean, an ultramafic mantle that is the source region for younger basalts ("maria") that fill impact basins, and a small metal core. A thick veneer of pulverized rock (regolith) covers the surface. Originally thought

Table 1.1 **Comparison of the terrestrial planets**

	Mercury	Venus	Earth	Mars	Moon
Semi-major axis	0.39 AU	0.72 AU	1.0 AU	1.50 AU	
Orbital period	0.24 yr	0.62 yr	1.0 yr	1.88 yr	27.3 d
Rotation period	58.6 d	−243 d	24 h	24.7 h	
Radius	2436 km	6051 km	6368 km	3390 km	1738 km
Mass	3.3×10^{23} kg	4.87×10^{24} kg	5.97×10^{24} kg	6.42×10^{23} kg	7.4×10^{22} kg
Mean density (ρ)	5.4 g/cm^3	5.3 g/cm^3	5.5 g/cm^3	3.9 g/cm^3	3.3 g/cm^3
Uncompressed ρ	5.3 g/cm^3	4.4 g/cm^3	4.4 g/cm^3	3.8 g/cm^3	3.2 g/cm^3
Atmosphere	~None	92 bar	1 bar	0.06 bar	None
Moons	0	0	1	2	

Figure 1.2 A family portrait of Mars rovers in the "Mars yard" at the Jet Propulsion Laboratory. Sojourner (*Mars Pathfinder*) in the foreground, Spirit and Opportunity (*Mars Exploration Rovers*) on the left, and Curiosity (*Mars Science Laboratory*) on the right. NASA and JPL image.

to be bone dry, recent data reveal traces of ice near the poles and of magmatic water in basalts.

The Soviet Union sent the first spacecraft to the Moon: *Luna 3* flew by the Moon in 1959 and sent back the first images of the far side. In 1966, *Luna 9* managed a controlled landing, and *Luna 10* first achieved lunar orbit. The USA also flew a number of unmanned *Surveyor* missions to different landing sites on the Moon. NASA's *Apollo* program conducted the first geologic exploration between 1968 and 1972: *Apollo 8* and *Apollo 10* were the first manned orbital flights, and six missions beginning with *Apollo 11* in 1969 landed astronauts on the surface. These manned missions returned the first lunar samples to Earth. Also during this time, the Soviet Union conducted additional unmanned *Luna* missions with rovers that collected and returned lunar soil samples to Earth. The locations of sampling sites, mostly in mare regions, are shown in Figure 1.4. It is not exaggerating to say that

Figure 1.3 Images of Mercury, Venus, Earth, and Mars (left to right), to scale. NASA image.

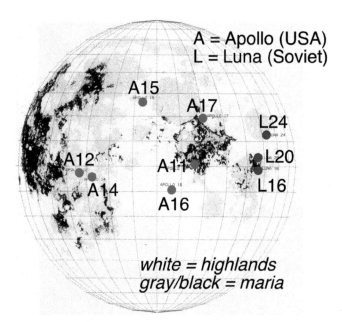

A = Apollo (USA)
L = Luna (Soviet)

white = highlands
gray/black = maria

Figure 1.4 Locations on the Moon from which samples have been returned to Earth.

samples of rocks and soils returned by these missions have revolutionized lunar science.

Major milestones in lunar exploration following *Apollo* have mostly employed orbiters. NASA's *Clementine* in 1994 and *Lunar Prospector* in 1998 mapped surface compositions and potential fields. ESA's *SMART-1* in 2003, JAXA's *Kaguya* (also called *SELENE*), CNSA's *Chang'e 1* in 2007, and ISRO's *Chandrayaan-1* in 2008 performed remote sensing measurements. NASA's *Lunar Reconnaissance Orbiter* and the *LCROSS* impactor in 2009 characterized the radiation environment and potential resources, including water. In 2011, the *GRAIL* mission, consisting of two orbiters named *Ebb* and *Flow*, refined our understanding of lunar gravity. In 2013, NASA's *LADEE* studied the atmosphere and airborne dust, and CNSA's *Chang'e 3* and *4* carried rovers to the lunar surface.

1.2.2 Mars

More spacecraft missions have been sent to Mars than to any other planet. Mars is divided into ancient, heavily cratered highlands in the southern hemisphere and younger, northern lowlands. The planet has gigantic volcanoes and extensive volcanic plains, a system of huge canyons, and layered deposits containing ices at the poles. It has a basaltic crust and a metallic core, and parts of the ancient crust are magnetized. Its climate has changed over time, and early Mars had liquid water and may possibly have been habitable. Clastic and chemical sediments are widespread, and surface soils are heaped into dunes. Analyses by orbiters and rovers, as well as martian

meteorites, have provided critical information on the planet's composition and history. Nowadays Mars is a windswept, dusty desert, although significant amounts of water in the form of permafrost occur in the subsurface at higher latitudes. Mars has Earth-like seasons and a thin atmosphere composed mostly of CO_2.

NASA's *Mariner 4* was the first successful flyby in 1964, and *Mariner 9* and the Soviet Union's *Mars 3* orbited Mars in 1971. NASA's *Viking 1* and *Viking 2* missions successfully deployed landers in 1975. The Viking probes analyzed soil and searched unsuccessfully for life. Beginning in 1996, NASA's *Mars Global Surveyor* mapped the composition of the planet's surface from orbit. *Mars Pathfinder* landed in the same year, deploying its Sojourner rover. NASA's *Mars Odyssey*, launched in 2001, and ESA's *Mars Express*, launched in 2003, obtained orbital images and spectra to understand surface geology and a radar sounder to probe the subsurface. The *Mars Exploration Rovers* Spirit and Opportunity landed in 2004 and conducted extensive science traverses for far longer than their designed lifetimes. In 2005, NASA's *Mars Reconnaissance Orbiter* began conducting remote-sensing measurements, and *Phoenix* landed near a pole and probed for ice. The *Mars Science Laboratory* Curiosity rover has explored the geology of an ancient lakebed since 2012. ISRO's *Mangalyaan* and NASA's *MAVEN*, both launched in 2013, are studying the martian atmosphere. The *Trace Gas Orbiter*, a collaboration between ESA and ROSCOSMOS, began atmospheric mapping in 2018, searching for methane and other minor gases. *InSight*, which landed in 2018, will study Mars' interior structure and heat flow. NASA's *Mars-2020* rover and ESA's *ExoMars* rover will land in 2020; *Mars-2020* will cache rock and soil samples for possible return to Earth as an international effort later in that decade, and *ExoMars* will search for organic compounds in the martian subsurface.

1.2.3 Venus

Venus is sometimes called Earth's sister because of its similar size and mass, but it rotates in the opposite direction from most planets. It has a dense atmosphere, mostly of CO_2, causing its surface to be blisteringly hot, with a mean temperature of 735 K. These hostile conditions and an obscuring shroud of clouds make exploration difficult. Radar imagery indicates that the surface is mostly smooth volcanic plains, but with two highlands regions. Volcanic features occur nearly everywhere and volcanism may be ongoing. The thick atmosphere screens out small impactors, and global resurfacing may have removed larger craters. Venus has a large metal core similar to Earth's, but no magnetic field. It has lost its

Figure 1.5 Images of Jupiter, Saturn, Uranus, and Neptune (left to right), with Earth (below Saturn) for scale. NASA images.

water, making its crust too strong to allow plate tectonics, which hampers heat loss.

NASA's *Mariner 2* in 1962, followed by *Mariner 5* in 1967, flew by Venus and probed the atmosphere and magnetic field. Between 1969 and 1983, the Soviet Union's *Venera* program sent numerous orbiters that analyzed the atmosphere and landers that imaged and measured the chemistry of surface rocks. Beginning in 1978, NASA's *Pioneer Venus Orbiter* conducted atmospheric experiments and made radar maps of the surface. Two *Vega* missions in 1984 continued the Soviet program of orbital and landed measurements. NASA's *Magellan* mission in 1990 provided global radar maps of the planet's surface. ESA's *Venus Express* operated from 2006 to 2014, observing atmospheric dynamics and the magnetic field.

1.2.4 Mercury

Mercury, the smallest and innermost planet, rotates three times for every two revolutions around the Sun – that is, it is in a spin–orbit resonance. Its high density indicates a huge metallic core that generates a magnetic field. Its surface is heavily cratered and covered with volcanic plains, making it appear almost lunar-like and indicating that geologic activity has ceased. Compression features reveal global shrinkage of the planet. Mercury's proximity to the Sun and lack of an atmosphere cause surface temperatures to vary wildly (between 100 and 700 K at the equator, daily).

Only two NASA spacecraft missions have made close observations of Mercury. *Mariner 10* flew past the planet three times in 1974 and 1975, and imaged less than half the surface. It also detected Mercury's magnetic field – a surprise since Mercury rotates so slowly. *MESSENGER* made passes in 2008 and 2009 before achieving orbit in 2011. It collected data until 2015, when it was allowed to crash into the surface. Its camera completed imaging of the whole surface and spectrometers characterized its chemical composition. In addition to studying its geologic history, the spacecraft quantified the size and state of the core and the magnetic field. *BepiColumbo* is actually two orbiters, one provided by ESA for surface imaging and one provided by JAXA for analyzing the magnetic field. It launched in 2018, although it will not reach Mercury's orbit until 2024.

1.3 Xenoplanets: Gas Giants and Ice Giants

The **giant planets** (Figure 1.5) are foreign to our geologic experience. They are commonly referred to as **gas giants**

Table 1.2 **Comparison of the giant planets**

	Jupiter	Saturn	Uranus	Neptune
Semi-major axis	5.2 AU	9.5 AU	19.2 AU	30.0 AU
Orbital period	11.9 yr	29.5 yr	84.0 yr	164.8 yr
Rotation period	10.0 h	10.2 h	17.2 h	16.1 h
Radius	71,400 km	60,270 km	25,600 km	24,750 km
Mass	1.90×10^{27} kg	5.68×10^{26} kg	8.68×10^{25} kg	1.02×10^{26} kg
Mean density (ρ)	1.33 g/cm^3	0.69 g/cm^3	1.27 g/cm^3	1.64 g/cm^3
Moons	69 + rings	62 + rings	27 + rings	14 + rings

(Jupiter and Saturn) and **ice giants** (Uranus and Neptune), depending on whether they are predominately composed of hydrogen and helium gas or of water, ammonia, and methane ices, respectively. The term "gas giant" is misleading, as their constituents are above the critical point and thus there is no distinction between gas and liquid. High pressures in the interiors of the ice giants transform ices into dense structures not seen elsewhere. Some orbital and global characteristics of the giant planets are given in Table 1.2.

1.3.1 Jupiter

The mass of Jupiter is about 2.5 times the mass of all the other planets combined. Its rotation, taking only about ten hours, is the fastest of the planets. It consists mostly of hydrogen and helium, and has no solid surface. The outer atmosphere is segregated into horizontal bands and has hurricane-like storms. Jupiter is thought to have a core of silicate rock and metal, surrounded by a mantle of dense metallic hydrogen extending out to 78 percent of the planet's radius. Above this is a layer of supercritical hydrogen and helium, which grades upward into the gaseous atmosphere. Jupiter has a strong magnetic field, generated by currents in the metallic hydrogen layer.

A number of spacecraft have flown by Jupiter, most en route to other targets, and obtained images and other data. NASA's *Pioneer 10* in 1973 and *Pioneer 11* in 1974 refined estimates of the planet's mass. NASA's iconic *Voyager 1 and 2* arrived five years later. These spacecraft focused on the geology of Jupiter's moons, and also discovered the existence of rings. Other flybys include *Ulysses* in 1992 and 2004, *Cassini* in 2000, and *New Horizons* in 2007.

The first mission to orbit Jupiter was NASA's *Galileo*, arriving in 1995 and operating for eight years. Besides investigating the Galilean moons on multiple passes and observing the impact of a comet onto Jupiter, it released a probe that parachuted through 150 km of Jupiter's atmosphere before it was crushed by the increasing pressure. NASA's *Juno* orbiter arrived in 2016 and is now measuring Jupiter's composition, as well as its gravity and magnetic fields.

1.3.2 Saturn

Like Jupiter, Saturn has a core of rock and metal, surrounded by metallic hydrogen, overlain by supercritical hydrogen and helium, and finally a gaseous atmosphere. Its pale yellow color is due to ammonia crystals in the atmosphere. Its magnetic field is much weaker than Jupiter's, and its mean density is less than that of water. Saturn's most prominent feature is, of course, its rings. These extend outward to 120,700 km from the equator, but are only about 20 m thick. The ring particles, ranging in size from dust to boulders, consist of water ice crystals with small amounts of organic and amorphous carbon.

Saturn was visited by *Pioneer 11* in 1979, *Voyager 1* in 1980, and *Voyager 2* in 1981. During flybys, these spacecraft imaged the rings and moons. The *Cassini* spacecraft, a collaboration between NASA and ESA, entered Saturn's orbit in 2004. This mission provided high-resolution images of the planet and its rings, and made significant discoveries about its moons. The mission ended in 2017, as the spacecraft plummeted into Saturn.

1.3.3 Uranus

Uranus is the least massive of the giant planets, although its diameter is slightly larger than Neptune's. It has a magnetic field, a ring system, and moons. It appears nearly featureless in visible light. Its rotation axis is tilted sideways, nearly into the ecliptic plane, so only a narrow strip near the equator experiences a day–night cycle and each pole receives 42 years of continuous sunlight followed by 42 years of darkness. Uranus is composed mostly of water, ammonia, and methane ices, along with other hydrocarbons. Underlying its icy mantle is a small core of silicate rock and metal. An overlying aquamarine-colored atmosphere is mostly hydrogen and helium, with methane as a coloring agent.

The only spacecraft to visit Uranus was *Voyager 2*, which flew by in 1986. It analyzed the composition and structure of the atmosphere, the magnetic field, and observed its moons and rings.

1.3.4 Neptune

Neptune is the densest giant planet. Its atmosphere is composed mostly of hydrogen and helium, with small

Figure 1.6 Images of the Galilean moons of Jupiter: Io, Europa, Ganymede, and Callisto (left to right), to scale. NASA images.

amounts of hydrocarbons and possible nitrogen. As for Uranus, traces of methane account for the planet's blue color. Unlike Uranus, however, its atmosphere has active weather patterns. The outer gas envelope grades downward into ices of water, ammonia, and methane. Below that is a core of silicate rock and metal. Neptune's magnetic field is strongly titled relative to its rotation axis.

Voyager 2 has been Neptune's only visitor, passing by in 1989 on its way out of the Solar System. The spacecraft imaged Neptune and its rings, measured the orientation of the magnetic field, and discovered a number of satellites. Because of its extreme distance and the dearth of exploration missions, much of what we know about Neptune (and Uranus) is based on observations using the Hubble Space Telescope and large Earth-based telescopes with adaptive optics.

1.4 The Most Interesting Moons

The giant planets themselves are not very amenable to geologic investigation, since they have no solid surfaces. However, their satellite systems are like miniature Solar Systems, and these moons can be studied using the same geologic tools that we apply to the terrestrial planets.

The numbers of moons for each giant planet are given in Table 1.2. There are far too many to describe here, and we know very little about most of them anyway. Instead, we will focus on the largest and most geologically interesting of the moons of Jupiter, Saturn, and Neptune; none of the moons of Uranus are particularly noteworthy, at least at our present level of ignorance.

1.4.1 Galilean Moons of Jupiter

The four largest satellites of Jupiter, collectively called **Galilean moons** after their discoverer, are Io, Europa, Ganymede, and Callisto (Figure 1.6). All are massive enough to have adopted nearly spherical shapes. The largest, Ganymede, has a diameter (5268 km) greater than Mercury. Io, Europa, and Ganymede are in a 4:2:1 orbital resonance with each other. All lie within the radiation and magnetic fields of Jupiter, making exploration difficult.

Io has the highest density ($3.5\,g/cm^3$) of the Galilean satellites, as the others have significant amounts of ice. It is a differentiated body composed of silicate rock, with a core of molten iron or iron sulfide. Its surface features more than 400 volcanoes, and so many of these are currently erupting that it qualifies as the most geologically active body in the Solar System.

Europa has a smooth (at global scale) and bright surface composed of ice. Reddish brown markings crisscross its tectonically active surface, and the occurrence of only a few craters testify to resurfacing. Below the icy crust is a salty ocean that has been hypothesized to be a possible abode for life, making Europa a high-priority target for further exploration. The interior of Europa is rock and likely a metallic core.

Ganymede is also believed to have a briny ocean, sandwiched between layers of ice. Below that is a mantle of rock and a core of iron. The icy surface of Ganymede is separated into dark, highly cratered (hence older) terrains and brighter, younger regions with grooves and ridges.

Callisto is the least dense ($1.83\,g/cm^3$) of the Galilean moons, with approximately equal amounts of ice and rock. It too may harbor a subsurface ocean. Its icy crust shows some tectonic features and numerous impact scars.

Pioneer 10 and 11 obtained low-resolution images of the moons when they flew past Jupiter in 1973 and 1974. In 1979, *Voyager 1 and 2* discovered volcanic eruptions on Io and a disrupted icy crust on Europa. The *Galileo* orbiter, arriving in 1995, made close approaches to the moons and found evidence for possible subsurface oceans on the outer three bodies. The *Cassini* probe in 2000 and *New Horizons* in 2007 flew by and made observations of the moons' orbital interactions with Jupiter. In 2016, *Juno*

Figure 1.7 Images of the largest moons of Saturn and Neptune: Titan, Enceladus, and Triton (top to bottom), not to scale. Titan is shown in an infrared image to see through its hazy atmosphere. Blue stripes on Enceladus are fractures from which jets erupt. Triton shows distinctive terrains. NASA images.

imaged the moons from above their orbital plane and made a movie of their motions.

1.4.2 Titan and Enceladus of Saturn

Titan (Figure 1.7), with a diameter of 5150 km, is the second largest moon in the Solar System, and perhaps the oddest. Its density (1.88 g/cm^3) suggests it is half ices and half silicate rock. This frigid world is characterized by a dense, hazy atmosphere composed of nitrogen, with clouds of methane and ethane and organic smog. Titan's surface is young and has only a few craters. Valleys and channels appear to have been cut by flowing organic liquids. Rocks and sand composed of frozen organic matter litter its surface. It also has lakes of organic liquids, and may have a subsurface ocean.

Enceladus (Figure 1.7) is small (508 km diameter) but nearly spherical. Its density (1.61 g/cm^3) indicates that it is a mixture of ice and rock. Its icy surface includes cratered terrains, as well as smooth areas with few craters. Long fracture systems are warm and emit jets of gas and dust. The eruptions may emanate from a subsurface ocean of water.

Voyager 1 and 2 in 1980 and 1981 obtained images of Titan and Enceladus. *Cassini*, arriving at Saturn in 2004, studied both bodies in detail. It made a number of Titan flybys, discovering hydrocarbon lakes in 2006. Cassini also released its *Huygens* probe, provided by ESA, which descended to the surface in 2005. The probe made measurements of the composition of Titan's atmosphere and took images of the surface. The *Cassini* orbiter also discovered the geyser-like eruptions on Enceladus.

1.4.3 Triton of Neptune

Triton (Figure 1.7), the largest moon of Neptune (2710 km diameter), has a retrograde orbit. Its surface is covered with a transparent layer of frozen nitrogen, over a crust of N_2, H_2O, and CO_2 ices, with an icy mantle and rocky core. Its mean density (2.06 g/cm^3) suggests ices comprise more than one-third of the body's mass. It has few impact craters, suggesting a young, active surface. A handful of geysers of nitrogen gas and dust have been observed erupting. The western hemisphere is called cantaloupe terrain, referring to its resemblance to that melon's skin. Other areas of the surface are furrowed plains formed by icy volcanic flows. Triton's retrograde motion and Pluto-like composition suggest that it is a captured body from the Kuiper belt (see Section 1.5).

Triton was deemed so interesting that *Voyager 2*'s trajectory was altered to allow its investigation. It imaged 40 percent of Triton's surface and measured its chemical composition.

1.5 Small Bodies, Big Rewards

The Solar System contains millions of smaller bodies orbiting the Sun; their small sizes belie their scientific importance. The largest of these are the **dwarf planets**. This term was adopted by the International Astronomical Union in 2006, and identifies objects that orbit the Sun, are massive enough for their gravity to have pulled them into a spheroidal shape, but have not cleared their orbital neighborhoods of other objects. Using this definition, Pluto was demoted from its prior designation as the ninth planet, and Ceres was promoted from its status as the largest asteroid. Three other currently recognized dwarf planets, located well beyond the orbit of Pluto, are Haumea, Makemake, and Eris. There are sure to be many more. The designation of dwarf planets remains somewhat controversial, and certainly does not recognize the role of geologic activity in planets.

Smaller bodies are designed as **asteroids** (the name means "star-like") and **comets** (from the Greek for "long hair," alluding to their prominent tails). Asteroids have conventionally been defined as rocky or metallic bodies, and comets are recognized by "cometary activity," referring to the release of gas and dust that form a bright coma and tail when ices are sublimated by solar heating. In recent years, the distinction between asteroids and comets has become blurred, because volatile eruptions have been seen on a few asteroids and some kinds of asteroids are known to contain or to have formerly contained ice and probably exhibited cometary activity in the distant past.

Before describing the exploration of these objects, we need to talk about their orbits. Most asteroids reside either in the **Main asteroid belt**, a band more than 30 million kilometers wide located between the orbits of Mars and Jupiter, or at **Lagrange points** around Jupiter (Figure 1.8). A Lagrange point is a position in the orbital configuration of two large bodies, in this case the Sun and Jupiter, where a small body, say an asteroid, can maintain a stable position. Asteroids orbiting Jupiter's L_4 and L_5 points (respectively located ahead and behind Jupiter in its orbit; Figure 1.8) are called Trojans. Centaurs are asteroids located between the outer planets, generally in unstable orbits.

Beyond Neptune lies the **Kuiper belt**, containing more than 1000 known objects and thought to have more than 100,000 objects larger than 100 km in size. The potential to discover so many planets in the Kuiper belt is what led to the designation of dwarf planets. Pluto is the innermost Kuiper belt object. Far beyond the Kuiper belt lies the **Oort cloud**, a theoretical, spherical collection of volatile-rich objects. Small fragments of Kuiper belt and

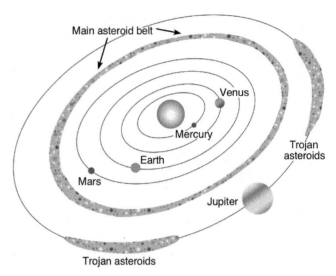

Figure 1.8 Birds-eye view of the Solar System, showing asteroids in the Main belt and Trojan asteroids in Lagrange points around Jupiter.

Oort cloud objects perturbed into the inner Solar System become comets.

1.5.1 Dwarf Planets: Ceres and Pluto

Ceres (Figure 1.9) is the first dwarf planet to be encountered by spacecraft, although Neptune's moon Triton (already described in Section 1.4) probably qualifies as a captured dwarf planet in terms of its size. Ceres is located within the Main asteroid belt, and was the first body discovered in that region. It has a diameter of 945 km and is partially differentiated, with a rocky core, altered rock mantle, and icy crust. The mineralogy of the surface consists of hydrated and ammoniated phyllosilicates and carbonates. Although most of its surface is ancient and heavily cratered, a cryovolcano formed recently, suggesting brines exist in the interior.

Pluto (Figure 1.9) is in orbital resonance with Neptune, but its orbit is inclined relative to the ecliptic. It has five moons, with Charon having a diameter just over half that of Pluto, whose diameter is 2380 km. Its density (1.86 g/cm^3) indicates a rocky core, and it likely hosts a subsurface ocean of liquid water. Its crust is composed mostly of nitrogen ice, with traces of methane and carbon monoxide. Its color varies from white, to orange, to black. The surface is virtually uncratered, indicating a very young age. It has mountains (possible volcanoes), and many landforms appear to be glacial in origin.

NASA's *Dawn* spacecraft studied Ceres from orbit during 2015–2018. Its spectrometers mapped the body's mineralogy and chemistry, its camera allowed studies of its geomorphology, and tracking of the spacecraft's orbit allowed the gravity and interior structure of Ceres to be

Figure 1.9 Dwarf planets visited by spacecraft: Ceres (top), showing bright spots of carbonate salts in a crater; and Pluto (bottom), showing terrains with distinctive coloration. Colors in both images are enhanced, and the images are not to scale. NASA images.

analyzed. NASA's *New Horizons* flew by Pluto in 2015 and returned images, chemical analyses, and other data. In 2019 it imaged a small icy body, and it is now en route to a body deeper within the Kuiper belt.

1.5.2 Asteroids

More than half of the mass of the Main belt is contained within the four largest bodies: Ceres (now a dwarf planet), Vesta, Pallas, and Hygia. Only a few large asteroids are thought to be intact, and many thousands of smaller asteroids are collisional rubble. Many of these are members of families that have similar orbital parameters and were once parts of the same larger object. Despite

popular movies that show spacecraft dodging asteroids as they race through the Main belt, the average spacing between asteroids is actually about three million kilometers.

Asteroid science benefits greatly from spectroscopy using ground-based telescopes, and from analyses of the thousands of asteroid samples that have fallen to Earth as meteorites. By comparing an asteroid's spectrum with laboratory spectra of meteorites, we can infer its mineral composition and thereby determine the kinds of rocks on its surface. Meteorite analyses also document the inorganic and organic chemistry of asteroids, and allow us to infer their geologic histories.

The best-studied asteroid is Vesta, a 500 km diameter rocky body with a large metal core. It is the parent body for more than 1000 meteorites: basalts, ultramafic rocks, and impact breccias formed from mixtures of those rocks. A huge and rather recent impact basin on Vesta is thought to have excavated these meteorites. Vesta's magmatic differentiation is apparent from its igneous rocks and high density (3.42 g/cm^3), indicating the presence of a core, and its heavily cratered surface reveals its ancient age.

A number of small asteroids have been imaged by spacecraft (some are shown in Figure 1.10). The irregular shapes of these bodies indicate their fragmentation during collisions. The only ones whose compositions have been well analyzed so far are Eros and tiny Itokawa, both ordinary chondrites.

Spacecraft missions that have flown by asteroids, beginning in 1991, include *Galileo*, *NEAR Shoemaker*, *Hayabusa*, *Rosetta*, and *Chang'e 2*. In 2000, NASA's *NEAR Shoemaker* became the first mission to orbit an asteroid (Eros), and it ended its mission by gently landing on the surface. JAXA's *Hayabusa* orbited asteroid Itokawa in 2005. It intended to collect a soil sample, but the sampling mechanism failed; nevertheless, dust particles on the spacecraft became the first returned asteroid sample. NASA's *Dawn* orbital exploration of Vesta in 2011 is the most exhaustive asteroid exploration to date. *Hayabusa2* orbited asteroid Ryugu in 2018. NASA's *OSIRIS-REx* arrived at asteroid Bennu in 2018 and is expected to return samples in 2023.

1.5.3 Comets

Comet **nuclei** (Figure 1.11) are popularly called dirty snowballs, reflecting the fact that they are mixtures of rock and ice (mostly water and carbon monoxide), along with abundant organic compounds. Their measured densities are 0.6 g/cm^3 or less. Inside their bright comae, the comet nuclei are as dark as black velvet. Their surfaces are coated with lag deposits of dust. The surfaces of irregularly shaped comets are constantly changing as solar warming causes

10 km

Figure 1.10 Rogues gallery of some small asteroids and martian Moons (thought to be captured asteroids) visited by spacecraft. Clockwise from upper left are Phobos, Eros, Ida, tiny Dactyl, Mathilde, and Deimos, with Gaspra in the center. Reprinted by permission from Cambridge University Press: Cosmochemistry, Harry Y. McSween Jr. and Gary R. Huss, Copyright (2010).

Figure 1.11 Images of some comet nuclei visited by spacecraft: Halley, Hartley 2, Wild 2 and Tempel 1, and Curyumov–Gerasimenko (top to bottom). Images are not to scale. NASA and ESA images.

ice to sublimate and sometimes form jets of gas. Some nuclei split apart, suggesting that they are fragile.

Comets are in highly elliptical orbits, and most appear to originate within the Kuiper belt or Oort cloud. The dust trails they leave behind can be sampled as interplanetary dust particles that intersect the Earth's orbit as meteorite showers.

ESA's *Giotto* encountered comet Halley in 1986, and was bombarded by particles that caused a temporary loss of signal. NASA's *Deep Space 1* flew by comet Borrelly in 2001. NASA's *Stardust* mission collected dust ejected from comet Wild 2 in 2004, and returned it to Earth for laboratory analysis. NASA's *Deep Impact* visited Tempel 1 in 2005 and delivered an impactor that created a crater on the comet's surface. ESA's *Rosetta* spacecraft orbited comet Churyumov–Gerasimenko in 2014, conducting the most extensive set of comet observations to date. Its *Philae* lander unfortunately was put down in shadow and keeled over, so it was unable to make many of the planned surface observations and measurements, although it detected organic compounds.

1.6 A Few Notes on Orbital Dynamics

Although this text focuses on planetary geoscience, it is useful to review a few principles of celestial mechanics. These concepts apply not only to orbiting planets and small bodies, but also to the spacecraft exploring them.

Each body in the Solar System orbits the system's barycenter (taken approximately as the Sun), tracing out elliptical paths with the barycenter at one focal point of the ellipse. At any point along the orbit, the sum of the kinetic and potential energy of the orbiting object remains constant. The potential energy decreases as the body approaches the point closest to the barycenter (called the "perihelion" in the case of objects orbiting the Sun; the farthest point is the "aphelion"), so its kinetic energy, i.e., its orbital velocity, increases.

Although the planets' orbital paths about the Sun are eccentric, they are nearly circular. The low eccentricity is believed to result from their gravitational interactions with the Sun, which has nearly circularized their orbits over time. Some asteroids and their smaller cousins (meteoroids) have been perturbed into highly elliptical orbits that cross the orbits of the inner planets. A few bodies, such as some comets, have hyperbolic orbits. Their velocity is greater than the Solar System escape velocity, so they curve around the Sun at closest approach, and then separate forever.

Planetary orbits lie within a few degrees of the same plane, the **ecliptic** – a characteristic inherited from the spinning disk of dust from which they grew. Smaller bodies, such as some asteroids, comets, and Pluto, have orbits highly inclined to the ecliptic.

So-called **orbital elements** are used to describe how bodies in space move relative to each other. In order to be able to predict an object's future position, we need to know its location and its velocity. Both of these quantities are vectors with three components – x, y, and z for location, and v_x, v_y, and v_z for velocity – so six values are required. However, these Cartesian coordinates change rapidly, making them very cumbersome to tabulate for any Solar System body. Planetary scientists instead have devised different ways of describing orbits that put most of the variability into a single parameter. Six orbital elements are still needed, but five of them only vary slowly.

Johannes Kepler showed in the early 1600s that planets move in elliptical orbits, with the Sun at one focus of the ellipse. The two parameters that describe the size and shape of the ellipse are the semi-major axis (a) and the eccentricity (e). The elliptical orbits of the different bodies in the Solar System are oriented differently in space. The inclination (i) describes the amount that the ellipse is tilted relative to the ecliptic. Two other parameters, the longitude of the ascending node (Ω) and the longitude of perihelion (ω), describe the orientation of the ellipse in space. Together, these five elements describe the orbit – the allowed positions of the body around the Sun. The sixth element, the true anomaly, is the angle between perihelion of the orbit and the object's position at any given time. In a perfect system with only two spherical bodies and no other gravitational effects, the first five elements would never change. Such an unchanging orbit is often referred to as Keplerian.

In reality, planets and moons are not only affected by the Sun's gravity, but also by other bodies. For example, the orbit of the Moon about our planet cannot be accurately described without considering the gravitational effects of both the Earth and the Sun. Orbits can also be affected by distortions in the gravity field, caused by non-spherical shapes or other irregularities in mass distribution – this requires periodic adjustments in orbiting spacecraft. Such effects cause a, e, i, ω, and Ω to change with time. This change is mostly very slow, so assumptions of Keplerian orbits are generally reliable over modest time scales (e.g., thousands to millions of years). However, some configurations can lead to relatively rapid changes in orbital elements.

Summary

BOX 1.2 **EARTH IS A PLANET TOO**

Understanding the workings of our own world has been markedly improved by Earth observations at the planetary scale from orbital vantage points. Moreover, from the study of other planets we can test the generality of the geologic processes we have worked so hard to understand on Earth. Other Solar System bodies also allow us to travel back in time to document the kinds of materials that accreted to form the Earth and to elucidate processes that must have affected it, but have been erased from the geologic record. A few examples of insights that derive from planetary exploration are:

- The early Earth, like the Moon and terrestrial planets, had a magma ocean, formed by heat from the decay of short-lived radionuclides and collisions with other bodies. Global-scale melting had profound implications for differentiation to form the core, mantle, and crust, and affected the partitioning of elements used by our civilization.

- Plate tectonics dominates Earth's geology, but its moving rocky plates are unique among Solar System bodies. Heat loss at plate boundaries differs from that on one-plate planets, affecting the mechanisms of melting and the duration of geologic evolution.
- Basalts, the most common volcanic rocks on Earth, are ubiquitous on other rocky worlds, but the pathways and extents of magma evolution differ, making granitic rocks virtually unrepresented elsewhere.
- Water, liquid or frozen, is not nearly as rare in the Solar System as we once thought. The Earth is not a unique indicator of the way water interacts with a lithosphere, of its effects as a geomorphic agent, and of its possible role in life.
- Impact cratering is the most significant geologic process on other planets, and must have been on the early Earth as well. Large impacts have had disastrous consequences for life, and unraveling this history has allowed us to realize that humans still live in the fast lane.

This travelogue is but a cursory overview of the many geologic wonders in the Solar System. We will return to many of the bodies introduced here in the following chapters, but will focus more on geologic processes and less on cataloging what's there. It is worth reflecting, though, on the mechanical (and rarely, human) explorers that have provided information in enough depth to make planets, moons, and smaller bodies into geologic objects.

We have not touched on the instruments carried on these spacecraft that allow us to address geologic questions. In the next several chapters, we will concentrate on the tools developed and used by planetary geoscientists, before delving into the processes that have shaped our Solar System.

Review Questions

1. Briefly describe the systemic challenges faced by the engineers who design interplanetary spacecraft and the scientists who design the instruments they carry.
2. What are uncompressed mean densities, and what can they tell us about planets?
3. What characteristics do the terrestrial planets share?
4. What characteristics do the giant planets share?
5. How do we subdivide/classify the small bodies of the Solar System?

SUGGESTION FOR FURTHER READING

Lodders, K., and Fegley, B. Jr. (1998) *The Planetary Scientist's Companion*. New York: Oxford University Press. A handy resource that tabulates a wealth of information, with appropriate references, on the physical and chemical properties of Solar System bodies.

REFERENCE

Lodders, K., and Fegley, B. Jr. (1998) *The Planetary Scientist's Companion*. New York: Oxford University Press.

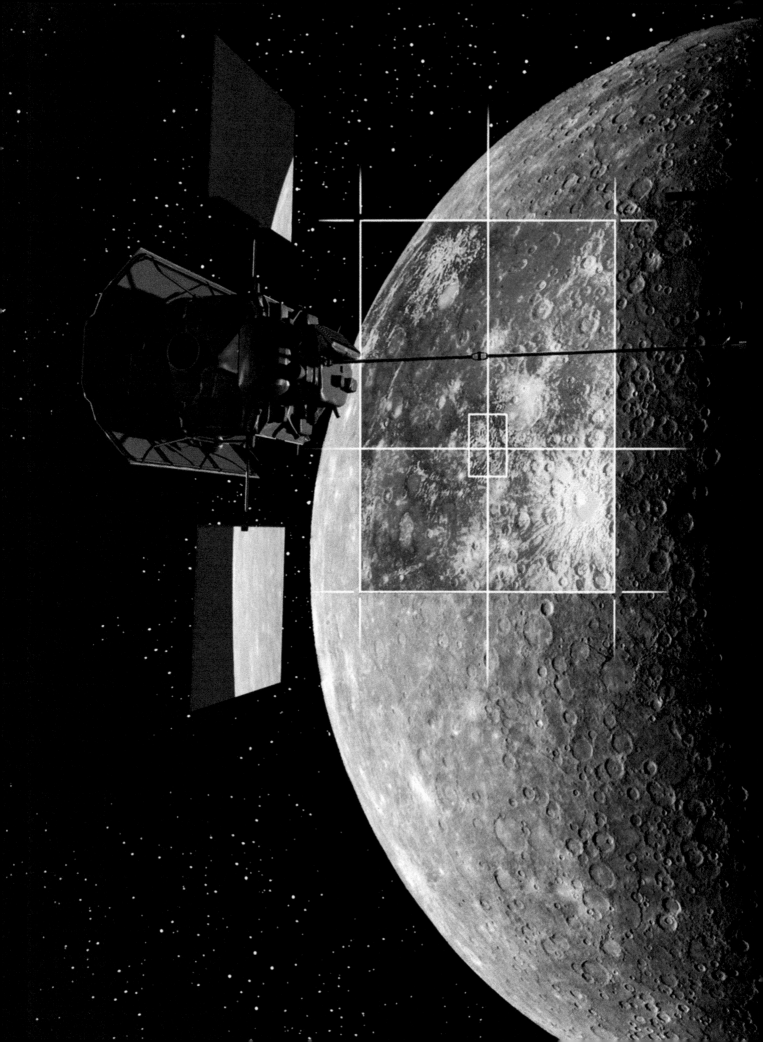

2

Toolkits for the Planetary Geoscientist: Spectroscopy and Imaging

With our preliminary survey of planetary bodies in the Solar System complete, we next turn our attention to developing an understanding of the tools that are used by planetary geologists to study these bodies. In this chapter, we will learn about some of the remote sensing techniques that are most commonly used in planetary geologic exploration. We will focus greatest attention on methods that employ light as the carrier of information about the remote target, but we will also consider some methods that employ the detection of other types of carriers. We will describe techniques that can be used from aircraft flying above the Earth, from spacecraft orbiting planets, and from landed and roving platforms on other worlds. We will examine both active and passive remote sensing techniques, provide guidance on how to select the right type of remote sensing method for the desired scientific outcome, and discuss the role of a successful ground campaign in the analysis of remote-sensing data when that option is available.

2.1 Sensing Remotely

While there is no substitute for *in situ* exploration of planetary surfaces as a method for understanding their geology, landed missions (both robotic and manned) are expensive and risky. Even high-priority planetary bodies, like the Moon and Mars, have only been explored with a handful of surface missions. The result is that there are relatively few places on other worlds that we have explored at ground level, and these sites do not adequately represent the full diversity of geologic features and environments found on those worlds. Fortunately, we can utilize the tools of remote sensing, particularly spectroscopic and imaging techniques, employed from

orbiting or flyby spacecraft missions passing overhead, to conduct reconnaissance of many planetary surfaces at global, regional, and even local scales.

Remote sensing, defined in the broadest sense, is any technique that allows one to characterize a target of interest without being in physical contact with the target. In the context of human "instruments," one's eyes and ears are remote sensing devices that rely on light and sound waves, respectively, as the carriers of information about the remote target. On planetary missions, the most common type of remote sensing device is the digital camera, but spectrometers are used nearly as often, particularly when developing an understanding of the composition of the target is a key priority.

In many situations, remote sensing is best thought of as an "efficiency enhancer" that may be employed when one wishes to study the geology of a new location. This is true whether one is studying the surface of another planet or a new field site on Earth. Because, by definition, remote-sensing tools are used at a distance, they typically can acquire large volumes of data covering vast expanses of a surface in relatively short amounts of time. For example, a camera in a high orbit around a planet can collect moderate-resolution images of an entire hemisphere of the planet in an instant. These data can be used to see a "big picture" that provides context and strategic guidance for further exploration at a more detailed scale, including landed missions to the surface. Even on landed missions, remote sensing experiments are often a key component of the instrument payload, typically used to pick the most scientifically valuable targets within a landed vehicle's reach for further study with experiments that require contact with samples, or to guide mobile landed vehicles (e.g., rovers) to new places where high-priority targets will be within reach of the contact instruments.

2.2 The Electromagnetic Spectrum

The most commonly employed remote sensing methods in planetary geology make use of **electromagnetic radiation**, otherwise known as light, as the carrier of information about the target's surface. Different methods rely on different sources of light – it can originate from the Sun, from the target surface itself, or even from the instrument making the observation. The smallest discrete packets of light, **photons**, have properties associated with both particles and waves. One fundamental property of a photon is its **wavelength**, defined as the distance between successive crests (or troughs) in the photon's waveform.

Our eyes are capable of sensing photons in a relatively restricted range of wavelengths called the **visible spectrum**. Within the visible spectrum, we perceive light of different wavelengths as different colors. Photons from the short-wavelength end of your detection range (around 390 nm) are perceived as the color violet, whereas those from the long-wavelength end (around 700 nm) are perceived as the color red. Figure 2.1 shows a map of the entire electromagnetic spectrum used in remote sensing. Whereas the wavelengths of visible photons only vary by about a factor of two, the range of wavelengths for all types of useful photons spans approximately 13 orders of magnitude! The visible spectrum is just one of many spectral regions that span different ranges of wavelengths over the electromagnetic spectrum (see Box 2.1).

Photons carry energy, and the energy of a single photon is given by the **Planck–Einstein relation**:

$$E = hc/\lambda \tag{2.1}$$

where E is the energy of the photon, h is the Planck constant (6.626×10^{-34} J/s), c is the speed of light in a vacuum (2.998×10^8 m/s), and λ is the wavelength of the photon. This relation may also be expressed in terms of the frequency of the photon ($v = c/\lambda$):

$$E = hv \tag{2.2}$$

The significance of the Planck–Einstein relation is that short-wavelength (high-frequency) photons, such as gamma rays or X-rays, are very energetic, whereas long-

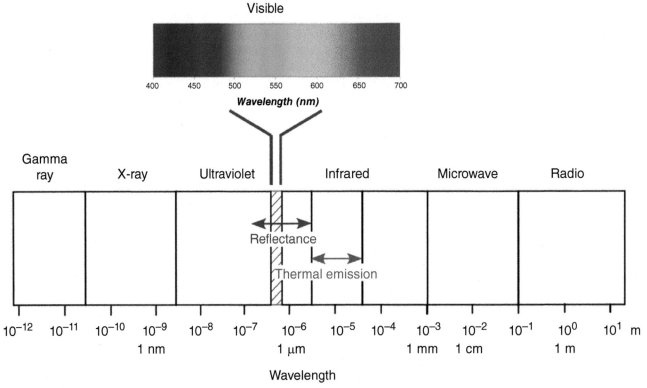

Figure 2.1 Map of the electromagnetic spectrum and its component spectral regions. Our eyes are sensitive to a narrow range of wavelengths, about 390–700 nm, called the visible spectrum. In planetary remote sensing, the long-wavelength end of the UV spectrum, the visible spectrum, and the short-wavelength end of the infrared spectrum (up to a wavelength of about 3000 nm, or 3 μm) is referred to as the reflectance part of the spectrum because the light available from planetary bodies at those wavelengths is dominated by reflected light that originates from the Sun. The thermal emission portion of the spectrum spans from about 3–50 μm, and is so called because light from planetary bodies at those wavelengths is dominated by self-emission of photons due to the fact that these bodies are warmer than absolute zero.

BOX 2.1 **SPECTRAL REGIONS AND UNIT CONVERSIONS**

Approximate ranges of photon wavelengths (λ) for different spectral regions are as follows:

- Gamma ray: $\lambda < 10\,\text{Å}$
- X-ray: $10\,\text{Å} < \lambda < 100\,\text{Å}$
- Ultraviolet (UV): $100\,\text{Å} < \lambda < 4000\,\text{Å}$
- Visible (Vis): $400\,\text{nm} < \lambda < 670\,\text{nm}$
- Near-infrared (NIR): $0.67\,\mu\text{m} < \lambda < 3.0\,\mu\text{m}$.
- Alternatively (typically used in terrestrial remote sensing):
  - Near-infrared (NIR): $0.67\,\mu\text{m} < \lambda < 1.0\,\mu\text{m}$
  - Shortwave infrared (SWIR): $1.0\,\mu\text{m} < \lambda < 3.0\,\mu\text{m}$
- Thermal infrared (TIR): $3\,\mu\text{m} < \lambda < 50\,\mu\text{m}$
- Microwaves, radio: $\lambda > 1\,\text{mm}$

Some useful unit conversions are:

$$1\,\text{Å} = 10^{-10}\,\text{m}$$
$$1\,\text{nm} = 10\,\text{Å} = 10^{-9}\,\text{m}$$
$$1\,\mu\text{m} = 10^{3}\,\text{nm} = 10^{-6}\,\text{m}$$
$$1\,\text{cm} = 10^{4}\,\mu\text{m} = 10^{-2}\,\text{m}$$

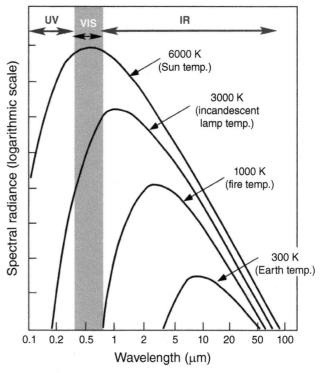

Figure 2.2 Blackbody spectra at temperatures corresponding to different familiar objects. Note the nonlinear plot axes.

wavelength (low-frequency) photons, such as radio waves, are very low in energy.

2.3 Blackbody Emission

Any object that is warmer than 0 K ("absolute zero") emits its own light. In everyday life, we observe light emitted from hot objects, such as the Sun or burning wood in a campfire, because these objects are warm enough to emit photons in the visible part of the spectrum, where our eyes are sensitive. Cooler objects, such as people or planetary surfaces, also emit their own light, but at longer wavelengths that the human eye cannot detect.

There is an idealized concept for self-emitting objects known as the **blackbody**. A blackbody is any substance that emits radiant energy (photons) at the maximum possible rate per unit area for any given temperature – in other words, a perfect emitter. Blackbodies also have the property of absorbing all of the photons that are incident upon them (thus the name). Blackbodies emit different amounts of photons at different wavelengths as a function of their temperatures, as given by the Planck equation:

$$L_{bb}(\lambda, T) = \frac{(10^{-6})\text{hc}}{\lambda^5 [e^{\text{hc}/\lambda kT} - 1]} \quad \frac{W}{m^2 \, \text{sr} \, \mu\text{m}} \qquad (2.3)$$

where $L_{\text{bb}}(\lambda, T)$ is the so-called **spectral radiance** of the blackbody, T is the temperature of the blackbody, and k is the Boltzmann constant (1.381×10^{-23} J/K) and wavelength λ is specified in microns. The physical units typically used for spectral radiance are watts per square meter per steradian per micron, which refer to the power emitted by the blackbody per unit area of its surface, per unit of solid angle into which it is emitted, per unit of wavelength along the spectrum. Note that the formulation of the Planck equation given above is specifically for calculating blackbody emission in terms of the wavelengths of photons; a different formulation of this equation must be used if photons are specified in terms of their frequencies.

Plotting Equation 2.3 at a fixed temperature yields a **blackbody spectrum** for that temperature – i.e., the radiance of a perfect emitter at that temperature as a function of wavelength. Figure 2.2 shows a set of blackbody spectra at temperatures corresponding to those of different familiar objects. Inspection of these spectra reveals a couple of intrinsic properties of blackbody spectra. First, the spectral radiance of a warmer object is greater than the spectral radiance of a cooler object at every wavelength – put another way, the curves corresponding to a family of blackbody spectra for the same target at different temperatures never cross. Another fundamental property of these spectra is that the wavelength of maximum emission for a blackbody at a given temperature is inversely proportional to that temperature – in other words, cooler blackbodies are brightest at longer wavelengths and warmer blackbodies are brightest at shorter wavelengths. This property is quantitatively characterized by a relationship called Wien's Law:

$$\lambda_{\max} = 2830/T \qquad (2.4)$$

where $\lambda_{\max}$ is the wavelength of peak emission in microns and T is in Kelvin. Wien's Law is the reason a horseshoe changes color as it heats up in a blacksmith's furnace. Initially, when the horseshoe is still cold, it emits light in the infrared and can only be seen by the human eye if it is illuminated by a source of visible light. But as it is heated, it eventually becomes warm enough that it emits its own visible light. First it will glow with a reddish hue corresponding to a $\lambda_{\max}$ at the long-wavelength end of the visible spectrum, but as it warms even more, it will begin glowing with a blueish-white color associated with a peak emission at shorter visible wavelengths.

Remote sensing of planetary surfaces can make use of reflected sunlight in the visible and near-infrared portion of the spectrum or light that is self-emitted by the surface in the thermal infrared portion of the spectrum. Figure 2.3 shows schematically how the sources of these two components compare for three different distances from the Sun. Note that there is a portion of the spectrum around 3–8 μm (depending on distance from the Sun) in which the contributions from reflected sunlight and self-emission are comparable. This is known as the "crossover region," and spectral measurements from this region are often difficult to interpret because the combination of the two sources acts to mute spectral absorption features (discussed in Section 2.4). For this reason, most spectral remote sensing of planetary surfaces tends to make use of reflected sunlight in the visible and near-infrared wavelengths between ~0.35 and 2.5 μm *or* self-emitted light from the surface in thermal infrared wavelengths between ~6 and 50 μm.

2.4 Emissivity and Reflectance Spectra

Most materials are not perfect emitters, so their emission spectra do not exactly match the theoretical ideal of a blackbody spectrum. Nevertheless, most natural materials have emission spectra with shapes that are fairly close to blackbodies except at certain wavelengths corresponding to **spectral absorption features**. At these wavelengths, emission of light from the material will be somewhat less than what a blackbody would emit. Different materials have different spectral absorption features. The Sun's emission spectrum (Figure 2.4), measured in space where it is not affected by the Earth's atmosphere, is essentially the same as a blackbody at a temperature slightly less than 6000 K, but with a deficit of emission at certain wavelengths caused by absorption of light by constituents of the solar atmosphere. These solar spectral absorption features are known as **Fraunhofer lines** and can be used to determine the Sun's chemical composition, as will be described in Section 4.3.

Most geologic materials also have spectral absorption features in the visible and/or infrared portions of the spectrum. Because the wavelengths and spectral shapes of these features are specific to different minerals, we can deduce the composition of an unknown target surface by

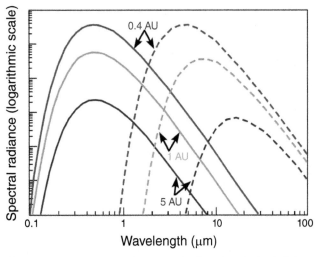

Figure 2.3 Comparison of spectra of reflected sunlight (solid lines) from a surface with a visual albedo of 0.1 and spectra of self-emitted thermal radiation (dashed lines) from the same surfaces with an IR emissivity of 1.00 for three different distances from the Sun, corresponding to the orbits of Mercury, Earth, and Jupiter. After Hapke (1993).

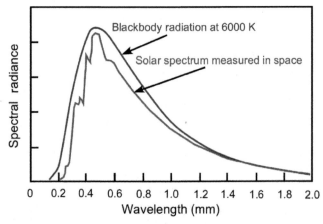

Figure 2.4 The Sun's emission spectrum is very close to that of a blackbody at a temperature slightly cooler than 6000 K, but with missing radiance at certain wavelengths corresponding to the spectral absorption features of gases in its atmosphere.

comparing the shape of its spectrum to a library of laboratory spectra of well-characterized samples. But to do this, we must first convert the measured radiance of the unknown target surface into a form that is directly comparable to that of the library spectra. The two solid lines in Figure 2.5a correspond to the thermal infrared radiance spectra of quartz at two different temperatures that are typical of the martian surface. These spectra have similar shapes, but are significantly different in amplitude because the warmer surface emits more thermal infrared light. The two dashed lines are blackbody spectra at the

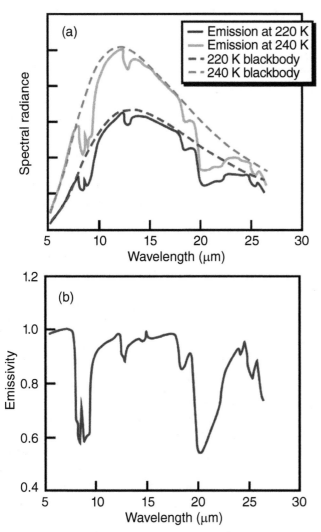

Figure 2.5 (a) Thermal infrared radiance spectra of quartz at two different temperatures appropriate for the surface of Mars (solid spectra), and theoretical blackbody spectra at the same two temperatures (dashed lines); (b) the emissivity spectrum of quartz, obtained by dividing either of the radiance spectra in (a) by their associated blackbody spectra. Conversion from the measured radiance spectra to emissivity spectra removes the temperature dependence of the curve and makes it directly comparable to the emissivity spectra of known samples measured in the laboratory at very different temperatures.

same temperatures, respectively. At the wavelengths where the solid and dashed lines are coincident, quartz is a perfect emitter. Spectral absorption features occur wherever the measured radiance spectra fall short of their corresponding blackbody spectra. To make the measured radiance spectra directly comparable to library spectra measured in the laboratory (possibly at very different temperatures), we divide them by their associated blackbody curves at each wavelength to create **emissivity** spectra:

$$\varepsilon(\lambda) \equiv \frac{L_{\text{measured}}(\lambda)}{L_{\text{bb}}(\lambda, T)} \tag{2.5}$$

where $\varepsilon(\lambda)$ is the emissivity spectrum, $L_{\text{measured}}(\lambda)$ is the measured spectral radiance of the target, and $L_{\text{bb}}(\lambda, T)$ is the spectral radiance of a theoretical blackbody at the same temperature as the target (from Equation 2.3). Conversion from a measured radiance spectrum to an emissivity spectrum is essentially a normalization step that removes the temperature dependence from the spectrum, leaving only a dependence on the composition of the surface as expressed in the form of spectral absorption features. Figure 2.5b shows the emissivity spectrum of quartz that is obtained by processing either of the two solid spectra in Figure 2.5a using Equation 2.5. Because the actual spectral radiance of the target cannot exceed that of a perfect emitter, $\varepsilon(\lambda)$ can only have values between 0 (a perfect absorber) and 1 (a perfect emitter) at any given wavelength. Library spectra of known samples measured in the laboratory are also converted to emissivity to facilitate comparison to the emissivity spectrum of the unknown target surface.

Emissivity spectra are usually used for spectral measurements made in the thermal infrared portion of the spectrum, where the photons have been self-emitted by the planetary surface being observed. A parallel concept, called **reflectance**, is used for spectral measurements of sunlight reflected off of planetary surfaces in the visible and near-infrared portions of the spectrum. Just as most real geologic materials are not perfect emitters at all wavelengths, they are also not perfect reflectors of sunlight at all wavelengths. Reflectance is defined as the ratio of the measured spectrum of reflected light coming off the surface to the spectrum of the incident light that is illuminating the surface:

$$R(\lambda) \equiv \frac{L_{\text{measured}}(\lambda)}{L_{\text{incident}}(\lambda)} \tag{2.6}$$

where $R(\lambda)$ is the reflectance spectrum, $L_{\text{measured}}(\lambda)$ is the radiance spectrum of reflected light coming off the surface, and $L_{\text{incident}}(\lambda)$ is the radiance spectrum of light that illuminates the surface. Converting a measured visible to near-infrared radiance spectrum to reflectance via Equation 2.6 has the desirable effect of removing the spectral shape of the illuminating light source from the spectrum, leaving only spectral absorption features that are associated with the composition of the surface. As with emissivity, reflectance values range from 0 (a perfect absorber) to 1 (a perfect reflector). The lamps used to illuminate known samples measured in laboratory spectrometers typically have a different spectrum than the Sun, so these spectra are also converted to reflectance to facilitate comparison to the spectra of the unknown target surface.

Occasionally, it may be necessary to compare a laboratory spectrum measured with reflected light to a remote sensing spectrum of an unknown surface measured with emitted light, or vice-versa. For example, early spectral libraries of known mineral samples in the thermal infrared region (e.g., Salisbury and D'Aria, 1992) were collected by cooling the samples so that their self-emission was minimized, and illuminating them with a lamp that shines in thermal infrared wavelengths. In this situation, the measured radiance spectra are first converted to reflectance using Equation 2.6, and then the resultant reflectance spectra are converted to emissivity using a fundamental relationship known as **Kirchhoff's Law**:

$$\varepsilon(\lambda) = 1 - R(\lambda) \tag{2.7}$$

Kirchhoff's Law states that emissivity is inversely related to reflectance at any given wavelength – in other words, the wavelengths at which a material is a poor emitter are also the wavelengths at which it is a good reflector. This is the reason that spectral remote sensing measurements made in the crossover region are so difficult to interpret – at any wavelength where there is a deficit of radiance in the emitted spectrum due to an emissivity minimum, there is also an abundance of reflected radiance due to a reflectance maximum. The net effect of these competing factors is that spectral features in the crossover region tend to cancel out, making compositional identifications challenging.

2.5 Making Spectra Useful: Information from Different Regions of the Electromagnetic Spectrum

The range of possible photon wavelengths is enormous, and, by the Planck–Einstein relation (Equation 2.1), the corresponding range of possible photon energies is also enormous. This large range of photon energies tends to translate into a large range of physical scales for the different microphysical processes that can emit or absorb photons. These different microphysical processes are associated with different compositional or physical

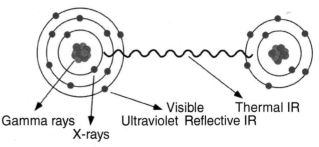

Figure 2.6 Schematic representation of the microphysical domains of the different energetic transitions that give rise to emission or absorption of photons in different spectral regions. The red and gray dots symbolize protons and neutrons in the nuclei of atoms, the blue dots represent electrons in orbits (black circles) around the nuclei, and the interatomic bond between two atoms in a mineral lattice is represented by the wavy black line. Higher-energy photons are associated with smaller-scale processes (e.g., gamma rays and nuclear transitions), and lower-energy photons are associated with larger-scale processes (e.g., thermal infrared photons and vibrational transitions in a mineral lattice). As a result, the type of information that can be gained about the composition of a planetary surface varies across the different spectral regions used for remote sensing.

properties of the material emitting or absorbing the photons. It is important to bear these associations in mind when selecting the specific wavelength region to use for a particular remote sensing study – different spectral regions convey different types of information about the surface.

Figure 2.6 shows a schematic graphical representation of the different types of energetic transitions that give rise to absorption or emission of photons from different spectral regions, as discussed below.

2.5.1 Gamma Rays

Gamma rays (γ) are the highest-energy photons, and as such the processes that can emit or absorb them involve the high-energy transitions associated with nuclear processes. For example, when a nucleus decays from an excited state to its ground state, one possible outcome is that the difference in energy between the excited and ground states is released in the form of a gamma ray. Every nuclear isotope has a set of allowable energetic transitions that are specific to that isotope. Thus, the gamma ray spectrum of an unknown planetary surface is diagnostic of the elemental and isotopic composition of the surface.

2.5.2 X-rays and Ultraviolet Photons

X-ray and UV radiations are the next-most energetic types of electromagnetic radiation, and they are emitted or absorbed by electron transitions that take place in the inner shells of the atom. For example, when an inner electron in an atom drops from an excited state to its ground state, it releases a photon with an energy equal to the difference between the two energy states of the electron. Different electron shell configurations of different elements have different allowable energy transitions. Thus, the X-ray and UV spectra of planetary surfaces are diagnostic of the elemental composition of the surface, but not its isotopic composition (because the allowable electron transitions are not sensitive to the number of neutrons in the nucleus).

2.5.3 Visible and Near-Infrared Photons

The energy differences associated with electronic transitions in outer-shell electrons correspond to the energies of visible photons and some of the shorter-wavelength near-infrared photons. As a result, visible and near-infrared remote sensing spectra of planetary surfaces are also sensitive to the elemental composition of the surface. However, there is also information about the mineral composition of the surface available at these wavelengths. The exact energies of electronic transitions in outer-shell electrons are influenced by the presence of other electrons around adjacent atoms in a mineral lattice. This effect varies from one mineral to another because different minerals have different elements arranged in different lattice structures. Minerals containing transition metals – iron is particularly important because of its high abundance – have spectral absorption features called **crystal field absorptions** that are caused by this effect. Analysis of these features can not only flag a surface as containing iron, but also the specific mineral host for that iron. Visible and near-infrared photons can also be emitted or absorbed by energetic transitions associated with electrons that are shared between atoms in the mineral lattice, creating spectral features known as **charge transfer features**.

2.5.4 Thermal Infrared Photons

In the longer-wavelength portion of the near-infrared region and in the thermal infrared region, photon energies are too low to be associated with electronic transitions. Instead, these photons are emitted or absorbed by **vibrational transitions**. These are energetic transitions associated with the excited and ground states of vibrations in mineral lattices that are warmer than absolute zero. The energies of these transitions are governed by the masses of the atoms in the lattice and the structure and strength of the interatomic bonds that hold the lattice together. Different minerals have different sets of allowable vibrational transitions. Thus, analysis of absorption features found in the longer-wavelength portion of the near-infrared and the thermal infrared spectra of

planetary surfaces provides information about the mineral composition of the surface. It is even possible to use thermal infrared spectra to distinguish different members of a solid solution series (e.g., within the forsterite–fayalite olivine series) because substitution of elements changes the masses in the lattice, which changes its fundamental vibrational frequencies, leading to a shift in the wavelength positions of the associated spectral absorption features.

Thermal infrared spectra of planetary surfaces may also be used to reveal information about grain sizes on a planetary surface. Recall that Equation 2.5 is typically used to convert measured radiance spectra into emissivity spectra when one wants to identify the mineral components of the surface. The other piece of information that may be extracted in this process (from the denominator on the right side of Equation 2.5) is the temperature of the surface at the time of measurement. **Thermal inertia** is the physical property of the surface that governs how readily the surface's temperature responds to changes in **insolation** (illumination by the Sun). A surface with low thermal inertia will exhibit larger diurnal temperature swings than a surface with a high thermal inertia. The thermal inertia of a surface is determined by a combination of the surface's specific heat capacity, bulk density, and bulk thermal conductivity. Of these, thermal conductivity is usually the dominant source of variability in thermal inertias of geologic surfaces with low moisture content. In turn, the primary factor controlling the thermal conductivity of a planetary surface is the size of the grains in the regolith. Surfaces composed of very fine, loose grains are much poorer conductors of heat than surfaces composed of large rocks. This means that surfaces composed of very fine, loose grains undergo significantly larger diurnal swings in temperature than surfaces composed of larger rocks. Thus, by extracting the temperature of a planetary surface from its remotely measured thermal infrared spectrum and knowing the local time when the measurement was made, one can infer the thermal inertia of the surface, which in turn is a proxy for the size of the grains on the surface. On Earth, especially in humid climates, this method can be complicated by the presence of soil moisture because water has an exceptionally high heat capacity that also influences the thermal inertia of the surface.

2.5.5 Microwave and Radio Photons

At longer wavelengths, in the microwave and radio portions of the electromagnetic spectrum, photon energies are even lower than the energies of mineral lattice vibrational transitions. Instead, the spectra of photons from planetary surfaces in these regions are shaped by scattering and conduction effects. Analyses of these spectra can reveal information about the roughness of a planetary surface and/or a property called the dielectric constant of the surface material, which governs how an electromagnetic wave propagates into the surface.

2.6 Example Spectra

In this section, we turn our attention from abstract discussion about the fundamental principles of geologic remote sensing to pragmatic information that will help the beginning analyst to interpret real data. We focus our attention on reflectance spectra, derived from measurements acquired in the visible and near-infrared portion of the spectrum, and emissivity spectra, derived from measurements acquired in the thermal infrared portion of the spectrum. These types of data contain some of the richest compositional information about planetary surfaces. Example spectra of selected geologic materials are presented, with key features noted. However, before we embark upon this survey, some discussion of relevant terminology is necessary.

In the context of remote sensing, a spectrum is a plot of some parameter as a function of wavelength (or, equivalently, in terms of photon frequency or energy for very short-wavelength photons). Usually the term is preceded by a modifier that indicates which parameter is being plotted – e.g., a radiance spectrum is a plot of spectral radiance as a function of wavelength. As described in Section 2.4, a reflectance spectrum of a planetary surface is typically derived by normalizing the measured visible and near-infrared spectral radiance reflected off the surface to the spectral radiance of sunlight incident upon the surface. An emissivity spectrum is derived by normalizing the measured thermal infrared spectral radiance emitted by the surface to the spectral radiance of an ideal blackbody at the same temperature.

Both reflectance and emissivity spectra have values that range between 0 and 1 as a function of wavelength. Typically, for most geologic materials, there are wide ranges of wavelengths for which the reflectance or emissivity are at or near 1.0. In a reflectance spectrum, a value of 1.0 at a given wavelength indicates complete reflectance of incident photons of that wavelength, and in an emissivity spectrum, a value of 1.0 at a given wavelength indicates the surface is emitting as efficiently as possible at that wavelength for its temperature. Wavelength regions in either type of spectrum that have values at or near 1.0 are referred to as being "on the **continuum**." Conversely, regions where the reflectance or emissivity is significantly less than 1.0 are called **spectral absorption features**, or sometimes just spectral features.

Spectral absorption features are caused by microphysical energy transitions. When a geologic material has an available transition (e.g., an electronic transition or a lattice vibrational transition) of a particular energy, it will tend to absorb photons at a wavelength corresponding (via the Planck–Einstein relation) to that energy. As a practical matter, the shapes and wavelength positions of spectral features from a target planetary surface may be compared to those found in laboratory spectra of materials with well-characterized compositions to make compositional identifications of the materials on the surface.

2.6.1 Visible/Near-Infrared Reflectance Spectra of Iron-Bearing Minerals

Figure 2.7a presents example spectra of four different iron-bearing minerals. As discussed in Section 2.5.3, the absorption features in these spectra associated with iron are caused by electronic charge transfer and crystal field effects. This means that although these absorptions are all due to iron, the exact positions and shapes of the features are influenced by the presence of other ions bonded to the iron in the mineral lattice, giving the spectrum of each iron-bearing mineral a different shape. Absorption features associated with charge transfer in iron typically manifest themselves in visible/near-infrared spectra as a steeply increasing slope in reflectance between $0.4\,\mu m$ and 1.0–$1.5\,\mu m$. The band center for this broad absorption feature actually falls within the UV portion of the spectrum; it appears in the visible/near-infrared region as a steep spectral slope because only the "right half" of the feature falls within this spectral range. This feature is particularly strong in iron oxide minerals such as hematite (Figure 2.7a). Visible/near-infrared reflectance spectra of iron-bearing pyroxene minerals such as augite (Figure 2.7a) have dual absorption features, one near $1.0\,\mu m$ and another near $2.0\,\mu m$, that produce a characteristic "W" shape in the spectrum. The exact positions and shapes of these bands provide clues as to which specific member of the pyroxene group is present. Iron-bearing olivine (Figure 2.7a) also displays a pronounced absorption feature near $1.0\,\mu m$, but can be distinguished from pyroxenes by the larger breadth of this feature and the absence of a second absorption near $2.0\,\mu m$. The spectrum of the iron-bearing carbonate mineral siderite (Figure 2.7a) is interesting because it displays not only electronic spectral absorption features caused by the iron in the 0.4–$1.5\,\mu m$ range, but also vibrational absorption features associated with the CO_3 structure around $2.3\,\mu m$ (see Section 2.6.2).

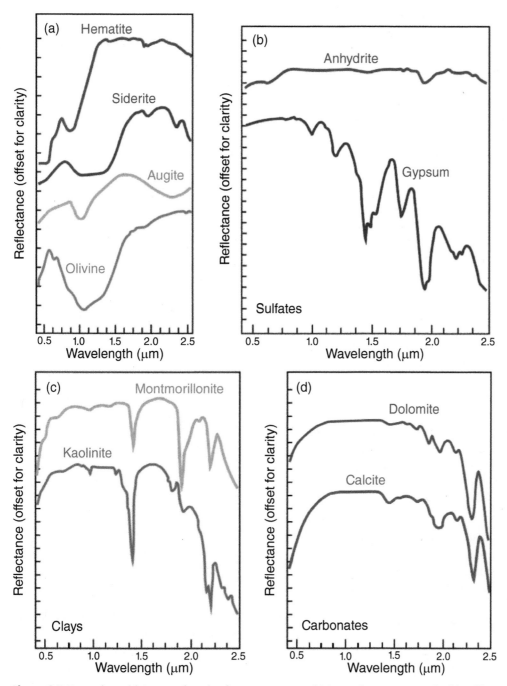

Figure 2.7 Example visible/near-infrared reflectance spectra of (a) iron-bearing minerals, (b) sulfates, (c) clay minerals, and (d) carbonates.

2.6.2 Vibrational Features in Near-Infrared Reflectance Spectra

Vibrational modes associated with the carbonate (CO_3), water (H_2O), and hydroxyl (OH) structures all give rise to spectral absorption features in the near-infrared. Thus, reflectance spectroscopy in this wavelength range is particularly well suited for mapping rock bodies that have experienced aqueous alteration. Comparing the spectra of gypsum ($CaSO_4 \cdot 2H_2O$) and anhydrite ($CaSO_4$) (Figure 2.7b) is instructive: the chemical difference

between these minerals is simply the addition of $2H_2O$ in gypsum, but this introduces a whole host of absorption features near the water vibrational bands at 1.4 and 1.9 μm, all of which are absent in the nearly featureless spectrum of anhydrite.

The presence of a hydroxyl ion in a mineral's structure gives rise to a spectral absorption feature near 2.2 μm that can be diagnostic for identifying phyllosilicates. Figure 2.7c shows the spectra of two clay minerals, montmorillonite and kaolinite. Note that, in addition to the aforementioned bands near 1.4 and 1.9 μm, these spectra also show feature(s) near 2.2 μm. The exact position and shape of this band can be used to distinguish different types of phyllosilicates. In the case of the kaolinite spectrum, the feature actually splits into a so-called "doublet" consisting of two narrow, adjacent bands that form an asymmetric "W" shape. In fact, the kaolinite doublet spectral feature is commonly used as a test of a spectrometer's spectral resolution – with sufficient resolution, the two minima are clearly separable as two distinct bands, but with a lower spectral resolution instrument, the two bands appear to merge into one band.

Carbonate minerals also have distinctive visible/near-infrared reflectance spectra due to the presence of a vibrational absorption feature near 2.3 μm (Figure 2.7d; also see siderite spectrum in Figure 2.7a). The exact wavelength position and shape of this band enable the identification of the specific type of carbonate. For example, in the dolomite ($CaMg(CO_3)_2$) spectrum in Figure 2.7d, the wavelength of the ~2.3 μm absorption feature is shifted to a slightly shorter wavelength than the equivalent feature in the calcite ($CaCO_3$) spectrum. This is because half of the Ca ions in the calcite have been replaced with Mg ions in the dolomite. The mass difference between these ions leads to slightly different vibrational frequencies for the two lattice structures, which in turn leads to spectral absorption features centered on slightly different wavelengths. It is worth mentioning that carbonates (and indeed many other minerals) also have diagnostic spectral absorption features at wavelengths longward of 2.5 μm. However, other than noting their existence, we do not discuss those features further because interpretation of spectral measurements made in the crossover region (Section 2.4) is rarely attempted.

2.6.3 Vibrational Features in Thermal Infrared Emissivity Spectra

Spectral remote sensing data acquired in the thermal infrared wavelength region are particularly useful for identifying silicate minerals. This is because the fundamental building block for silicates, the silica tetrahedron, has fundamental vibrational modes within itself and in its bonds with the rest of the silicate lattice with energies that fall in the thermal infrared portion of the electromagnetic spectrum. Figure 2.8 shows an example set of thermal infrared emissivity spectra for silicate minerals with different lattice structures, ranging from silica tetrahedra tightly bound in a framework silicate structure (quartz) to a loosely bound, isolated tetrahedra structure (olivine). The set of spectral features on the short-wavelength ends of these spectra are collectively and colloquially known as "the 10-μm silicate absorption feature," but in reality the

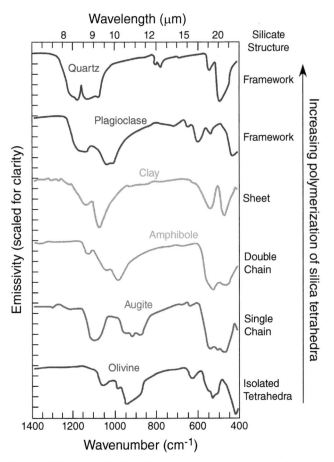

Figure 2.8 Thermal infrared emissivity spectra of example minerals with different silicate lattice structures, ranging from a framework silicate lattice structure to an isolated tetrahedral silicate lattice structure. The x-axis is plotted in both wavelength (top axis) and wavenumber (bottom axis), which is a form of reporting light frequencies often used by spectroscopists. Figure courtesy of P. R. Christensen.

wavelength position of this feature shifts from ~8.5 μm in framework silicates up to ~10.5 μm for the isolated tetrahedral lattice structure, with the wavelength of the feature strongly inverse-correlated with the degree of polymerization of the silica tetrahedra.

Minerals from other important groups, including carbonates, sulfates, and oxides, also have diagnostic absorption features in the thermal infrared portion of the spectrum.

2.6.4 Complicating Factors in Making Spectral Identifications

The observant student will have noticed that all of the preceding example spectra are of individual minerals, whereas most planetary surfaces that one might want to map with remote sensing consist of multiple compositions. How do spectra of multi-mineralic rocks compare to the spectra of the individual minerals that make up those rocks? The answer depends on the region of the spectrum in which one is working. Thermal infrared emissivity spectra of multi-mineralic rocks are very closely approximated by a weighted average of the spectra of the individual constituent minerals, in which the weightings are proportional to the abundances of each phase. If one has the thermal IR emissivity spectrum of an unknown multi-mineralic target and a library of spectra of individual minerals (including those in the target), a mathematical operation known as linear unmixing may be used to determine which minerals are present in the target, and in what abundances (Ramsey and Christensen, 1998). Knowing the abundances of constituent minerals from thermal IR spectra can be very useful in constraining potential rock types on the surface. Deriving abundances of constituent minerals from the spectrum of an unknown mixed target is much more difficult when working with reflectance spectra acquired in the visible/near-infrared. This is because the whole-rock spectrum is not well-approximated by a linear mixture of its constituent spectra. Determination of constituent mineral abundances from visible/near-infrared reflectance spectra is not impossible, but it does require use of computationally intensive light scattering models (e.g., Shkuratov et al., 1999).

We must also bear in mind that not all minerals have absorption features in all regions of the electromagnetic spectrum. Compositional "blind spots" can lead to spurious interpretations of spectral remote sensing data if this bias is not taken into account. For example, neither quartz nor (most types of) feldspar have spectral absorption features in the visible/near-infrared portion of the spectrum, so trying to map the composition of a granite outcrop with visible/near-infrared reflectance data will present significant difficulties. In some cases, combining visible/near-infrared reflectance data with thermal infrared emissivity data will eliminate relevant blind spots, since most minerals that lack spectral features in one region will have them in the other. However, some minerals lack features in both spectral ranges, making them exceptionally difficult to map. Halite is perhaps the best example of a common mineral that has no spectral features in the visible/near-infrared or the thermal infrared. The best one can do when attempting to compositionally map an area suspected to have extensive halite is to look for spectrally featureless regions that provide other contextual clues about candidate compositions – e.g., a spectrally featureless surface that is low-lying, flat, and embays a possible shoreline would be a candidate halite surface.

Finally, if the target surface is on a planet with an atmosphere, some surface spectral features may be obscured by overlapping spectral features of gases or aerosols in the atmosphere. For example, the Earth's atmosphere contains water vapor, which absorbs incoming sunlight near $1.4\,\mu m$ and $1.9\,\mu m$. Because light at these wavelengths does not make it through the atmosphere to illuminate the surface, no light at the same wavelengths is reflected off the surface. Reflectance spectra near these wavelengths can be very noisy, making it impossible to identify surface absorption features. Many remote sensing instruments are designed with these atmospheric absorptions in mind and do not acquire data in wavelengths where the surface is not illuminated (Section 2.7.4).

2.7 Remote Sensing Instrumentation and Observational Considerations

A fundamental challenge in mapping the compositions of planetary surfaces with spectral remote sensing techniques is that, typically, the data required are intrinsically three-dimensional. The basic data product used in this type of remote sensing analysis is called a **multispectral** or **hyperspectral image cube**. The distinction between these two types of products is simply the number of wavelengths they capture; to be considered "hyperspectral" the number of spectral channels must be sufficient to oversample the width of a typical spectral absorption feature. In practice, multispectral image cubes typically have no more than a dozen or so wavelengths, whereas a hyperspectral image cube normally has hundreds of wavelengths. Figure 2.9 presents a schematic representation of a multi- or hyperspectral image cube. This product can be thought of as a stack of images, all spatially registered to each other, taken at tens (for multispectral) or hundreds (for hyperspectral) of wavelengths. If a single pixel is extracted from the scene and the brightness of that pixel from each of the wavelengths in the stack is plotted, the result is the spectrum of the location of that pixel. The reason three-dimensional data products are challenging to acquire is that most detectors that operate in these spectral regions are only capable of collecting either one or two dimensions of the data at a time. As discussed below, different classes of instruments solve this problem with different approaches.

2.7.1 Framing Cameras

One of the most straightforward classes of instruments for acquiring multispectral image cubes is the **framing camera** (Figure 2.10a). The detector element for these instruments consists of a two-dimensional photosensitive array, not unlike those found in the cameras on mobile phones. The **dispersing element** (the portion of the

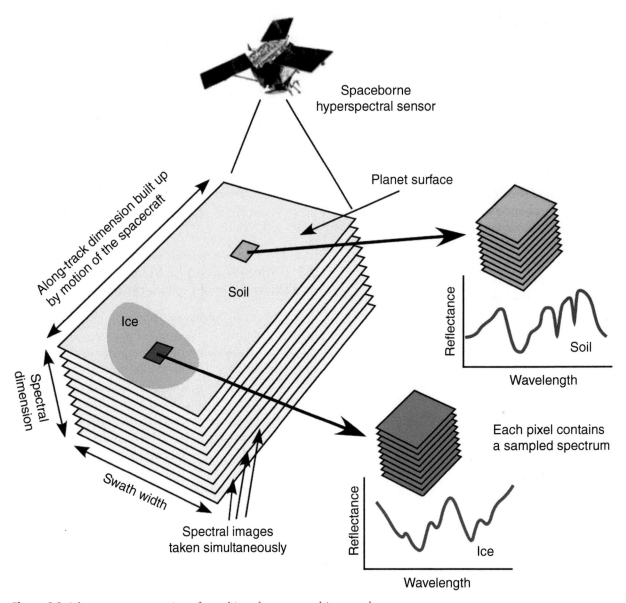

Figure 2.9 Schematic representation of a multi- or hyperspectral image cube.

instrument that breaks the light up into different wavelengths) in this class of instrument consists of a set of filters that can be placed in front of the imaging array, each of which only passes light within a certain wavelength range to the array. Typically, the filters are mounted in a ring on a wheel, which can be rotated between exposures to capture the scene through the different filters. Thus, when a framing camera is used to acquire a multispectral image cube, the two spatial dimensions of the cube are captured simultaneously with each exposure of the array through a given filter, but the wavelength dimension of the cube is acquired over time, as the filter wheel rotates through its positions.

The benefits of framing cameras include their relatively simple instrument design and the fact that all pixels at a given wavelength are acquired simultaneously, which can help when viewing conditions are variable. Also, the resultant multispectral image cubes have good geometric control for the two spatial dimensions.

The biggest drawback to using a framing camera approach is that they can only acquire a modest number of wavelengths, so while multispectral image cube collection is possible, hyperspectral image cube collection is not. Typically, these instruments are mounted on a satellite or an aircraft that moves down-track at a constant speed (Figure 2.10a). This sets up a competition between the time it takes for a given location on the ground to pass down-track from one end of the camera's field of view to the other and the time it takes for the camera to acquire images through all the filters on the filter wheel. If

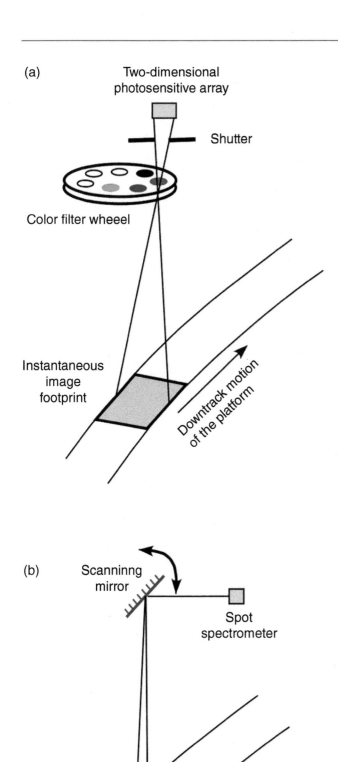

(a) Two-dimensional photosensitive array

Shutter

Color filter wheeel

Instantaneous image footprint

Downtrack motion of the platform

(b) Scanninng mirror

Spot spectrometer

Downtrack motion of the platform

the system does not have enough time to acquire images through each of the filters before the scene has passed underneath, then no pixel in the scene will have complete wavelength coverage. In practice, this typically limits the number of filters possible with a framing camera system on a moving platform to no more than approximately ten wavelength ranges, give or take. For framing cameras in fixed positions (e.g., on a rover mast), the number of filters is limited by other factors, such as the size and mass of the filter wheel, and how narrow a range of wavelengths the filters can be manufactured to transmit.

2.7.2 Scanning Systems

The simplest class of instrument for the collection of hyperspectral image cubes is called a **scanning system** (Figure 2.10b). In this approach, the detector element is a single "spot" spectrometer, and its sensing footprint on the ground is swept back and forth across the scene using a pivoting scan mirror. Down-track pointing is taken care of by the forward progress of the platform, be it an aircraft or a satellite. The dispersing element in this type of system can be a dispersion grating, a prism, or any other optical element used inside a spectrometer to break up light into its constituent wavelengths.

The benefits of scanning systems include the inherent simplicity of the detector itself (the spectrometer), and the fact that this type of detector makes it easy to acquire a large number of spectral channels. On the negative side, the constant motion of the detector's footprint, via a combination of the scanning mirror and down-track motion, means that the amount of time the detector spends on any given footprint (called "dwell time") is short. This means fewer photons can be collected, which in turn can limit the signal-to-noise ratio of the measurement (Section 2.7.5). Additionally, any uncertainties in the positioning of the spectrometer's footprint (e.g., as a result of turbulence on an aircraft platform) can reduce the geometric fidelity of the resultant image cube in its spatial dimensions. Finally, because the spectra composing the pixels of the image cube are collected one at a

Figure 2.10 (a) Schematic representation of a framing camera for acquiring multispectral image cubes. Images in different wavelength ranges are acquired through a set of color filters that may be rotated in front of the imaging array. (b) Schematic representation of a scanning system for acquisition of hyperspectral image cubes. The image cube is constructed by a spectrometer that records the spectrum of each pixel individually. A scanning mirror is used to sweep the footprint of the spectrometer back and forth in the cross-track direction, and the down-track motion of the platform is used to build successive rows of pixels.

time, in series, there can be issues with instrument calibration drifting or observing conditions changing (e.g., water vapor varying) between the beginning and end of image acquisition. While some early hyperspectral imaging experiments employed the scanning system approach, today it has been largely superseded by hyperspectral push-broom imaging.

2.7.3 Hyperspectral Push-Broom Imagers

Although they are more complex, **hyperspectral push-broom imagers** (Figure 2.11) offer the spectral resolution benefits of a scanning system with fewer of the downsides associated with low pixel dwell times. The detector element in this class of instrument is a two-dimensional photosensitive array, similar to that used in framing cameras. However, rather than have both dimensions of

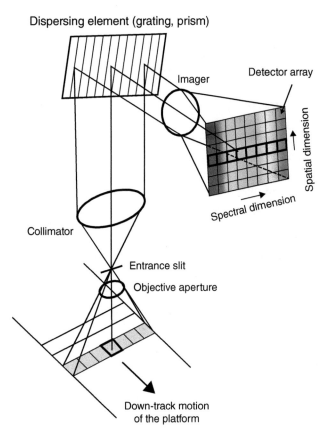

Figure 2.11 Schematic representation of a hyperspectral push-broom imager. Light from a cross-track strip of the surface passes through the entrance slit, is collimated, and then is projected onto a dispersing element that spreads the light out into its constituent wavelengths. Light leaving the dispersing element is projected onto a two-dimensional photosensitive array, with the cross-track spatial dimension mapped onto one direction of the array and the constituent wavelengths of light spread out in the other direction of the array. A hyperspectral image cube is produced one "vertical slice" at a time via the forward motion of the instrument platform.

the array mapped to spatial dimensions on the ground as a framing camera would, one dimension on the hyperspectral push-broom imager's array is used spatially for the cross-track dimension and the other dimension on the array is used to capture spectral information. Light coming into the instrument must pass through an entrance slit oriented in the cross-track direction. This restricts the instrument to "seeing" only one cross-track row of pixels at a time. The long, narrow beam of light is passed to a dispersing element (often a diffraction grating) that breaks the light down into its component wavelengths. The dispersion element is oriented in such a way that the wavelengths of light are spread out perpendicular to the long axis of the slit. Light coming out of the dispersion element is projected onto the detector array with cross-track pixels spanning one dimension of the array, and the constituent wavelengths of light for each cross-track pixel spanning the other dimension of the array. Thus, in any given moment, complete spectra are acquired for every pixel in a cross-track row of the image cube, and successive rows of pixels are acquired by the instrument platform moving down-track and taking new exposures (which explains the "push-broom" moniker). Essentially, this class of instrument builds the three dimensions of a hyperspectral image cube one spatial row of spectra at a time, something like assembling a loaf of bread by stacking slices next to each other

While hyperspectral push-broom imagers are significantly more complex to design and build than framing cameras or scanning systems, they enjoy several advantages. First, they have no moving parts, which enhances reliability – a key quality in an instrument that must operate without human intervention on a robotic spacecraft. Hyperspectral push-broom imagers also offer longer dwell times per pixel than comparable scanning systems, which improves the signal-to-noise ratio, and they can easily acquire far more wavelength channels than a framing camera that relies on a filter wheel. Geometric fidelity is good in the cross-track direction, though turbulence and/or other factors can still introduce geometric distortion between image rows.

2.7.4 Band Placement and Atmospheric Transmission

Most hyperspectral instruments have their spectral channels spaced at equal and contiguous intervals over the entire spectral range (e.g., visible/near-infrared) within which they operate, and it is understood that some of the channels will be useless for remote sensing of the surface because certain wavelengths of light are absorbed or scattered by the atmosphere. This is acceptable because the channels that are rendered useless by the atmosphere are a small fraction of the hundreds of spectral channels these instruments acquire. Multispectral instruments are different because the number of spectral bands is limited and their positions and widths must be chosen carefully for the science the experiment is meant to accomplish.

Every planetary atmosphere has a unique combination of gases and aerosols, many of which absorb or scatter (or both) certain wavelengths of light. On Earth, the primary atmospheric components that do this are H_2O, CO_2, and O_3. A "spectral map" showing the net effect of these components on light traveling through the Earth's atmosphere is shown in the top portion of Figure 2.12. The combined effects of absorption and scattering in the atmosphere are called **extinction** – a term that refers to the total amount of light at any given wavelength that is removed from a beam passing through the atmosphere. In this plot, wavelengths at which light passes through the atmosphere relatively unhindered have values near 0 percent extinction; these spectral regions are referred to as "telluric windows." Wavelengths where no light makes it to the surface have values near 100 percent extinction. Note, for example, the 100 percent values at 1.4 μm and 1.9 μm – these are the water vapor bands mentioned in Section 2.6.4. Similar maps have been made for

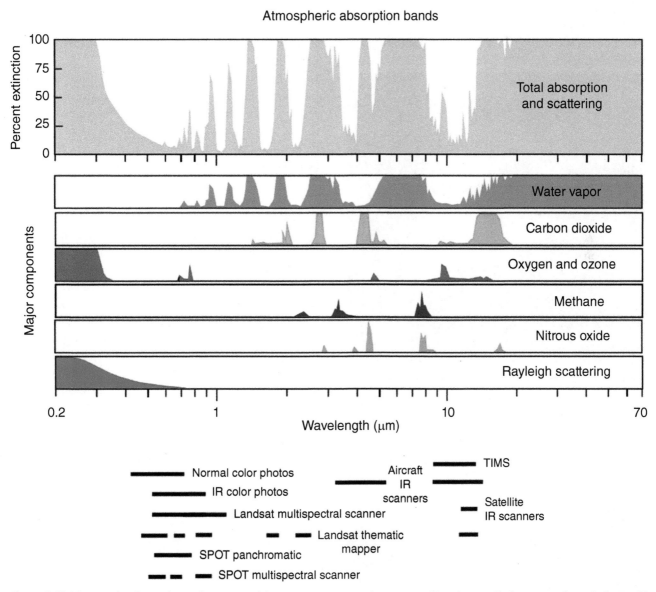

Figure 2.12 The top plot shows the total extinction (absorption + scattering) experienced by a beam of light passing through the Earth's atmosphere, expressed as a percentage of the initial intensity at each wavelength that is removed from the beam. Below this are plots showing the individual contributions to this total from different components of the atmosphere. At the bottom, horizontal bars show the wavelength ranges used by several well-known terrestrial multi- or hyperspectral remote sensing instruments. Note that these instruments tend to avoid using wavelengths where no light makes it through the atmosphere.

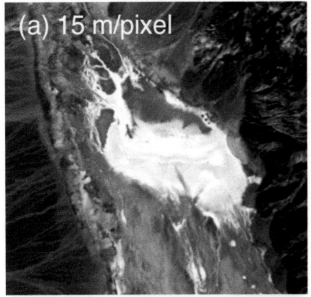

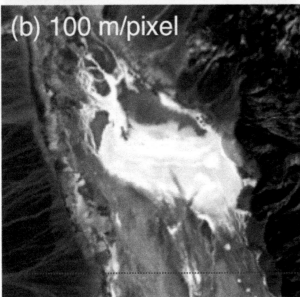

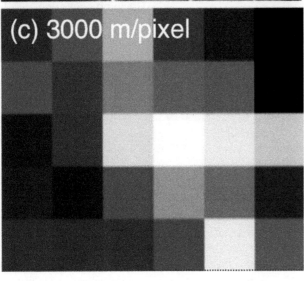

atmospheric extinction on other planets. They are not the same as the Earth's because of the different atmospheric components present and different pressure and temperature conditions.

Multispectral instruments designed for mapping surface compositions avoid wavelengths with high atmospheric extinction because little, if any, light from the ground at these wavelengths would make it through the atmosphere and into the instrument. The bottom portion of Figure 2.12 shows the wavelength positions of the spectral bands used by several well-known terrestrial multispectral remote sensing instruments. Note that the positions of these bands fall within the telluric windows and avoid spectral ranges with high extinction.

2.7.5 Other Instrumental/Experimental Considerations

In addition to the class of instrument and the wavelength positions and widths of the spectral channels, there are several other top-level instrumental specifications that should be considered when choosing a remote sensing dataset for a particular scientific goal. These can broadly be grouped into spatial, spectral, radiometric, and "logistical" categories.

Spatial characteristics of a remote sensing instrument include the field of regard (the spatial dimensions of the full scene), the **spatial resolution** (the scale of the smallest feature resolvable by the instrument), and the total image size in terms of rows and columns of pixels. The first two of these are actually determined by a combination of the instrument itself and the range between the instrument and the target – the same instrument raised to twice the height above the ground will have a field of regard that is twice as large in both dimensions and a spatial resolution that is two times worse in each dimension. Obviously, for most applications, higher resolution is desirable, but sometimes competing factors, such as a need to minimize the size of the instrument's optics or a desire for greater global coverage, will weigh against this. Figure 2.13 shows a false-color composite made from the same multispectral image of Badwater Basin, an evaporitic playa in Death

Figure 2.13 False-color composite image of Badwater Basin, an evaporite playa in Death Valley, seen (a) at its native resolution of 15 m/pixel, (b) at 100 m/pixel, and (c) at 3000 m/pixel. These resolutions roughly correspond to those of three multi- and hyperspectral instruments that have been sent to map the surface of Mars from orbit. Terrestrial analog sites, such as this one, inform us of the necessary capabilities for remote sensing instruments we send to other planets.

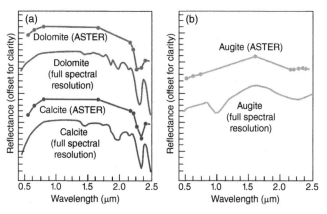

Figure 2.14 Example (a) carbonate spectra and (b) an augite spectrum seen at hyperspectral (5 nm) resolution and through the multispectral bands of ASTER (spectra with dots indicating band centers).

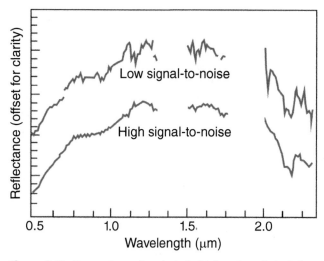

Figure 2.15 Comparison of a relatively high and a relatively low signal-to-noise spectrum of the same target, kaolinite. In the low signal-to-noise spectrum, the magnitude of the noise is comparable to that of the kaolinite spectral doublet near 2.2 μm, making the presence of the mineral difficult to identify. Missing data around 1.4 μm and 1.9 μm are caused by atmospheric absorptions.

Valley, California, at three different representative spatial resolutions. The three resolutions were chosen to match those of instruments that have been sent to Mars on orbiters to look for similar types of geologic features.

Spectral characteristics of a multi- or hyperspectral remote sensing instrument include the spectral range within which the instrument operates (visible/near-infrared, thermal infrared, etc.), the **spectral resolution**, and the number of spectral bands. These parameters can "make or break" an instrument's utility for a given scientific goal. Examples of visible/near-infrared spectra of two carbonate minerals (Figure 2.14a) and the mineral augite (Figure 2.14b) at 5 nm spectral resolution (comparable to many hyperspectral instruments) are also shown as they would be observed by the Advanced Spaceborne Thermal Emission and Reflection Radiometer (ASTER), a highly successful satellite-based terrestrial multispectral imager with nine spectral bands in the visible/near-infrared portion of the spectrum. Notice how the clever placement of two of ASTER's spectral bands, with one centered at 2.330 μm and another at 2.395 μm, allows for discrimination between the dolomite and calcite spectra. The dolomite spectrum has its absorption feature at a slightly shorter wavelength than the calcite spectrum, with the result being that the two bands show comparably low reflectance for the dolomite. By contrast, the calcite spectrum displays significantly less absorption in the shorter-wavelength band than the longer-wavelength band. Because of this, ASTER is well suited to mapping carbonate deposits where some diagenesis converting limestone to dolostone has occurred. The augite spectrum illustrates a counterpoint, which is that multispectral instruments can have "blind spots" for important spectral features. The 5 nm resolution spectrum of augite shows the classic pyroxene "W" spectral shape, with features near 1.0 μm

and 2.0 μm. However, the absence of any ASTER spectral bands near 1.0 μm means that the augite 1.0 μm feature cannot be detected by that instrument.

Radiometric characteristics include anything related to the instrument's ability to measure the amount of light it receives. One of the most important radiometric parameters is the signal-to-noise ratio of the instrument. All instruments have some amount of noise in the data they report, and the "signal" (i.e., the true spectrum of the target) must be significantly stronger than this noise if spectral features are to be recognizable. Figure 2.15 compares two kaolinite spectra, one a "clean" spectrum with high signal-to-noise ratio and one a "dirty" spectrum with poor signal-to-noise ratio. The key spectral feature that marks this target as kaolinite is the aforementioned doublet absorption feature near 2.2 μm. Note that it is easy to pick this feature out in the clean spectrum, but in the dirty spectrum the noise is of comparable magnitude to the absorption feature, making it difficult to tell whether kaolinite is actually present. Many different strategies are available to instrument designers for improving the signal-to-noise ratio. Cooling the detector and associated electronics can bring down the noise levels inherent to the system, but this advantage comes at the expense of weight, size, and complexity. Many options exist for increasing the signal (i.e., raising the number of photons hitting the detector elements), such as using wider spectral channels, using a larger aperture, and/or increasing the dwell time.

BOX 2.2 SPECTRAL CLASSIFICATION OF ASTEROIDS

Satellites, aircraft, and drones are not the only types of platforms employed for remote sensing experiments – Earth-based telescopes also qualify! Spectrometers mounted on telescopes have been used since the early nineteenth century to attempt to characterize the composition of the Sun, planets, and other heavenly objects. Today, telescopic visible/near-infrared reflectance spectra provide the basis for remote sensing classification of asteroids. These studies have a long history, and the current taxonomies have been well described in a variety of publications (e.g., Tholen and Barucci, 1989; Bus et al., 2002). Letters are used to identify the various asteroid classes, which are qualitatively distinguished based on spectral shapes, as shown in Figure 2.16a.

The absorption bands that dominate these spectra are mostly from iron-bearing minerals like pyroxene and olivine (e.g., the 1 and 2 µm bands in V and S asteroid spectra (Figure 2.16a)), although OH vibrational features due to phyllosilicates play a role in C asteroid spectra. The distinctive slope in M asteroids (Figure 2.16a) reflects spectral reddening caused by iron metal. Other phases are not usually abundant enough to be recognized in asteroid spectra, although phases such as carbonates, organic matter, and ice have been identified in visible/near-infrared spectra of Ceres from the orbiting *Dawn* spacecraft.

Further insights into the mineralogy of asteroids can be gleaned from comparisons with laboratory spectra of meteorites (Figure 2.16b). The meteorite classes in this figure are igneous samples (howardite, eucrite, brachinite, aubrite), iron and stony iron (pallasite) meteorites, or chondrites, which will be explained in Box 4.1. Some of the comparisons are good enough to correlate meteorite classes to asteroid classes, although a specific parent body for meteorites can only be identified confidently for Vesta, the largest V class asteroid.

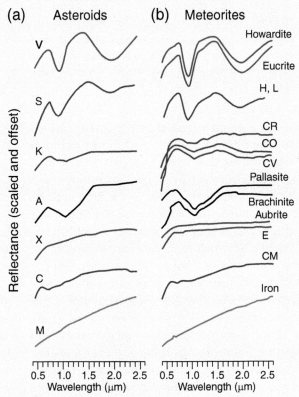

Figure 2.16 (a) Telescopic visible/near-infrared spectra of asteroids are used to classify them; asteroid taxonomy by Bus et al. (2002). Comparison with laboratory spectra of meteorites (b) provides a means of assessing likely asteroidal parent bodies.

Finally, logistical considerations often play a key role in determining which remote sensing dataset is most appropriate for a given scientific goal. For example, datasets from instruments deployed on aircraft often have relatively limited global coverage, whereas those from instruments on satellites typically have access to nearly the entire planetary surface. There are trade-offs, though – because satellites "fly" higher, the same instrument deployed in space will have significantly worse spatial resolution than it would from an aircraft. Aircraft-based instruments must also deal with the geometric distortions that come from turbulence in flight, whereas satellites typically provide rock-steady observation platforms. On the other hand, aircraft-based instruments are easier to maintain and upgrade, whereas many satellite-based instruments can never be serviced after they are deployed. Recently, drones have also started to become available as platforms for collecting remote sensing data. Drones offer huge savings in deployment costs and, because they tend to fly low, often offer vastly superior spatial resolution over other platforms. The major limitation on drone-based remote sensing is instrument mass relative to the payload capacity of the drone, but this is gradually becoming less of a problem as instruments are miniaturized. Another logistical consideration for potential users of remote sensing data is cost. Many government-funded remote sensing datasets (including those from other planets) are free, but there are also commercial providers of terrestrial remote sensing data that charge thousands of dollars per scene.

2.8 Analysis of Multi- and Hyperspectral Image Cubes

As mentioned at the beginning of the chapter, multi- and hyperspectral instruments generate huge volumes of data. Reducing these data to a useful form and then extracting the information one wants from them can be a non-trivial exercise. Typically, the first step in this process is to perform an atmospheric correction on the data. For some datasets, this is already done for the users by the team that operates the instrument, but in other cases the users must obtain the appropriate computer code and do it themselves. The purpose of the atmospheric correction is to turn the "radiance at sensor" values reported by the instrument into "radiance leaving the ground" – i.e., the spectrum one would measure if the instrument had no interaction with the atmosphere after leaving the ground on the way to the instrument. For visible/near-infrared data, this step often also incorporates a normalization to a modeled spectrum of incident sunlight on the surface, thereby producing an image cube in units of reflectance (via Equation 2.6). For thermal infrared data, the atmospherically corrected radiance leaving the surface is divided by the spectrum of a blackbody at the same temperature as the surface to produce an image cube in units of emissivity (via Equation 2.5).

Once the data have been corrected and recast in either reflectance or emissivity values, the remaining analysis steps depend on what is needed from the data. In some cases, there is a particular mineral of interest (e.g., tracer minerals for ore deposits), and the goal is simply to map the geographic occurrence and/or abundance of that mineral. In other cases, the goal may be to map the regional geology of a new site, with compositional maps of all detectable minerals used to complement ground surveys. In the following discussion, descriptions of analysis methods refer to working with reflectance values, but the same methods work if the image cube is of emissivity values from thermal infrared observations.

If the goal is to map a particular mineral of interest, one of the simplest approaches is to devise a **spectral index**, which is a parameter created by mathematically manipulating the reflectance values for each pixel in such a way that the results highlight the presence of the mineral. The most basic type of spectral index is a band ratio. Band ratios are created by dividing the reflectance from a band at a wavelength on the continuum (away from any absorption features) by the reflectance from a band centered on a spectral absorption feature that is diagnostic of the mineral of interest. This ratio goes up if the absorption feature is present in a given pixel's spectrum, and the deeper the absorption feature, the higher the ratio value. A new image is created in which the value of each pixel is given by this ratio. An example is a project to map the occurrence of evaporitic calcite in Badwater Basin in Death Valley, California (Figure 2.17a). A band ratio map of calcite is shown in Figure 2.17b. Calcite has a strong absorption feature at $2.35\,\mu m$ (green arrow in Figure 2.17c), and in most locations within the scene no minerals have spectral absorptions at $2.10\,\mu m$ (i.e., this wavelength is on the continuum). A band ratio image, created by dividing the reflectance in the continuum band by the reflectance in the calcite absorption feature band, reveals that calcite is deposited all around the outer margin of the playa in a classic "bathtub ring" evaporite facies pattern.

If the goal of a remote sensing project is to make a more general map of all detectable compositions, a much more sophisticated approach that takes advantage of all the information available in the image cube is necessary. One of the most commonly employed approaches is to use an automated cluster analysis on the spectral shapes displayed in the scene.

Each pixel in a multi- or hyperspectral image cube contains a spectrum. One way of representing each of these spectra is to treat them as points in "n-dimensional" plot space, where n is the number of spectral bands. In the simple case of a regular color digital camera image, each pixel has a red, green, and blue brightness value. Thus, such an image is essentially a three-band multispectral image cube, and each pixel has a three-point RGB spectrum. The brightness in each of the three color channels may be used as x, y, and z coordinates for points representing the spectra of each pixel in a three-dimensional plot space. Points from pixels in the image

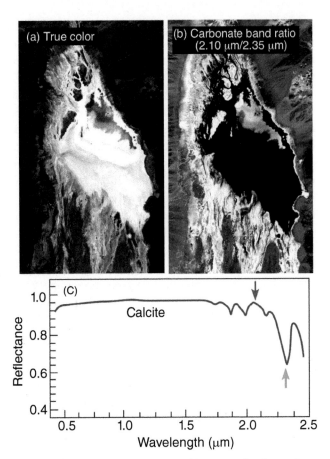

Figure 2.17 (a) Approximate true-color image of Badwater Basin in Death Valley, California, created from red, green, and blue spectral bands in a hyperspectral image cube acquired by the Airborne Visible/Infrared Imaging Spectrometer (AVIRIS). (b) A band ratio image, created by dividing the reflectance at 2.10 μm by the reflectance at 2.35 μm for each pixel, highlights the presence of calcite around the margin of the playa. (c) Identification is made possible because few compositions in the scene have an absorption feature at 2.10 μm (red arrow), whereas calcite has a deep absorption feature at 2.35 μm (green arrow).

that have similar colors will cluster together in the plot. The same concept may be used with any three bands from a multispectral scene (Figure 2.18).

With the help of powerful, number-crunching computers, this concept is easily extended to image cubes with more than three bands. While humans have a hard time envisioning plots with more than three dimensions, computers with sufficient processing power have no trouble thinking in terms of plots with hundreds of dimensions. The more spectral channels in an image cube, the more dimensions used to plot the points, which tends to provide more nuanced clustering of the resulting data cloud

in n-dimensional space. Computer algorithms can perform statistical analyses on this clustering to determine how many different spectral shapes (statistically separable clusters) are present in the scene and derive a representative spectrum for each of the clusters. The analyst then compares these so-called **end member spectra** from the scene to libraries of spectra from samples of known compositions measured in the laboratory, so as to make compositional assignments to the end member spectra. Finally, another computer algorithm can examine every pixel in the scene and classify it, based on the shape of its spectrum, as belonging to one of the end member spectral classes, resulting in a spectral classification map (e.g., Figure 2.19). Other mapping algorithms may also be applied, depending on the site and the science goals of the project. For example, if it is thought that many of the pixels in the scene contain mixtures of spectral end members, a spectral unmixing program may be used to estimate the abundances of every compositional end member in each pixel. Compositional maps made from multi- or hyperspectral data may also be combined with other data to highlight relationships of interest. For example, draping a color spectral classification map over a three-dimensional digital topographic model can sometimes highlight structural relationships between lithologic units that aren't obvious in flat, two-dimensional map representations.

2.9 Ground Truthing

Spectral classification maps from remote sensing data, such as the one in Figure 2.19, present patches of different colors representing different compositions, making them visually similar, at least at first glance, to a geologic map. However, the two types of maps should not be confused. Geologic maps make use of field observations that go far beyond compositional identifications, such as textures and structural relationships (see Section 3.2 for additional details). Also, spectral classification maps only identify compositions exposed in the very near-surface (down to a depth equal to a few times the wavelength being used to observe), so any thin veneer of material that might be easily ignored as exogenous by the geologic field mapper will dominate the classification of affected pixels. Also, as previously mentioned, some highly important minerals lack spectral features in large portions of the spectrum, leaving compositional "blind spots" in classification maps that would never be missed by someone mapping from the

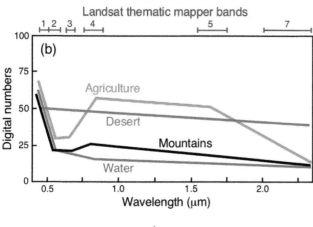

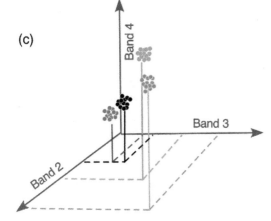

Figure 2.18 (a) A single band image from the Landsat Thematic Mapper (TM) multispectral imager, taken over the Salton Sea (a) and Imperial Valley of California, which includes several distinct surface spectral types. (b) Seven-band spectra extracted from the multispectral scene for each of the major surface compositions. (c) An *n*-dimensional plot of pixels from each of the four surface types – in this figure, only three bands (TM bands 2, 3, and 4) are used as coordinates in a three-axis plot, but the computer algorithm can work with all of the bands in a seven-dimensional statistical analysis.

ground – many of the black unclassified pixels in Figure 2.19 contain halite, for example. These potential pitfalls, and others, dictate that every remote sensing project should be accompanied by a ground-truthing campaign if at all possible. In the context of planetary exploration, this can be costly, but it is important. In fact, it is fair to say that the twin *Mars Exploration Rovers*, Opportunity and Spirit, and the *Mars Science Laboratory* rover, Curiosity, are missions sent to conduct ground-truthing campaigns for hypotheses about regional water-related geology that were generated from analyses of orbital remote sensing experiments on previous missions (described in Chapter 17).

Most ground-truthing campaigns serve three distinct purposes: *calibration*, *validation*, and *augmentation* of the remote sensing analyses. "Calibration" refers to collecting data in the field that will improve the fidelity of the remote sensing data. For example, on Earth a portable field spectrometer may be used to collect spectra at locations within the scene from the ground that are compositionally homogeneous and large compared to the scale of a single pixel in the remote scene. Because there is virtually no atmosphere between the field spectrometer and the ground, these field spectra may be used to "anchor" the atmospheric correction of the remote scene to surface spectra acquired *in situ* at key locations. This then improves the quality of spectra from everywhere else in the scene, perhaps making it easier to assign compositional identifications to spectral end members.

"Validation" refers to tasks aimed at checking the veracity of interpretations made in the remote analysis. For example, samples of surface material from locations displaying exceptionally pure spectral end members may be collected and returned to the laboratory for petrographic and mineralogic analyses to see if the compositional assignments made by comparing the end member spectra to library spectra were correct. Such work will also tend to reveal compositions missed in the remote-sensing analysis because of spectral "blind spots."

"Augmentation" refers to new data collected in the field that provide geologic insights that would be unavailable from the remote sensing analysis alone. For example, this could include observations of structural relationships, small-scale textures, and index fossils. Finally, it is worth noting that while the ground-truthing campaign will improve the quality of the remote sensing analysis, the benefits run in both directions. Performing a preliminary analysis of remote sensing data for a new field area

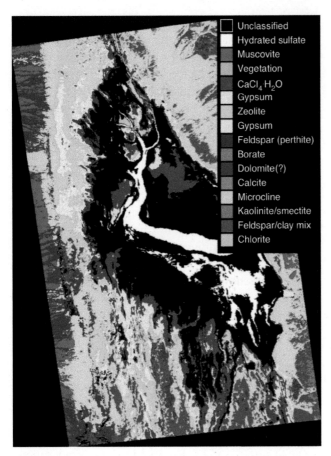

Unclassified
Hydrated sulfate
Muscovite
Vegetation
$CaCl_4 H_2O$
Gypsum
Zeolite
Gypsum
Feldspar (perthite)
Borate
Dolomite(?)
Calcite
Microcline
Kaolinite/smectite
Feldspar/clay mix
Chlorite

Figure 2.19 Spectral classification map for the AVIRIS scene of Badwater Basin shown in Figure 2.17. Reflectance spectra from every pixel, taken in over 200 spectral channels, were processed in an *n*-dimensional analysis that identified 15 different spectral end members. Mean spectra from each of the end members were compared to library spectra of known samples to make the compositional identifications listed in the color key. Then, the spectra associated with every pixel in the scene were compared to the 15 end member spectra in an attempt to classify them and assign colors for the map. In some cases, the candidate pixel had no close matches and was left as unclassified (black pixels). Ground truthing revealed that most unclassified pixels near the middle of the scene were dominated by halite, which is unsurprising considering halite has no visible/near-infrared absorption features.

will inevitably enhance the efficiency of any ground-based expedition by focusing attention on locations of key contacts and other high-value targets.

2.10 Nuclear Remote Sensing

Gamma ray and neutron remote sensing (collectively referred to here as "nuclear remote sensing") experiments are probably the second most common type of remote-sensing approach used in planetary exploration, after the optical (visible/near-infrared and thermal infrared) techniques just described. This may come as a surprise to some, because these techniques have very limited utility in remote sensing studies of the Earth. However, on planetary bodies with little or no atmosphere, there is a wealth of geologic information to be gained from them. As discussed in Section 2.5, gamma rays carry the signatures of energetic transitions in the nuclei of atoms, so gamma ray spectra of planetary surfaces allow one to map elemental and isotopic compositions. And, as discussed in Section 2.10.2, neutron spectroscopy of planetary surfaces is particularly sensitive to the presence of hydrogen.

As with other types of electromagnetic remote sensing, the energy of each photon (given by the Planck–Einstein relation, Equation 2.1) emitted by the surface is the most important quantity to consider in gamma ray spectroscopy. For neutron remote sensing, the analogous quantity is the kinetic energy of the neutron, $E_K = \frac{1}{2}mv^2$, where m is the mass of the neutron and v is the speed at which it is traveling. Energies for gamma rays and neutrons are usually expressed in units of electron-volts (eV, with $1\ eV = 1.60218 \times 10^{-19}$ J), and typically spectra are plotted using energy on the x-axis rather than wavelength. Typical energies of interest can range from so-called "thermal" energies of ~0.02 eV, all the way up to "relativistic" energies of ~10 GeV. A critical concept for all nuclear transitions is that the total energy of the system is conserved – in other words, the sum of the energies of all the reactants must equal the sum of the energies of all the products. For example, consider this reaction:

$$^{16}O(n, n\gamma)^{16}O$$

In this form of nuclear reaction notation, the reactants appear on the left side of the comma and the products appear on the right. So this reaction denotes the interaction of an ^{16}O target nucleus with an incoming neutron, with the result being the ejection of a neutron and emission of a gamma ray. This reaction can only work if the kinetic energy of the inbound neutron is equal to the sum of the kinetic energy of the outbound neutron plus the energy of the emitted gamma ray.

2.10.1 Gamma Rays

The ultimate stimulus for the production of most gamma rays in the surfaces of thin- or no-atmosphere planetary bodies is the bombardment of the surface by **galactic cosmic rays** (GCRs) from space. Figure 2.20 illustrates the process. Galactic cosmic rays are mostly highly energetic **protons** (*not* **ph**otons) that are emitted by the Sun or by other sources in the galaxy. On planets with relatively thick atmospheres (including the Earth),

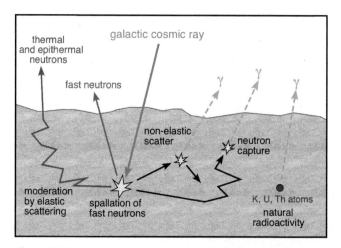

Figure 2.20 Production of gamma rays and neutrons in the regolith of a thin- or no-atmosphere planet by galactic cosmic rays and natural radioactive decay.

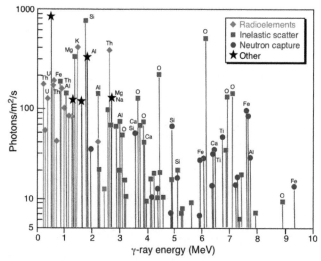

Figure 2.21 A spectrum of gamma rays emitted by the lunar surface. The type of nuclear reaction that gives rise to the photons in each emission line is denoted by the plot symbol at the apex of the line. This spectrum was generated by a computer model; actual gamma ray spectra from planetary bodies also include background counts.

almost all GCRs are absorbed in the atmosphere and never make it to the surface. However, on planetary bodies with thin atmospheres (such as Mars) or no atmosphere (such as the Moon, Mercury, or asteroids), GCRs are able to reach the surface and interact with the nuclei there. A typical rate of bombardment for GCRs on an airless planetary body is around 1.5 particles per square centimeter every second. Typical penetration depths for GCRs in planetary regoliths are on the order of tens of centimeters, so the compositional information retrieved from nuclear remote sensing techniques represents a significantly deeper surface layer than that obtained from, for example, visible/near-infrared or thermal infrared remote sensing.

When a GCR impinges upon a nucleus in the regolith, the nucleus undergoes a process called **spallation**, in which it ejects an average of about nine high-energy (i.e., fast-moving) neutrons. These neutrons "rattle around" in the surface regolith, interacting with the nuclei there. When interactions with these neutrons excite target nuclei, those nuclei can revert to their lower-energy states by giving up a gamma ray photon with an energy equal to the energy difference between the excited and de-excited states of the nucleus. Sometimes the nuclear excitation happens because the nucleus actually absorbs the incident neutron in a process called **neutron capture**. Geologically relevant elements that typically produce gamma rays through neutron capture include H, Al, Si, Cl, Ca, Ti, Cr, Fe, and Ni. In other cases, the interaction only serves to slow the neutron down by transferring a portion of its kinetic energy to the nucleus, in which case the process is referred to as **non-elastic scatter**. Geologically relevant elements that

typically produce gamma rays through non-elastic scattering include C, O, Mg, Al, Si, S, Ca, Ti, and Fe. If the target nucleus remains the same isotope after a non-elastic scattering event, the event is considered a subcategory of non-elastic scattering called "inelastic scatter." If, on the other hand, the target nucleus is transmuted into a different element or different isotope as a result of the event, that reaction is simply referred to as "other non-elastic scattering."

An additional process that produces gamma rays is the decay of naturally occurring radioactive isotopes. However U, K, and Th are the only elements with naturally occurring radioactive isotopes that are present in sufficient quantities to be useful for gamma ray remote sensing of most planetary surfaces. In fact, these are the only elements that may be mapped in terrestrial applications of gamma ray remote sensing because they do not rely on the impingement of GCRs in the surface to produce gamma rays.

Unlike the vibrational and electronic transitions that give rise to spectral features in the visible and infrared, which are spread out over a range of wavelengths, nuclear reactions that produce gamma rays have very specific energies. Thus, the spectrum of gamma rays leaving a planetary surface (Figure 2.21) looks more like a "forest" of gamma ray emission lines with different intensities, rather than a continuum with broad absorptions. The abundance of each element in the regolith may be

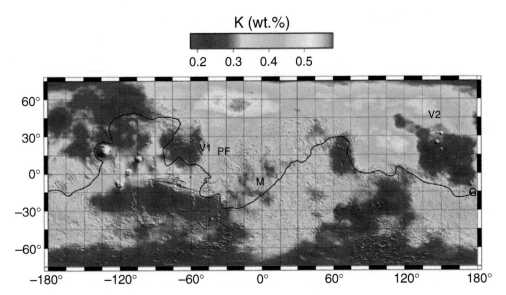

Figure 2.22 Global potassium abundance map of Mars, obtained by the Gamma Ray Spectrometer (GRS) instrument on the *Mars Odyssey* orbiter. The letters refer to spacecraft landing sites: V = *Viking*; PF = *Pathfinder*; M = *Meridiani*; and G = *Gusev*. From Boynton et al. (2008), with permission.

determined from the intensity of that element's gamma ray lines using computer models. An example of the type of map that may be produced for each element appears in Figure 2.22.

2.10.2 Neutrons

Not all neutrons produced by GCR-induced spallation are absorbed by nuclei in the regolith. Some fraction of the neutrons actually "leak" out of the surface after many interactions with nuclei in the regolith. Collisions that do not convert kinetic energy to any other form (and thus do not result in the production of a gamma ray) are called **elastic collisions**. The energies of leaked neutrons may still be relatively high, or in some cases elastic collisions with nuclei in the regolith may act to slow down, or moderate their energies. The key factor that determines the energetic outcome for the neutrons in these collisions is the masses of the nuclei they collide with before leaking out. If the regolith is almost entirely composed of nuclei that are significantly more massive than a neutron, the neutrons will give up very little of their kinetic energy with each collision and will still be moving quickly as they leak from the surface. A macroscopic analogy would be a table tennis ball shot at a bowling ball; because of the significant mass difference between the projectile and the target, the target recoils very little and the projectile bounces off, going almost as fast as immediately prior to contact. On the other hand, if a substantial fraction of

the regolith is composed of nuclei with masses comparable to the mass of a neutron, the neutron will give up an average of half of its kinetic energy with each collision, and by the time it leaks out of the surface it will be slowed (moderated) significantly. A macroscopic analogy for this would be a cue ball shot into a collection of other pool balls. Because the projectile and the target have the same mass, there is significant recoil of the target that takes away kinetic energy from the projectile, and after a few collisions the projectile and targets are all moving at similarly slow speeds.

Of course, only one type of nucleus – hydrogen – has a mass comparable to the mass of a neutron. The next-most massive elements are helium and lithium, which have masses four times and seven times larger than the mass of a neutron, respectively. Thus, the energies of neutrons leaking from a planetary surface are very sensitive to the abundance of hydrogen in the surface because hydrogen is, by far, the most efficient moderator of neutron energies. Hydrogen abundances derived from neutron remote sensing are usually expressed in terms of "water-equivalent hydrogen" (WEH), which is the percentage (by weight) abundance of water that would be present if all of the detected hydrogen were hosted in H_2O. It is important to recognize that neutron remote sensing is not sensitive to the host molecule for the hydrogen, so a given WEH value may not literally mean there is that

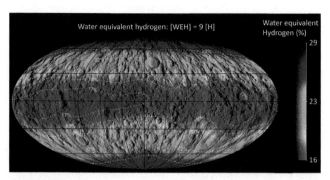

Water equivalent hydrogen: [WEH] = 9 [H]

Water equivalent Hydrogen (%)

29

23

16

Figure 2.23 Map of hydrogen concentration, expressed as water-equivalent hydrogen, on dwarf planet Ceres, obtained from neutron measurements made by the *Dawn* orbiter. Adapted from Prettyman et al. (2017).

percentage of water present in the regolith – it may also be present in other hosts, such as hydroxyl ions tied up in clay minerals. The more hydrogen present, the more neutron moderation takes place, up to a WEH value of around 20 wt%. Above that abundance, the method saturates because all of the neutrons have already been fully moderated. Figure 2.23 shows an example of hydrogen mapped on the dwarf planet Ceres using data from the Gamma Ray and Neutron Detector (GRaND) experiment on the *Dawn* orbiter (Prettyman et al., 2017). The WEH values increase toward the poles because ice approaches the surface at high latitudes.

2.10.3 Observational Considerations in Nuclear Remote Sensing

Nuclear remote sensing differs from optical remote sensing in more ways than just the type of compositional information it can provide. In optical remote sensing, photons are typically very abundant, allowing for good signal-to-noise characteristics in a single exposure. By contrast, gamma rays are emitted relatively slowly, and each gamma ray photon is detected by the instrument as a single event or "count." The only way to obtain enough photons to provide useful levels of uncertainty in elemental abundances derived from the measurements is to count gamma rays for a long time. Of course, this is complicated if the instrument is on a moving platform, such as an orbiter. Unless one simply wishes to obtain a globally averaged abundance for the element of interest, the locations over which each gamma ray is detected must be retained, and over the course of many, many orbits, enough photons will have been counted over a given

location to determine the abundance of the associated element. For some elements with strong gamma ray lines, useful numbers of counts may be obtained in a matter of weeks or months, but for elements with weaker lines, it can take years of collecting counts in orbit to make a useful map.

Another oddity about nuclear remote sensing is that most of the detector types available are so-called "4π steradian detectors," which means they count gamma rays (and/or neutrons) hitting the detector from any direction and cannot discern the direction from which the count came. This fundamentally limits the spatial resolution of an elemental map that can be made from such an instrument mounted on an orbiter because the gamma rays (or neutrons) counted at any given position in orbit could have originated anywhere from the part of the planet's surface directly under the spacecraft all the way to the local horizon. It happens to work out that the spatial resolution of maps that can be made from orbit is roughly comparable to the height of the orbit above the surface, which is why many of the maps made with nuclear remote sensing look so "fuzzy." The easiest way to improve upon this situation is to bring the detector much closer to the surface, and in fact this is exactly the strategy employed by the neutron detector on the Curiosity rover, which has a horizontal spatial resolution of about 1 m in the profiles of subsurface hydrogen it makes along the rover's traverse route.

2.11 Radar Remote Sensing

All of the methods discussed in depth so far in this chapter could be characterized as **passive remote sensing** techniques – i.e., they rely on naturally occurring photons (or neutrons) that are emitted by or reflected off of a planetary surface. By contrast, an **active remote sensing** technique is one in which the source of illumination is provided by the experiment. The most common type of active remote sensing in planetary exploration is radar.

Originally developed in the early twentieth century for military applications, RADAR (as it was originally written) is an acronym derived from RAdio Direction And Ranging. The basic method of radar is that a transmitter is used to generate radio waves (i.e., very long-wavelength photons), which are directed at a remote target by an antenna. These waves reflect off the target and are returned to the antenna. The timing and intensity of the returned signal are used to

Figure 2.24 Radar image of a portion of Venus' surface showing impact craters in a fractured crust. The largest crater is ~50 km across. The ejecta surrounding the craters is rough and thus radar-bright. Image from the *Magellan* spacecraft, courtesy of NASA.

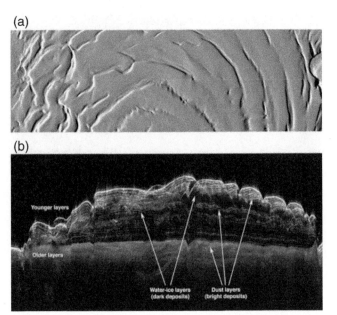

Figure 2.25 Image of (a) the martian north pole and (b) subsurface ice layers obtained by the Mars SHAllow RADar sounder (SHARAD) on the *Mars Reconnaissance Orbiter* (Stuurman et al., 2016). The dielectric constant is consistent with a mixture of dust and water ice. Image courtesy of NASA.

determine properties of the target. Timing of the returned signal gives the position of the target, and the intensity of the returned signal carries information about the physical state of the material – e.g., how rough the target is relative to the wavelength of the radar, and the **dielectric constant**, which is a material property that controls how radio waves propagate into or off of the material. Radar can be used in a configuration known as **side-looking radar** (SLR) to map surface topography and scattering properties. Another application, referred to as **ground penetrating radar** (GPR), employs longer wavelength radio waves to map subsurface compositional heterogeneities at depths ranging from a few centimeters to kilometers.

Side-looking radar has been used from orbiting spacecraft to map planetary surfaces on cloud-covered worlds, such as Venus and Titan. This works because radio waves (unlike shorter-wavelength photons) are able to penetrate through atmospheric gases and aerosols. A pulse of radio waves is generated in the transmitter on the spacecraft and directed with an antenna obliquely at the surface in an azimuthal direction that is perpendicular to the orbit track. The intensity of the returned signal is recorded with very high time resolution because the radio waves travel at the speed of light. Some portions of the ground are closer to the transmitter than others, and the two-way travel time for the signal to return to the spacecraft will therefore vary with the range between each portion of surface and the spacecraft. The intensity of the returned pulse from any given point on the ground is controlled by surface geometry, roughness, and the aforementioned dielectric constant. Generally speaking, a surface that is smooth on the scale of the wavelength of light being used will provide a low-intensity return unless that surface happens to be aligned perpendicular to the incoming beam. This is because smooth surfaces tend to reflect the beam like a mirror, so unless the surface is oriented in such a way as to reflect the beam directly back at the spacecraft, the surface will look dark to the instrument. Surfaces that are rough on a scale compared to the wavelength of the radar will tend to scatter the incoming beam in all directions, including the direction of the spacecraft. Thus, these surfaces tend to register at the instrument with relatively high intensities. The dielectric constant of the surface material also contributes to determining the intensity of the returned signal. Some materials, such as water, have dielectric constants that are ten times higher (or more) than most natural, dry surface materials, making their returned signal intensities very high. This is why experiments using radar on radio telescopes have been used to good effect in mapping polar ice deposits on Mercury and the Moon.

The final SLR image is generated by measuring the intensity of a returned pulse as a function of time after the pulse, and plotting these data as a row of pixels. Then, as the spacecraft continues flying down-track, the radar sends out another pulse, which is used to construct the next row of the final image. This is repeated until a complete image has been constructed (Figure 2.24).

Ground penetrating radar is typically used to look for subsurface strata and other buried features, such as ice lenses or (on Earth) archeological structures. It may be used from orbit (Figure 2.25) or, as planned for the upcoming *Mars-2020* rover mission, from a moving platform on the surface. The depth that can be sensed is correlated with the wavelength of light used (longer wavelengths go deeper), but as the wavelength increases, it becomes harder to resolve fine-scale vertical details. Unlike SLR, GPR directs its signal vertically downward, using the timing of the returned signal as a proxy for depth. This yields a column of pixels for the final image that show returned intensity versus depth. As the platform moves over the surface, additional sounding columns are acquired and assembled into a continuous profile. The intensity of the returned signal is governed by a material property called permittivity, which (like the dielectric constant) governs how much of the beam gets transmitted into the material versus being scattered at its interface with other materials. Ground penetrating radar is somewhat similar to active seismic profiling, except that it employs electromagnetic waves instead of acoustic waves.

Summary

Remote sensing techniques offer the opportunity to study the geology of planetary bodies at a distance, which is important because landed missions for *in situ* exploration are expensive and risky. Most remote sensing techniques used in planetary exploration rely on light as the carrier of information about the surface. Photons from different regions of the spectrum correspond to different scales of physical processes. Nuclear transitions produce gamma rays that can be used to characterize the surface's chemical and isotopic composition; electronic transitions absorb or emit X-ray, UV, visible, and near-infrared photons which allow for the characterization of elemental and mineralogic composition; and vibrational transitions absorb near-infrared and thermal infrared photons, which also enable the characterization of mineralogic composition.

In optical remote sensing (visible/near-infrared and thermal infrared), different minerals have different absorption bands in their spectra that are diagnostic of their composition. Analysts assign compositional labels to the spectra of unknown surfaces by comparing the spectral shapes in the unknown surface spectrum to libraries of spectra of known minerals acquired in the laboratory. For many optical remote sensing experiments, the basic data product is a multi- or hyperspectral image cube, which consists of a stack of spatially co-registered images of the surface taken at different wavelengths. Each pixel in such an image cube contains a complete spectrum for that location on the ground. Different classes of instruments are available to acquire these image cubes, and each class has its strengths and weaknesses. The approach used in analyzing multi- or hyperspectral image cubes depends on the science goals that are sought, and can range from creating simple band ratio images that map the location and abundance of a particular mineral of interest, to sophisticated "*n*-dimensional" statistical techniques that attempt to classify the composition of every pixel in the image cube. Ground truthing of these analytical approaches is highly desirable when possible, as it can improve the calibration of the remote scene, validate the preliminary interpretations made in the analysis of the remote scene, and augment what is learned from the remote scene with field observations of properties like small-scale textures and structural relationships.

Gamma ray and neutron remote sensing rely on galactic cosmic rays from space that bombard planetary surfaces where there is little or no atmosphere to stop them or on natural radioactivity. The cosmic rays stimulate the release of fast neutrons from regolith nuclei via spallation. The fast neutrons interact with nuclei in the surface in a variety of ways, some of which cause the nuclei to produce gamma rays, which can be sensed by an instrument above the surface and are used to map the abundance of the different isotopes that produced them. Likewise, the energies of neutrons that leak out of the surface may be used to infer the abundance of subsurface hydrogen.

Radar remote sensing is an active technique in which radio waves are directed at a planetary surface. The timing and intensity of the reflected signal are used to determine the geometry and material properties of the surface. Side-looking radar is used to map surface topography, roughness, and an aspect of composition called the dielectric constant. Ground penetrating radar aims the beam downward into the surface and uses the returned signal to map vertical profiles of subsurface layering and other buried features.

In the next chapter, we will explore other kinds of tools used in planetary exploration.

Review Questions

1. Why do you suppose our eyes evolved to be sensitive to visible wavelengths of light?
2. What is the difference between a radiance spectrum, an emissivity spectrum, and a reflectance spectrum?
3. What is meant by the term "hyperspectral cube"?
4. How does n-dimensional cluster analysis work in hyperspectral mapping?
5. How can combining analyses of different remote sensing instruments (with different spectral ranges, spatial resolutions, and other characteristics) provide a more complete picture of the geology of a given site than is possible with a single instrument?
6. What are the mechanisms by which gamma rays are produced in planetary surfaces?
7. Why are the energies of neutrons that leak from planetary surfaces so strongly dependent on the amount of hydrogen in the surface?

SUGGESTIONS FOR FURTHER READING

Bishop, J., Bell, J. F. III, and Moersch, J. E. (eds.) (2019) *Remote Compositional Analysis: Techniques for Understanding Spectroscopy, Mineralogy, and Geochemistry of Planetary Surfaces*. Cambridge: Cambridge University Press. A volume of chapters contributed by leading experts that focuses specifically on planetary geologic remote sensing.

Elachi, C., and van Zyl, J. (2006) *Introduction to the Physics and Techniques of Remote Sensing*, 2nd edition. Hoboken, NJ: John Wiley & Sons. An authoritative text that is particularly good at explaining the physical underpinnings of remote sensing.

Sabins, F. F. (2007) *Remote Sensing: Principles and Interpretation*, 3rd edition. Long Grove, IL: Waveland Press, Inc. A re-issue of a classic textbook on remote sensing, usually used in graduate courses.

REFERENCES

Boynton, W. V., Taylor, G. J., Karunatillake, S., et al. (2008) Elemental abundances determined via the Mars Odyssey GRS. In *The Martian Surface: Composition, Mineralogy, and Physical Properties*, ed. Bell, J. Cambridge: Cambridge University Press, pp. 105–124.

Bus, S. J., Vilas, F., and Barrucci, M. A. (2002) Visible-wavelength spectroscopy of asteroids. In *Asteroids III*, eds. Bottke, W. F., Cellino, A., Paolicchi, P., and Binzel, R. P. Tucson, AZ: University of Arizona Press, pp. 169–182.

Hapke, B. (1993) *Theory of Reflectance and Emittance Spectroscopy*. Cambridge: Cambridge University Press.

Prettyman, T. H., Yamashita, N., Toplis, M. J., et al. (2017) Extensive water ice within Ceres' aqueously altered regolith: evidence from nuclear spectroscopy. *Science*, **355**, 55–59.

Ramsey, M. S., and Christensen, P. R. (1998) Mineral abundance determination: quantitative deconvolution of thermal emission spectra. *Journal of Geophysical Research: Solid Earth* **103**, 577–596.

Salisbury, J. W., and D'Aria, D. M. (1992) Emissivity of terrestrial materials in the 8–14 µm atmospheric window. *Remote Sensing of Environment*, **42**: 83–106.

Shkuratov, Y., Starukhina, L., Hoffman, H., et al. (1999) A model of spectral albedo of particulate surfaces: Implications for optical properties of the Moon. *Icarus*, **137**: 235–246.

Stuurman, C. M., Osinski, G. R., Holt, J. W., et al. (2016) SHARAD detection and characterization of subsurface water ice deposits in Utopia Planitia, Mars. *Geophysical Research Letters*, **43**, 9484–9491.

Tholen, D. J., and Barucci, M. A. (1989) Asteroid taxonomy. In *Asteroids II*, eds. Bottke, W. F., Cellino, A., Paolicchi, P., and Binzel, R. P. Tucson, AZ: University of Arizona Press, pp. 298–315.

3

More Toolkits for the Planetary Geoscientist:
Chronology, Mapping, Geophysics, and Laboratory Analysis

In addition to spectroscopy, planetary geoscience uses some other tools familiar to most geologists, and some tools that are either unique or involve new twists in how they are employed. We explain how stratigraphic principles are adapted for planets (using strata produced by impacts), how the density of craters can be quantified to derive relative ages of geologic units, and how radioisotope measurements on samples, where available, give absolute ages. We explain how images from orbiting and landed spacecraft are used, along with chronologic and remote-sensing data, to make planetary geologic maps at different scales. We consider various geophysical techniques that are used on spacecraft to obtain information about planetary potential fields, interior structure, and surface topography. We summarize the kinds of extraterrestrial materials that are available for laboratory investigations, and briefly describe the analytical techniques used to characterize their mineralogy, petrology, and geochemistry. We also examine some techniques that are adapted as remote sensing tools for analyses of rocks and soils on planetary surfaces.

3.1 Geochronology

Geologic time is a critical parameter in understanding planetary evolution. Several different techniques are employed to determine relative and absolute time in planetary geoscience.

3.1.1 Planetary Stratigraphy
Stratigraphy is a familiar tool for geologists. Volcanic strata are common features on planets, but how can the

principle of superposition be applied to planets and small bodies where water and wind are not agents for creating strata? To answer this question, we should examine the first planetary geologic map based on stratigraphic principles, which was published by Eugene Shoemaker (1962) in a landmark telescopic study of the region surrounding Copernicus crater on the Moon. The stratigraphic units recognized in this map (Figure 3.1) were mostly ejecta blankets from impact craters. This novel definition of strata as the materials excavated by impacts was respectful of the geologic linkage of rocks and time, and has since been used on every heavily cratered body. Some of the issues in mapping strata as viewed from above are described in Section 3.2. Following publication of this map, Shoemaker and Hackman (1962) proposed a lunar timescale divided into periods delineated by cataclysmic impacts, with time-rock units (systems) defined as all the rocks formed during periods and major rock units (formations) defined as the ejecta from the largest impact basins. Figure 3.2 shows these rock, time-rock, and time units for the time–stratigraphic system now used for the Moon.

The stratigraphic principles of crosscutting relationships and lateral continuity can likewise be used on planets. Faults in planetary imagery are recognized by offsets that can be used to gauge relative time, and deformation bands must necessarily cut across older units. Figure 3.3 shows a map of part of Ganymede, with older terrain truncated by several generations of younger grooved terrains and superposed craters. Correlation of units can sometimes be difficult using only images from orbiting spacecraft, but separated parts of units can be

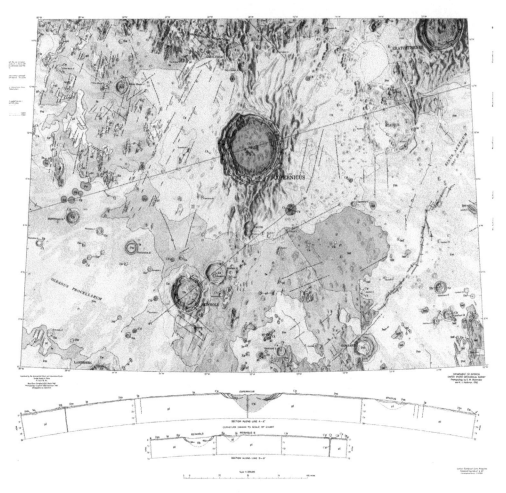

Figure 3.1 Geologic map of the Copernicus crater region of the Moon by Eugene Shoemaker, the first planetary application of stratigraphic principles. See Shoemaker (1962) for explanation of the units. Courtesy of the U.S. Geological Survey.

recognized if they have similar cratering histories, as described below.

3.1.2 Crater Size–Frequency Distribution as a Chronometer

The density of impact craters on some portion of a planetary surface is proportional to the length of time that surface has been exposed to meteor impacts. Simple inspection of crater density is sufficient to distinguish surface units having very different ages, such as the heavily cratered highlands and sparsely cratered maria of the Moon. However, quantifying the **crater size–frequency distribution** is often required to recognize smaller differences in age.

Crater densities are quantified in the following way: First, we select an area believed to have a homogeneous geologic history; it is important not to mix terrains having different ages. The diameters (D) of visibly recognizable craters of all sizes are measured in a terrain of known area, and the diameters are placed into size bins. These bins commonly increase in size between D and $2D$, or D and the square root of $2D$, rather than at fixed size intervals. We then sum the

number of craters in each bin. Finally, the cumulative number of craters above some size (N_{cum}) per unit area is plotted versus crater diameter D. This is normally done using a log–log plot, as illustrated in Figure 3.4a.

The number of craters per unit area with D greater than or equal to a given diameter approximates a power function of D, and the plot has a negative slope because small craters are produced in greater abundance than large craters. The cumulative size–frequency distribution curve shifts to the right as time goes on and more craters accumulate, so the relative positions of the curves indicate the relative ages of units (t_2 is older than t_1 in Figure 3.4a).

In practical applications, computer tools are available to obtain undistorted crater diameters from oblique images, and software packages can analyze crater statistics. Platz et al. (2013) provide a very useful summary of Mars crater counting, describing how areas to be counted are defined, how measured crater populations are validated, and how crater density data are interpreted.

A potential problem with this method is that very old surfaces can become saturated with craters, so that each new

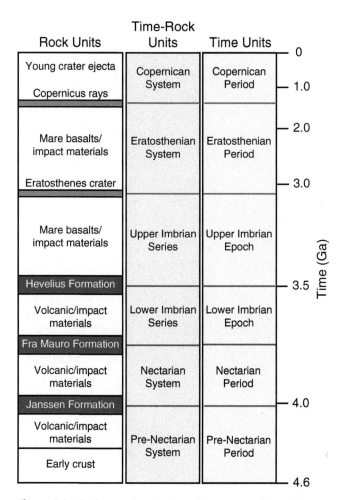

Rock Units	Time-Rock Units	Time Units
Young crater ejecta Copernicus rays	Copernican System	Copernican Period
Mare basalts/ impact materials Eratosthenes crater	Eratosthenian System	Eratosthenian Period
Mare basalts/ impact materials Hevelius Formation	Upper Imbrian Series	Upper Imbrian Epoch
Volcanic/impact materials Fra Mauro Formation	Lower Imbrian Series	Lower Imbrian Epoch
Volcanic/impact materials Janssen Formation	Nectarian System	Nectarian Period
Volcanic/impact materials Early crust	Pre-Nectarian System	Pre-Nectarian Period

Figure 3.2 The lunar time–stratigraphic system, adapted from Wilhelms (1987). Time units (periods) and time-rock units (systems) are delineated by rock units (formations), which are the ejecta excavated from large impact basins. The absolute time markers on the right (Ga = billions of years before present) are based on radiometric analyses of returned lunar rocks.

crater obliterates, on average, one older crater. This situation is illustrated in Figure 3.4b, plotting the production of craters per unit area versus time. Geometric saturation is never actually achieved, because nature does not precisely fit small craters into all the spaces between larger craters. Instead, some lower "equilibrium" value is reached, as shown by the observed curve in this figure. The consequence of reaching equilibrium is illustrated in Figure 3.4c. Because small craters can reach equilibrium before large craters, the small-diameter craters exhibit the equilibrium slope, which is lower than the slope of the crater production curve, whereas the large craters can still define the steeper production curve. This produces a kink in the observed cumulative crater distribution. In this case, only the large-diameter craters give actual age information.

Another problem occurs if **secondary craters** are counted. Large impacts can produce ejected blocks that form their own secondary craters, recognized as lines or arcs of small craters emanating from a larger crater. These secondary craters do not indicate additional time, and thus give erroneous ages if included in crater density measurements.

Although crater size–frequency distributions give only relative ages, they can be used to estimate absolute ages if the rate of impacts through time is known. This rate, the **crater production function**, is not linear – impact rates were higher in the early Solar System and have declined over time. The crater production function has been quantitatively determined only for the Moon (Neukum et al., 2001), where the crater densities of formations have been compared with the measured radiometric ages of rocks from those same units that were returned to Earth by *Apollo* astronauts. The crater production functions for other bodies are sometimes approximated as multiples or fractions of the lunar production function. In other cases, they are based on models for collisions of bodies within the asteroid belt, the presumed source for most impactors. The crater production function for asteroids is an order of magnitude higher than that for the Moon (O'Brien et al., 2006).

3.1.3 Radioactive Isotopes as a Chronometer

Radiometric ages can only be measured, of course, on rocks – either lunar samples returned to Earth by astronauts, or meteorites from known source bodies. A complete discussion of radiometric dating is beyond the scope of this book, but we can illustrate the determination of absolute ages using several commonly used isotopic systems.

Radioactive ^{87}Rb decays to radiogenic (produced by radioactive decay) ^{87}Sr at a known, constant rate. Strontium has two isotopes, ^{87}Sr and nonradiogenic ^{86}Sr, and the newly produced ^{87}Sr is slowly added to ^{87}Sr already in the rock. The amount of ^{87}Sr in the rock initially is an unknown quantity. We can get around this problem by use of an **isochron** diagram, plotting ^{87}Sr/^{86}Sr versus ^{87}Rb/^{86}Sr (Figure 3.5). At the time the rock forms (let's say it's a crystallized lava), every mineral will have the ^{87}Sr/^{86}Sr ratio of the magma; however, each mineral will have a different ^{87}Rb/^{86}Sr ratio, because rubidium and strontium have different geochemical behaviors. As time goes on, ^{87}Rb decays, and each ^{87}Rb atom produces one ^{87}Sr atom, thereby increasing ^{87}Sr/^{86}Sr and decreasing ^{87}Rb/^{86}Sr, as shown by the diagonal arrows in Figure 3.5. As more time passes, the line defined by the minerals gets steeper. The elapsed time is defined by the slope of the line, called an isochron, and can be calculated knowing the decay rate.

Uranium has two radioactive isotopes, ^{235}U and ^{238}U, which decay into ^{207}Pb and ^{206}Pb, respectively. Because these radioisotopes decay at different rates, they are effectively two independent clocks that, by definition, must give the same age. The isotopic measurements are

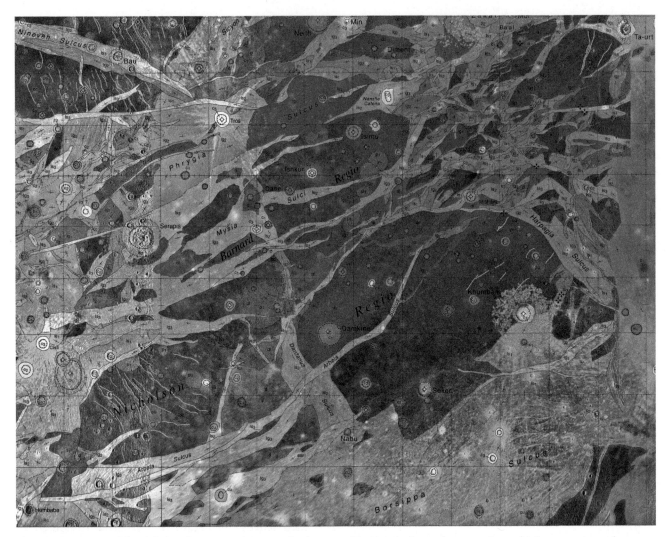

Figure 3.3 Portion of the global geologic map of Ganymede, showing older terrain (brown) crosscut by multiple generations of younger grooved terrain and superimposed craters. See Collins et al. (2013) for explanation of the units.

made on a mineral like zircon that incorporates uranium but contains no lead initially. Data from the U–Pb system can be treated in several ways. Here, we focus on the "Pb–Pb" diagram, plotting the ratios of radiogenic ^{207}Pb and ^{206}Pb to a stable, nonradiogenic isotope of lead, ^{204}Pb (Figure 3.6). This method offers the advantage of not having to measure the uranium isotopes. Different samples may initially contain the same mixture of lead isotopes but vary in their uranium abundance (U/Pb ratio). Samples with varying U/Pb will evolve along different "growth curves" (Figure 3.6), but at any time will align to form a Pb–Pb isochron. As in the Rb–Sr system, the age of the rock can be calculated from the slope of the isochron, which in this case becomes less steep with time. There are other variations on the Pb–Pb isochron diagram (for example, we will see a plot of ^{207}Pb/^{206}Pb versus ^{204}P/^{206}Pb in a later chapter), but the principle remains the same. This is an extremely accurate chronometer that has found wide use in planetary science.

The Rb–Sr and U–Pb chronometers require mineral separations and careful geochemical manipulations in the laboratory. A mass spectrometer measures the isotope abundances in these chemical separates. Attempts are underway to devise methods to measure radiometric ages by remote sensing on a planet's surface, but only one such measurement has been accomplished to date, using the decay of radioactive ^{40}K to ^{40}Ar. By measuring, in a powdered Mars rock sample, potassium using the APXS instrument and ^{40}Ar using the SAM instrument on the Curiosity rover, Farley et al. (2014) obtained an age of ~4.2 billion years (Ga). This age is interpreted as either a mixture of the ages of detrital components in this sedimentary rock or the minimum age of deposition of this sediment.

The isotope systems we have just described are long-lived, so that only small amounts of the radioactive isotopes have decayed away. However, some radioactive isotopes decay so rapidly that they were present only

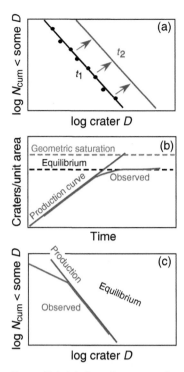

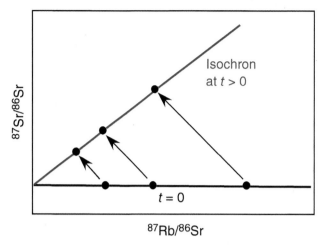

Figure 3.5 Rb–Sr isochron diagram, illustrating the isotopic evolution of three minerals in a rock with time. At the time the rock formed ($t = 0$), all the minerals have the same $^{87}Sr/^{86}Sr$ ratio but different $^{87}Rb/^{86}Sr$ ratios. The isotopic composition of each mineral moves along a line of slope –1 as ^{87}Rb decays to ^{87}Sr, the length of the arrow determined by the amount of radioactive ^{87}Rb in each mineral. After some time has elapsed ($t > 0$), the minerals define an isochron whose slope becomes steeper with time. The age of the rock can be calculated from the slope of the isochron.

Figure 3.4 (a) Cumulative size–frequency distribution of craters, showing the number of craters per unit area with $D >$ a given diameter. The curve at time t_1 shifts to the right with increasing age (t_2). (b) Diagram illustrating how crater saturation causes the crater density to level off over time. Idealized geometric saturation is never reached, and the crater density eventually reaches some lower equilibrium value. (c) Crater saturation can produce a kink in the observed crater distribution, because small craters reach saturation before large craters. In this case, only the large craters give age information.

fleetingly in the early Solar System and are now virtually extinct. A few of these **short-lived radioisotopes** can provide very precise measurements of the relative ages of early Solar System materials. The most widely used are ^{26}Al, which decayed to ^{26}Mg with a half-life of only 0.7 million years (Ma), and ^{182}Hf, which decayed to ^{182}W with a half-life of 9 Ma. Formation ages differing by only one million years or so can be distinguished, if the events occurred while the short-lived isotope was still undergoing radioactive decay.

Extinct radioisotopes give only relative ages, so they must be linked to **long-lived radioisotopes** to determine absolute ages. For example, ^{26}Al measurements might indicate that two igneous meteorites from the same parent body crystallized 2 Ma apart, but a U–Pb measurement on one of the samples would be necessary to say that the events occurred at 4.552 and 4.550 Ga.

Radiogenic isotopes are produced by radioactive decay, whereas **cosmogenic isotopes** form by interaction of atoms with cosmic rays. Some examples of cosmogenic nuclides are ^{3}He, ^{10}Be, ^{14}C, ^{21}Ne, ^{26}Al (so it's not

completely extinct in the Solar System after all), ^{36}Cl, and ^{36}Ar. The reactions that form these isotopes are inefficient, so the duration of irradiation is important. These cosmogenic isotopes are radioactive, and measurements of their abundances give cosmic-ray exposure ages. Because cosmic rays can only penetrate a fraction of a meter into rock, exposure ages define the times spent on or very near a planet's surface or, in the case of meteorites, the time spent as a small rock orbiting in space.

3.2 Geologic Mapping

The purpose of a geologic map is to illustrate an interpretation of the surface geology. If the map has associated cross-sections, it also illustrates an interpretation of the subsurface geology. The map is based on observations that can be reproducibly collected either from particular locations or from mosaics of images across a region. The geologic history is represented by an associated sequence of map units or, ideally, by a column of stratigraphic units that represents the relative or the absolute ages of the units outcropping in a region.

The quality of the map is dependent on the number and types of geologic observations that are available. A local map with many observations to constrain interpretations should be of higher quality than a regional or global map (Figure 3.8) with a limited number of observations. Still, if a regional or global map is needed to

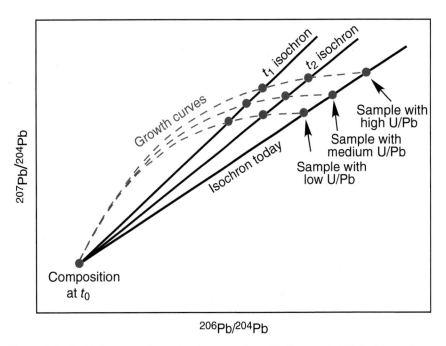

Figure 3.6 Pb–Pb diagram, illustrating how samples with the same initial lead isotopic composition but different U/Pb ratios at time t_0 evolve over time. Although the samples follow different lead isotopic growth curves, at any later time t_1, t_2, or today they define an isochron. The age can be calculated from the slope of the isochron.

address a geologic problem, then that map should be constructed and allowance made for uncertainties due to the limitations in data. After all, geologic interpretations are based on incomplete datasets, so defining uncertainty should be part of the mindset for creating geologic maps.

When constructing geologic maps on the Earth, we are accustomed to literally walking the ground, be it for the compilation of the Earth's oldest known geologic map (Turin Papyrus Map), the first modern map (William Smith's Geologic Map of England, Wales and parts of Scotland), or a new geologic map in the Himalayas. While traversing the map area, we collect observations, gather useful rock samples, and compile data onto a topographic map, which now is typically digital and in a geographic information system (GIS). The samples may be subsequently analyzed to provide other position-specific geologic observations. Also, the field-derived data may be collated during or after the field campaign with remote-sensing data or subsurface geophysical data to locally validate or improve the interpretations.

Herein lies one of the greatest differences between geologic mapping on Earth versus on other Solar System bodies. With the exception of a few astronauts on the Moon and a few robotic rovers on the Moon and Mars, geologic maps beyond Earth have not involved data generated with sample collection by observers traversing a region. (The Mars traverse map in Figure 3.9 involved rock analyses by rover instruments, and projection of those identified rock types farther afield using spectral measurements.) This inability to "ground truth" the geology of other bodies is a limitation.

So, how do we compile a geologic map, if we cannot "walk the ground"? The good news is that spacecraft missions, particularly orbital missions, have utilized a variety of electromagnetic frequencies (discussed in Chapter 2) to image most major rocky or icy bodies in the Solar System. Since the early 1990s, these datasets have been digital, which is advantageous for image processing, analysis, and referencing. The type of electromagnetic radiation, the tool for generating information about the radiation, and the method for gathering the data enable the collection of information about surface appearance, elevation, and composition. An example is the radar map of Venus shown in Figure 3.10; this image has been rotated to provide a perspective view. Successful analysis of various kinds of imagery is the key to creating useful geologic maps for other Solar System bodies.

3.2.1 Imagery

Unless geologists are involved in image acquisition and processing during a spacecraft mission, they will typically submit a data-request proposal to the organization that is the repository for mission data. These organizations include the NASA Planetary Data System (PDS) and USGS Astrogeology Science Center. These data typically contain named features based on the International

BOX 3.1 PLANETARY CHRONOLOGY: WORKS IN PROGRESS

The geologic timescale for the Earth is a testimony to the efforts of many geologists during the past two centuries. It continues to evolve slowly as more discoveries are made. The timescale of Earth is compared with those of the Moon and terrestrial planets (Figure 3.7), based on the lunar stratigraphic techniques, measurements of crater size–frequency distributions, and radiometric ages of the limited available samples. The Earth's timescale is obviously much more detailed than this illustration and only the broadest divisions (eons and eras) are shown, because the inclusion of periods, epochs, and ages would make it too crowded. The time units defined for the other bodies are called periods, although they are longer than periods in our planet's timescale. The absolute ages of their boundaries are only approximations.

One interesting observation is that most of the subdivisions in the Earth's timescale are in recent times, whereas planetary timescales are more finely divided prior to ~3 Ga. This reflects the different criteria used: Planetary timescales are based mostly on terrains formed when large impacts were prevalent; in contrast, the Earth's early history has mostly been erased by plate tectonics, and its Phanerozoic chronostratigraphy is based on fossils. An exception is Venus, for which the early geologic record has been obscured by global-scale volcanism.

Geologic timescales have also been formulated for asteroid Vesta and dwarf planet Ceres, the only small bodies for which sufficiently detailed stratigraphy is available.

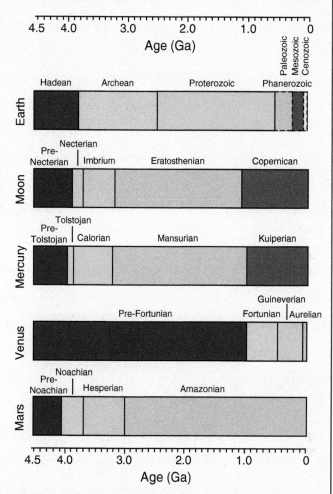

Figure 3.7 Geologic timescales for the Earth, Moon, and terrestrial planets.

Astronomical Union (IAU) Gazetteer of Planetary Nomenclature, and conventions for additional naming are governed by the IAU.

Planetary geologists use software such as the Integrated Software for Imagers and Spectrometers (ISIS) to compile and organize the needed image data. Where datum and elevation data are not available and appropriate overlapping images with appropriate emission angles are available, a user may employ software such as the BAE Systems SOCET SET (Softcopy Exploitation Toolkit) to create a digital elevation map (DEM).

3.2.2 Definition of Map Units
The geologist then needs to define map units, identify structures, and establish a map unit sequence. With

imagery for an entire study area, the geologist can consider the entire picture before diving into the local details of the area. Given this advantage, it is recommended that she/he consider the large areas of similar appearance and most obvious linear features, whether curvilinear or straight. An example of this approach would be to consider all of Figure 3.11a rather than starting by considering only 5 percent of the image at the outset. Focusing on the details of a small area increases the likelihood that important regional patterns and characteristics will not be recognized and that the work will become mired in details. Also, it is important to remember that large-scale features commonly relate to larger-scale processes and behaviors, so it is important to recognize them first. Later, detail may be added and information refined if the resolution of the imagery allows.

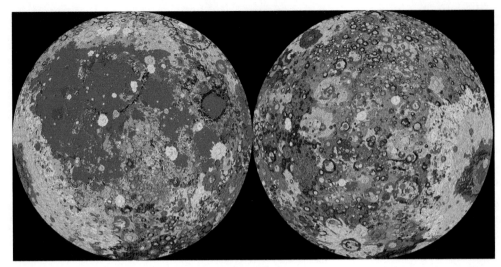

Figure 3.8 Global geological map of the near (left) and far (right) sides of the Moon. Image courtesy of P. D. Spudis, www.spudis lunarresources.com.

Figure 3.9 A portion of the geologic traverse map in Gusev crater by Spirit rover, with five images of rock outcrops encountered during the traverse. Circular features in outcrops are rover analysis sites. Traverse map modified from Crumpler et al. (2011). NASA images.

Figure 3.10 Stereoscopic Synthetic Aperture Radar image (with vertical exaggeration) of Venus showing Maat Mons. *Magellan* image, JPL.

As with Earth-based geologic mapping, a map unit contains a set of similarities that identify and distinguish it (e.g., Figure 3.11b). For planetary imagery, the type of similarity depends on the image type, and typically involves identifying similar surface features (e.g., smaller conical volcanoes on Venus, linear topographic features with less topography and less vertical curvature on Europa, etc.). The characteristic features must be morphologically definable in a manner that enables reproducible identification by another geologist. While developing ideas about the interpretation of the origin and history of a unit during this unit-identification process is appropriate and useful, those interpretations

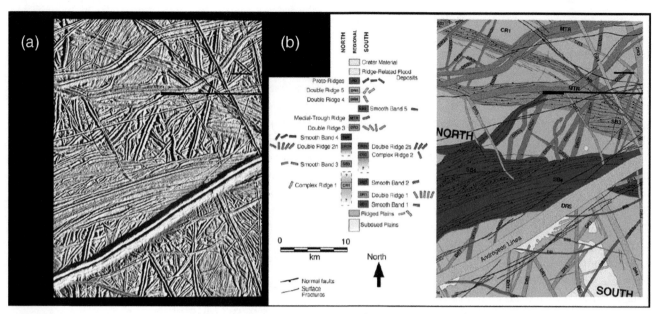

Figure 3.11 (a) Image mosaic of the Bright Plains region of Europa at ~20 m/pixel. (b) Geologic interpretation of Bright Plains region. Modified from Kattenhorn (2002).

should not define the physical characteristics used to identify the unit. For example, a unit could be defined by a dark **albedo** and high crater density count, without surface cones, and by being overlain by all the surrounding units. The interpretation might be, "the oldest basaltic lava flow without a visible extrusion source," but that interpretation may be subject to change. Well-defined units should not change through time, whereas interpretations are likely to change as greater knowledge and understanding are accumulated.

An important attribute of a map unit is its contacts, which normally separate units having different characteristics, such as color, albedo, surface morphological features, or abundance of particular types of structures. The form of the contact may also be important for defining the difference between adjacent map units. Possibilities include: Is the contact straight (e.g., Figure 3.11a) or curvilinear with embayments (Figure 3.9), associated with a vertical change in topography, or associated with a change from well-defined features to features that are less well-defined because they are partially obscured by an overlying unit/deposit? Correspondingly, contacts may be sharp so positioning the boundary is simple, or gradational where the boundary position is approximate and a judgment call.

Thus, as a geologic map is constructed, it should become an array of map units, where some of the contacts may be related to geologic structures (Figure 3.9 versus Figure 3.11b). Presently, the typical illustration of the geologic units, unit boundaries and structures for a geologic map will be done in layers overlaying the imagery in software such as Adobe Illustrator or ESRI ArcGIS.

3.2.3 Relative Age Determination of Units

During geologic mapping, an important activity is determining the relative ages of map units. The principles of superposition and crosscutting relationships are described in Section 3.1.1, but we consider their usage here. For superposition, a contact is examined to determine which map unit rests on which other map unit. Such an analysis is much easier where the units and topographic information are superimposed, as topographically higher units are generally considered to be younger than the lower units. For crosscutting relationships, the rule of thumb is that a unit or feature must have been present to be "cut," so that feature or unit is older than the feature or unit that creates the cut (e.g., younger DB5 cuts and offsets older SB4 in the middle of Figure 3.11b).

While the procedures of determining crater size–frequency distributions are described in Section 3.1.2, knowing when to apply these procedures is necessary to bring value to the relative dating of units for geologic maps. Relative-age dating with crater density is more effective when the number of craters is abundant, the range of crater sizes is large, and the crater abundance correlates to changes in map units. For example, Io has a sufficiently geologically active surface that very few craters are preserved and crater density is not a useful tool for determining the relative age. Another situation is illustrated by Venus, where the dense atmosphere only allows large impactors to reach the surface, typically forming craters with diameters greater than 2 km. These craters are fewer than 1000 in number. They can be used to determine an average age for the entire planetary surface, but they are not sufficient to determine the relative ages of regional-scale map units. In contrast, a number of bodies such as the Moon, Mercury, Mars, or Callisto have sufficient numbers of craters of varying sizes that crater densities do change with map units, so relative ages can be determined and compared to relative-age determinations from the other two approaches.

3.2.4 Rock (or Ice) Units and Rock (or Ice)-Time Units

The map units identified on planets are rock units (or, on some bodies, ice units) that are equivalent to **lithostratigraphic units** for Earth-based geologic maps, although the units on planets typically lack compositional or textural data because they have not been observed at the outcrop- and micro-scale. Still, the rock (or ice) units have defining characteristics and relative ages, so they can be used to construct rock (or ice)-time units, which are systems or sequences of units across the map or larger areas with the same relative ages. These rock-time or ice-time units are equivalent to **chronostratigraphic units** on Earth.

One useful ways to identify rock-time or ice-time units is by their crater densities. Thus, different rock or ice units in a region with the same crater density are assumed to have the same relative age and constitute the same rock-/ice-time unit. A second way is to distinguish units that formed rapidly. Rock or ice units that are bounded at the top and bottom by the same units formed by rapid geologic events are considered to belong to the same rock-/ice-time unit. Examples of units due to rapid geologic events include crater ejecta, individual lava flows, or extensive deposits due to a catastrophic sedimentary event. So, rock/ice units sharing an overlying unit that is a single ejecta blanket and underlain by a single extensive lava flow would be treated as a single rock-/ice-time unit.

3.2.5 Mapping Tectonic Structures

The construction of some geologic maps will not involve the presence of tectonic structures (e.g., Figure 3.9), whereas other maps would be essentially impossible to construct and interpret if tectonic structures were not considered (e.g., Figure 3.11b). Common structures are open dilational fractures, extensional normal faults with scarp faces that may occur as pairs bounding rift valleys, contractional wrinkle ridges, and folds. All of these features are linear to curvilinear, may be persistent across areas, and have topographic characteristics that are well illuminated by oblique electromagnetic radiation. These features may be restricted to within single map units, may occur at the boundaries of map units, or may be

abundant and transect all map units equally, and as such have different possible roles for defining map units. Younger structures may crosscut and offset older structures, unit boundaries, impact craters, or volcanic cones. Conversely, older structures may have superimposed younger craters and ejecta blankets, lava flows, sedimentary deposits, or volcanic cones. Thus, structures can be quite important for building the time sequence of map units or even rock-/ice-time units.

In summary, constructing a planetary geologic map offers some challenges not faced in terrestrial geology mapping. The identification and interpretation of map units, map unit contacts, and relative ages of units and tectonic structures is an interactive process in which rules matter, but intuition will be needed.

3.3 Geophysical Methods

In planetary science, geophysical methods, with rare exceptions, have mostly been employed from orbiting spacecraft platforms. Here we consider some of the most important geophysical tools for characterizing planets and small bodies. The intent of most geophysical investigations is to develop a global picture of the hidden interiors of these objects. Oftentimes, interpretation of geophysical measurements in geologic terms requires complex numerical modeling. Figure 3.12 shows the best current interpretation of the interior structure of the planet Mercury, compared to Earth. Development of this model required the combined use of almost all of the geophysical techniques described below.

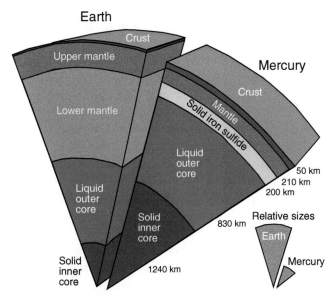

Figure 3.12 Comparison models of the internal structure of Mercury and Earth, based on geophysical data. Mercury model adapted from Smith et al. (2012).

3.3.1 Topography

A fundamental property of a planet is its shape or, at higher resolution, its surface **topography**. Topography can be obtained most accurately using a laser or radar altimeter on an orbiting spacecraft. By measuring the two-way travel time for a pulse of light or a microwave to be reflected from a planet's surface, we can calculate its distance. Basic altimetry data consist of spots on the surface for which average heights are measured. When sufficient density of spots is achieved, topography is often interpolated to a regular grid for ease of use. The height accuracy is commonly 10–30 cm for planetary topography. Multiple measurements can be expressed as deviation in elevation from the mean planetary radius. Color-coded topographic maps are essential tools for planetary exploration. In digital form, these are called **digital terrain models** (DTMs). Laser altimeters have measured the topography of Mercury, the Moon, Mars, and several asteroids, and radar altimeters have measured elevations on Venus and Titan.

Under proper conditions, images can also be used to compute topography. In **stereo imaging**, two images are taken of the same region under similar illumination conditions, but with different viewing geometries. The apparent relative locations of the same features in the two images enable computation of relative distances from the spacecraft, in all three dimensions. If only a single image is available, heights can be computed from the lengths of shadows cast by topographic features. This requires the illumination to be from a shallower angle than the topographic slope. This trigonometric method uses the tangent function:

$$h = L/\tan\theta \tag{3.1}$$

where h is height, L is the shadow length, and θ is the solar incidence angle. There are more complex formulations, but the principle is the same. If the albedo of the surface is fairly constant, the brightness of an image will vary with the topographic slope. This change in brightness across a single image can therefore be inverted to compute topographic slopes. The benefits of using imagery are that images generally have higher spatial resolution than laser or radar altimetry spots and images provide continuous coverage. The drawback is that it is extremely difficult to apply these techniques to global datasets. Images of the limb of a body can also be used to measure heights and slopes of topographic features that are on the limb at the time of the image. Limb images are also commonly used to constrain global shape, particularly for irregularly shaped bodies, such as asteroids.

Orbital imagery is generally more effective for other planets than for the Earth, because terrains on other bodies are generally not blocked from view by oceans or plant life. Venus and Saturn's moon Titan, however, are special cases. Their thick, hazy atmospheres preclude direct images at visible wavelengths, but radar imaging has proven effective in topographic mapping. Radar sounding instruments can also be used to penetrate tens of meters to kilometers below a planetary surface to image subsurface structures (see Section 2.11).

Quantitative physical measurements of topographic features, such as ridges, volcanoes, and fractures, can provide important constraints on geodynamic processes in the interior. For example, measurements of shortening on the global network of lobate fault scarps on Mercury demonstrate that the planet has undergone global contraction which, in turn, provides a constraint on models of the size and cooling history of Mercury's massive core.

3.3.2 Gravity

The force of gravity (F) that attracts an object is given by

$$F = (GM_1M_2)/d^2 \tag{3.2}$$

where G is the universal gravitational constant (6.673×10^{-11} Nm^2/kg^2), M_1 and M_2 are the masses of the two objects, and d is the distance between the centers of the two objects. Gravity can be measured by observing the path of a body orbiting or passing near a planet. This can be a moon or a spacecraft. The more massive the planet, the more strongly it attracts a second object, and the faster it moves. (This velocity increase finds practical usage in gravity-assist maneuvers that accelerate spacecraft during close approaches to planets, thereby saving fuel.) By carefully tracking the trajectory and velocity of an orbiting spacecraft from its radio signals, it is even possible to map changes in gravity over different portions of the planet. During times when engines are turned off so that the spacecraft is influenced only by gravity, changes in spacecraft position are measured through direct radio contact with Earth. In the case of the *GRAIL* mission to the Moon, two spacecraft orbited in tandem and communicated with each other as well as Earth, vastly improving the precision of gravity measurements. Local surface gravity is affected by differences in topography, rock density, and buried tectonic structures, producing gravity anomalies. Maps of **gravity anomalies** provide information on local mass variations in the subsurface.

The spatial resolution achievable through gravity measurements depends on the type of trajectory (single flyby, multiple flybys, orbit) and the height above the surface (gravity field perturbations decrease rapidly in strength relative to the background field with distance from the surface). Single flybys generally provide a measurement of the mass and possibly some information on hemispheric-scale lateral density variations. Multiple fly-bys can constrain the radial distribution of density, if at least one flyby occurs over the equatorial region and one over the polar region. Orbital measurements provide the most accurate gravity measurements.

If a planet's internal distribution of mass differs from that of a spherically symmetrical body, measurements of surface gravity can be modeled to deduce the nature of the interior. Gravity determinations of a planet's mass are also essential for calculating its mean density. We will see examples of these applications in Chapter 8.

3.3.3 Magnetics

A planet's magnetic field is generated by a dynamo powered by electric currents in layers of conductive material, i.e., molten metals in planetary interiors. The strength and orientation of a magnetic field can be measured by a magnetometer on an orbiting spacecraft. These devices must be mounted on booms that extend far from the body of the spacecraft, in order to isolate the magnetometer from electromagnetic noise of the spacecraft and other remote-sensing instruments. Plasma and ion mass spectrometers measure compositions and trajectories of charged particles trapped in magnetic fields. Radio and plasma wave instruments measure electromagnetic waves that are produced by the interactions of particles with planetary magnetic fields.

Maps of local **magnetic anomalies** obtained during close approaches can provide information on compositional variations in the crust. Planetary fields define dipoles at the surface (a pole is a point where the magnetic field is vertical). Past reversals in the polarity of a magnetic field are recorded in igneous rocks, as seen in the Earth's ocean floor and possibly in magnetic anomalies in the ancient crust of Mars. The most direct measurement indicating that Europa has a subsurface ocean comes from mapping the effect that Europa has on Jupiter's magnetic field.

A magnetic field can deflect most of the **solar wind**, protecting the atmosphere from being stripped away over time. The magnetosphere can also protect the surface from cosmic rays, which could be harmful to life, and thus its occurrence may have astrobiological significance.

Paleomagnetism can also be measured in the laboratory, if samples are available. Remanent magnetism provides evidence for a conducting, molten core, and thus constrains the thermal history of planets and smaller differentiated bodies.

3.3.4 Seismicity

Seismic measurements have been extremely successful in mapping the internal structure of the Earth. Such measurements, however, require a network of seismometers on the surface. Seismometers measure motion of the ground, and because these movements are small the instruments must be deployed on a planet's surface. *Apollo* missions carried both active and passive seismic experiments to the Moon. Several of the passive stations remained operational until 1977 and provided very useful information on the near-surface structure and deep interior of the Moon. Unfortunately, no seismic measurements have yet been made for any other planetary body. The *Philae* lander on the European Space Agency's ROSETTA mission to a comet had several seismic instruments, but the lander ended up in shadow and did not function properly. NASA's *InSight* mission to Mars, launched in 2018, includes a seismometer and should detect marsquakes and impacts of falling meteors.

The operating principle can be described by a weight suspended on a spring, both connected to the ground on a frame. During an earthquake the ground moves, and the force used to hold the weight stationary is the output of the seismometer. Instruments usually measure movements along three axes: vertical and two horizontal directions.

The analysis of seismic wave paths and travel times provides the most rigorous information on planetary interiors. On a gross scale, seismic data constrain the size of a core and the depth to critical boundaries, such as that between the mantle and crust, which have contrasts in density. Seismic data also provide indirect information on crust, mantle, and core compositions. These uses are discussed further in Chapter 7.

3.3.5 Radiometry

Quantitative measurements of the heat flowing outward from a planet's surface are important constraints on the processes creating and transporting heat in the interiors. On Earth, **heat flux** is commonly determined by measuring the temperature at different depths in the crust. This method was also used by *Apollo* astronauts on the Moon, but it has not yet been possible on any other body. The *InSight* mission will measure heat flow on Mars during surface operations in 2019.

Heat flux can also be determined remotely through measurements of thermal radiation emitted by the surface. Orbital instruments used for this purpose include bolometers, multispectral thermal imagers, and infrared spectrometers. Bolometers collect thermal radiation over a very broad wavelength range in a single channel to measure the total emitted energy, without regard for the spectral distribution. Multispectral thermal imagers observe the surface through several narrow infrared filters and combine the measured fluxes into an estimate of the total emitted flux. Infrared spectrometers measure the thermal emission spectrum of the surface, preferably over a fairly broad wavelength range. As with multispectral thermal imagers, the flux is integrated over wavelength to estimate the total emitted flux.

Surfaces are heated by sunlight as well as internal sources. In order to isolate the heat flux coming from the interior, the contribution from solar heating must be modeled and removed.

3.4 Analysis of Planetary Materials

The analysis of samples, even if their sources cannot be confidently identified, in laboratories on Earth provides mineralogic, petrologic, geochemical, and geochronological information that cannot be otherwise obtained.

3.4.1 Available Extraterrestrial Samples

Meteorites constitute the most abundant kinds of extraterrestrial materials available for laboratory analysis. **Chondrites**, the most common type of meteorite, are ultramafic rocks – basically cosmic sediments composed of an assortment of solid materials accreted from the solar nebula. Chondrites have never been geologically processed by melting and differentiation. They are samples of asteroids or, in some cases, possibly comets. **Achondrites** are either igneous rocks or the solid residues after partial melting (the latter are sometimes called "primitive achondrites," although the term doesn't make much sense). The magmatic achondrites can be either volcanic or plutonic rocks, and may represent melt compositions or cumulates formed by the accumulation of crystals. Most achondrites are samples of melted asteroids, but a few achondrites are martian or lunar rocks. **Irons** are composed of iron–nickel alloys, and represent samples of asteroidal cores. **Stony-irons** are mixtures of metal and olivine or silicate rock, the olivine-bearing meteorites (pallasites) likely representing samples from core–mantle boundaries and the other stony-irons (mesosiderites) representing impact-generated mixtures. Meteorites can fall anywhere, but thousands have been collected in Antarctica and in some deserts like the Sahara, where they accumulate over time under conditions with minimal weathering.

Interplanetary dust particles (IDPs in the vernacular of planetary scientists) once roamed the regions between the planets. They are small lithic fragments of asteroids or the mineral dust expelled from comets as they approach the Sun (Flynn et al., 2016). IDPs generally resemble chondrites and are collected in the stratosphere by high-flying aircraft, although they can accumulate on any clean surface on the ground, such as Antarctic ice.

Many chondrites contain small amounts of **presolar grains** – tiny mineral particles that formed around other stars and were incorporated as dust into the early solar nebula (Nguyen and Messenger, 2011). Their presolar origin is recognized by their anomalous isotopic compositions, relative to normal Solar System matter.

Astronauts on six historic voyages (NASA's *Apollo* missions) collected and returned to Earth 382 kg of rocks and soils from both the highlands and the maria. These precious lunar samples are curated at the Johnson Space Center in Houston, Texas, and allocated to scientists for research. The Soviet unmanned *Luna* missions also scooped up small amounts of soil and brought those samples back to Earth. A few achondritic meteorites are also from the Moon, and are important because they sample other locations; the landed spacecraft only sampled the lunar nearside.

Several unmanned spacecraft missions have also collected samples and brought them home. The *Stardust* mission sampled **cometary particles** as it passed through the dust cloud surrounding the nucleus of comet Wild 2 in 2004 (Brownlee et al., 2006). The particles were trapped in aerogel, a highly porous silica foam, necessitating careful extraction and manipulation. The *Genesis* mission collected solar wind particles for several years, returning them to Earth in 2004. Although the spacecraft crashed on return, shattering many of the sample collectors, careful work has allowed the recovery of the expected science (Burnett, 2013). The JAXA *Hayabusa* mission obtained a tiny amount of soil particles from the Itokowa near-Earth asteroid. These particles are samples of chondrite.

3.4.2 Laboratory Analysis Techniques

The techniques (see Table 3.1) used for characterization of extraterrestrial materials are, in many cases, the same as those used for analysis of terrestrial samples. However, the small quantities and extremely small grain sizes of some extraterrestrial samples, such as *Stardust* comet and *Genesis* solar wind particles, IDPs, and especially presolar grains, pose daunting challenges.

Thin sections of meteorites and lunar rocks are examined with the petrographic microscope, allowing the optical identification of minerals and textures. Mineral structures and compositions are examined using electron-beam techniques. The electron microprobe measures element abundances in minerals from

Table 3.1 **Instruments used to analyze rocks and soils at spacecraft landing sites**

Alpha particle X-ray spectrometers (APXS) on *MER and MSL*: measures elemental chemistry of rocks and soils using interactions of alpha particles with the target.

Alpha proton X-ray spectrometer (APXS) on *MPF*: measured elemental chemistry of rocks and soils using interactions of alpha particles and protons with the target.

ChemCam on *MSL*: fires a laser and analyzes the element abundances of vaporized areas on rocks and soils a few meters away.

ChemMin on *MSL*: powder X-ray diffraction instrument used to identify minerals.

Gas chromatograph/mass spectrometer (GCMS) on *Viking*: instruments that analyzed chemical compounds in soils.

Dynamic Albedo of Neutrons (DAN) on *MSL*: neutron detector for sensing water.

Imager for Mars Pathfinder (IMP) on *MPF*: lander-mounted digital imager for stereo color images and visible/near-infrared reflectance spectra of minerals.

Mars Hand Lens Imager (MAHLI) on *MSL*: camera that provides close-up views of the textures of rocks and soils.

Mast Camera (MASTCAM) on *MSL*: digital imaging system for stereo color images and reflectance spectra of minerals.

Microscopic Imager (MI) on *MER*: high-resolution camera used to image textures of rocks and soils.

Microscopy, Electrochemistry, and Conductivity Analyzer (MECA) on *Phoenix*: included wet chemistry laboratory, optical and atomic force microscopes, and thermal and electrical conductivity probe.

Miniature Thermal Emission Spectrometers (MINI-TES) on *MER*: identified minerals via thermal infrared spectral characteristics.

Mössbauer spectrometer (MB) on *MER*: identifies iron-bearing minerals and distribution of iron oxidation states by measuring scattered gamma rays.

Panoramic Camera (PANCAM) on *MER*: digital imaging system for stereo color images and reflectance spectra of minerals.

Rock Abrasion Tool (RAT) on *MER*: brushes or grinds rock surface to reveal interiors.

SAM on *MSL*: suite of three instruments (mass spectrometer, gas chromatograph, tunable laser spectrometer) used to identify carbon compounds and to analyze hydrogen, oxygen, and nitrogen isotopes.

Sampling System (SA/SPaH) on *MSL*: includes a drill, brush, soil scoop, and sample processing device.

Surface Stereo Imager (SSI) on *Phoenix*: digital imaging system for stereo color images and reflectance spectra of minerals.

Thermal and Evolved Gas Analyzer (TEGA) on *Phoenix*: furnace and mass spectrometer to analyze ice and soil.

X-ray fluorescence spectrometer (XRFS) on *Viking*: XRF instrument to analyze the elemental composition of soils.

characteristic X-rays generated by electrons. The scanning electron microscope (SEM) utilizes an electron beam to construct an image of the surface of a sample; SEMs are also often equipped with X-ray detectors for chemical analysis. The transmission electron microscope (TEM) uses electrons to probe the structure and composition of very thin samples. X-ray diffraction (XRD), well known to mineralogy students, uses the scattering of X-rays from crystal structures to identify minerals. X-rays emitted from high-energy synchrotron electrons are used to study tiny samples; some applications include microscopic X-ray fluorescence (XRF), scanning transmission X-ray microscopy (STXM), and X-ray absorption near-edge spectroscopy (XANES). Another somewhat exotic technique is proton-induced X-ray emission (PIXE), which determines element abundances from characteristic X-rays produced by a proton beam.

Mass spectrometers are used primarily for measuring isotopic compositions, although some kinds are increasingly utilized to determine element abundances. Most of the time, the mass spectrometer is identified by its source of ions. Gas-source spectrometers introduce extracted elements as a gas that is ionized by a beam of electrons, and are used primarily for measuring the stable isotopes of hydrogen, carbon, nitrogen, oxygen, and noble gases. Thermal-ionization mass spectrometers (TIMS) use a hot filament to ionize an already purified sample. TIMS can measure a wide variety of radioactive and radiogenic isotopes used in geochronology. Inductively coupled-plasma mass spectrometers (ICPMS) use a plasma source, which makes most of the periodic table accessible for measurement. Both single- and multi-collector machines are used. Trace elements are commonly analyzed using a single-collector ICPMS coupled to a laser. Multi-collector ICPMS instruments are used for high-precision isotope analyses.

Ion microprobes use magnetic sector mass spectrometers with focused ion beams that can analyze trace elements and isotopes in micron-size spots on individual mineral grains. The NanoSIMS is a multi-collector mass spectrometer that can be focused to less than 100 nm, originally adapted to study presolar grains.

The characterization of organic molecules in extraterrestrial materials is an emerging priority. Organic contamination on Earth is a big problem – simply holding a meteorite in your fingers can transfer organic matter. The analytical techniques used for organic analyses distinguish compounds by their elemental composition, structure, reactive groups, and mass, and are unfamiliar to most geologists. Isotopic compositions are also measured by mass spectrometry. Some extraterrestrial organic matter is readily extractable from rocks and consists of hydrocarbons having chain (aliphatic) and ring (aromatic) structures, often with nitrogen, oxygen, and sulfur. However, most organic matter occurs as complex macromolecules that can only be studied by demineralizing the sample with harsh acid treatments.

3.4.3 Geochemical, Mineralogical, and Geophysical Instruments Adapted for Landed Operations

A variety of analytical instruments on landers and rovers, originally developed for laboratory use, have been used to process and analyze rocks and soils on the martian surface. These instruments (Table 3.1) were flown on the *Viking* and *Phoenix* landers, the *Mars Pathfinder* (MPF) Sojourner rover, the *Mars Exploration Rovers* (MERs) Spirit and Opportunity, and the *Mars Science Laboratory* (MSL) Curiosity rover. The missions and their instruments are commonly (and sometimes confusingly) identified by their acronyms.

Summary

The ages of units on planetary surfaces are determined from stratigraphy (mostly ejecta deposits from large impact craters) and the densities of craters (crater size–frequency distributions). These relative ages can be calibrated with radiometric ages if samples are available. Planetary science makes use of both long-lived and short-lived radioisotope systems.

Planetary geologic mapping utilizes visible images, combined with spectroscopic data, to define units, recognize structures, determine ages, and interpret origins. Mapping allows the geologic histories of bodies dominated by rock and by ice to be constrained.

Laser or radar altimeters are used to map topography, and gravity anomalies are determined from precise tracking of the trajectories of nearby spacecraft. Magnetic anomalies require magnetometer measurements. Seismometers to measure seismic activity have been carried on only a few missions. Heat flux can be measured by instruments on orbiting spacecraft.

Extraterrestrial materials that can be analyzed in terrestrial laboratories include samples that arrive naturally on Earth (meteorites, interplanetary dust particles, and presolar grains) and those brought back by astronauts (lunar samples) and collected by spacecraft (comet and asteroid particles, solar wind). A number of analytical techniques developed for laboratory analyses have been adapted for remote sensing purposes.

Review Questions

1. How are stratigraphic principles modified for use in planetary geologic mapping and development of planetary stratigraphic timescales?
2. How are crater populations quantified, and how can crater plots be used to tell relative time?
3. What is an equilibrium population of craters, and why is recognizing it important?
4. Distinguish between geochronologies based on long-lived and extinct radioisotopes.
5. For the lower (front) half of Figure 3.10, define two map units. State their characteristics and the characteristics of their contact. Draw a contact on the image between the two units (e.g., use tracing or thin paper, or screen capture the image and place in image-handling software and use a pen tool).
6. In Figure 3.11b, identify and state the observations for why the unit DR5 is almost younger than every other unit around it and why the Subdued Plains are older than every other unit around it.
7. What would be the most difficult step for you to undertake if you had the opportunity to use Figure 3.11a to create a geologic map? Why? Discuss with your classmates.
8. Which geophysical techniques might have been used to construct Figure 3.12 (for both Earth and Mercury)?

SUGGESTIONS FOR FURTHER READING

McSween, H. Y. and Huss, G. R. (2010) *Cosmochemistry.* Cambridge: Cambridge University Press. The appendix of this book describes the analytical techniques used for extraterrestrial materials in detail.

Platz, T., Michael, G., Tanaka, K. L., et al. (2013) Crater-based dating of geological units on Mars: methods and application for the new global geological map. *Icarus,* **225,** 806–827. This paper is an up-to-date reference on the methodology of crater counting, and illustrates how this method was used to determine the ages of mapped units on Mars.

Tanaka, K. L., Anderson, R., Dohm, J. M., et al. (2010) Planetary structural mapping. In *Planetary Tectonics,* eds. Watters, T. R., and Schultz, R. A. Cambridge: Cambridge Press, pp. 351–396. This chapter considers tectonic structures on geologic maps, including implications for interpretations of geologic histories.

Tanaka, K. L., Skinner, J. A., and Hare, T. M. (2011) *Planetary Geologic Mapping Handbook.* Flagstaff, AZ: United States Geological Center, Astrogeology Science Center.

Wilhelms, D. E. (1990) Geological mapping. In *Planetary Mapping,* eds. Greeley, R., and Batson, R. M. Cambridge: Cambridge University Press, pp. 208–260. This chapter gives a complete explanation of geologic map creation, although the data compilation techniques do not reflect current digital tools.

Wilhelms, D. E. (1987) The geologic history of the Moon. *U.S. Geological Survey Professional Paper* 1348. Washington, DC: USGS. This comprehensive work summarizes everything you might want to know about lunar stratigraphy and chronology.

REFERENCES

Brownlee, D. E., Tsou, P., Aleon, J., et al. (2006) Comet 81P/Wild2 under a microscope. *Science,* **314,** 1711–1716.

Burnett, D. S. (2013) The Genesis solar wind sample return mission: Past, present, and future. *Meteoritics & Planetary Science,* **48,** 2351–2370.

Collins, G. C., Patterson, G. W., Head, J. W., et al. (2013) Global geologic map of Ganymede. U.S. Geological Survey Investigations Map 3237. DOI: 10.3133/sim3237.

Crumpler, L. S., Arvidson, R. E., Squyres, S. W., et al. (2011) Field reconnaissance geologic mapping of the Columbia Hills, Mars, based on Mars Exploration Rover *Spirit* and MRO HiRISE observations. *Journal of Geophysical Research,* **116,** E00F24.

Farley, K. A., Malespin, C., Mahaffy, P., et al. (2014) In situ radiometric and exposure age dating of the martian surface. *Science,* **343,** 124–166.

Flynn, G. J., Nittler, L. R., and Engrand, C. (2016) Composition of cosmic dust: sources and implications for the early Solar System. *Elements,* **12,** 177–183.

Kattenhorn, S. A. (2002) Nonsynchronous rotation evidence and fracture history in the Bright Plains region, Europa. *Icarus,* **157,** 490–506.

Neukum, G., Ivanov, B. A., and Hartmann, W. K. (2001) Cratering records in the early Solar System in relation to the lunar reference system. *Space Science Reviews,* **96,** 55–86.

Nguyen, A. N., and Messenger, S. (2011) Presolar history recorded in extraterrestrial materials. *Elements,* **7,** 17–22.

O'Brien, D. P., Greenberg, R., and Richardson, J. E. (2006) Craters on asteroids: reconciling diverse impact records with a common impacting population. *Icarus,* **183,** 79–92.

Platz, T., Michael, G., Tanaka, K. L., et al. (2013) Crater-based dating of geological units on Mars: methods and application for the new global geological map. *Icarus,* **225,** 806–827.

Shoemaker, E. M. (1962) Interpretation of lunar craters. In *Physics and Astronomy of the Moon,* ed. Kopal, Z. New York: Academic Press, pp. 283–359.

Shoemaker, E. M., and Hackman, R. J. (1962) Stratigraphic basis for a lunar time scale. In *The Moon,* eds. Kopal, Z., and Miklhalov, S. K. London: Academic Press, pp. 289–300.

Smith, D. E., Zuber, M. T., Phillips, R. J., et al. (2012) Gravity field and internal structure of Mercury from MESSENGER. *Science,* **336,** 214–217.

Wilhelms, D. E. (1987) The geologic history of the Moon. U.S. Geological Survey Professional Paper 1348. Washington, DC: USGS.

4

Solar System Raw Materials

We explain nucleosynthesis in evolving stars and use this foundation to understand the chemical composition of our own star and of the Solar System. Element abundances are determined from the Sun's spectrum, and from laboratory measurements of the solar wind and chondritic meteorites. The metal-rich Solar System composition reflects the recycling of elements formed in earlier generations of stars. Condensation models of a cooling nebular gas having this composition produced the minerals found in refractory inclusions in chondrites. The deuterium enrichment in organic matter in chondrites suggests that hydrocarbons formed at low temperatures in molecular clouds and were subsequently processed into complex molecules in the solar nebula and in parent bodies. Ices condensed far from the Sun and were incorporated into the giant planets and comets. Element fractionations in the nebula were largely controlled by element volatility or by the physical sorting of solid grains. Separation of isotopes by mass was common in the nebula, although oxygen shows mass-independent fractionation.

4.1 Adding Cosmo to Chemistry

Cosmochemistry is the name given to the study of the chemical composition of the Universe and the processes that produced that composition. In this chapter, we use the methods of cosmochemistry to describe the raw materials – the elements and the compounds formed from their combinations – from which the Sun and its retinue of planets and satellites were fashioned.

The Sun comprises 99.8 percent of the mass of the Solar System, so to a first approximation the solar composition is that of the Solar System. What are the relative abundances of elements in the Sun, and how did that mixture come to be? Surrounding the infant Sun was a **solar nebula**, a swirling cocoon of gas and dust that

eventually became our planetary system. How was the nebula processed to form the minerals that are found in surviving nebular samples from that early period? The initial chemistry of the nebula must have been the same as the solar composition, but the differing compositions of planets indicate chemical fractionation processes that we also need to understand.

To address these problems, cosmochemistry utilizes laboratory analyses of extraterrestrial samples, supplemented by remote sensing (spectroscopic) measurements. The types of materials in hand (meteorites, interplanetary dust particles, samples returned by spacecraft missions, presolar grains) and the principles of spectroscopy were previously described in Chapters 2 and 3. Chemical and mineralogical analyses can be used to decipher the composition of the Solar System and to reveal the processes that produced its raw materials.

4.2 Origin of the Elements

Elements are synthesized primarily in stars, so the raw materials for the formation of planets and the origin of life are stardust. Let's see how elements are made.

4.2.1 Stellar (and Solar) Formation and Evolution

Some astronomical perspective is useful in understanding the composition of our neighboring star. Clusters of stars form together in giant **molecular clouds**, concentrations of gas and dust in interstellar space. An example is shown in Figure 4.1. The molecular clouds are mostly hydrogen (which occurs as H_2 molecules, hence the name) and some helium, with oxygen, carbon, and nitrogen contributing about 1 percent by mass and all the other elements occurring mostly as tiny dust motes. In astronomical jargon, everything but hydrogen and helium are "metals." The newly formed stars inherit the elemental composition of their parent molecular cloud.

Figure 4.1 Hubble Space Telescope image of a star-forming region in the Carina Nebula. Concentrations of gas and dust in this molecular cloud are collapsing to form new stars. Stellar winds compress the cloud into pillars. One new star at the tip of the longest pillar has bipolar outflows. NASA image.

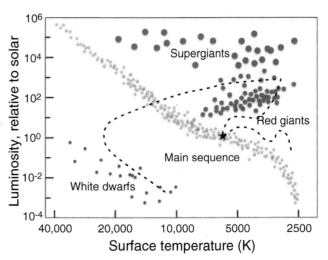

Figure 4.2 Herzsprung–Russell (H-R) diagram showing the distribution of stars in terms of luminosity and temperature. The black star is our Sun. Also shown by dashed lines are the Sun's evolutionary path to the main sequence, and thence to the red giant stage and finally to the white dwarf stage.

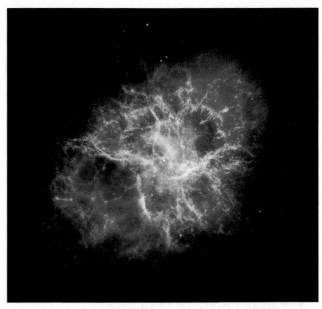

Figure 4.3 The Crab Nebula, the remnant of a supernova that exploded a thousand years ago. X-ray, visible, and infrared images from the Chandra, Hubble, and Spitzer Space Telescopes were combined to produce this composite. Combined NASA and ESA images.

Stars can be classified using the **Herzsprung–Russell (H-R) diagram**, plotting stellar luminosity (brightness is a proxy for mass) versus surface temperature (Figure 4.2). The most massive stars are situated at the top of the diagram, and the hottest stars plot on the left. The main sequence, a band stretching diagonally from upper left to lower right, contains about 90 percent of all stars. Our Sun lies on the main sequence; its position is marked with a black star. A new star appears on the main sequence when its temperature increases to the point where fusion of hydrogen, its main constituent, is triggered. Massive, hot stars near the top of the main sequence evolve at a furious rate and have lifetimes of only a few million years, whereas cooler, dimmer stars last for billions of years. Once the hydrogen fuel in the core of a massive star is exhausted, it expands and cools to form a red giant or a supergiant. Fusion of heavier elements continues within the interiors of red giants, as described in Section 4.2.2. Stars with masses greater than ten times that of the Sun

are the progenitors of supergiants, which end their lives in gigantic explosions called **supernovae** (Figure 4.3). These explosions can leave behind stellar corpses highly compressed by gravity – black holes or neutron stars. Smaller stars like the Sun form red giants that increase in temperature, and eventually slough off their outer layers and migrate across the H-R diagram to form white dwarfs (the Sun's path in Figure 4.2). White dwarfs, too, can undergo supernovae, if they merge with matter from a neighboring star.

4.2.2 Nucleosynthesis, Slow and Fast

The transmutation of hydrogen into helium occurs by fusion in the cores of main sequence stars. (Although we talk about elements in stars, a more appropriate term is "nuclides" because multiple isotopes of the elements are formed.) Hydrogen fusion is often called hydrogen "burning." Although the hydrogen is fuel, fusion in stars is a nuclear rather than chemical reaction, and no combination with oxygen occurs as in conventional burning. Once hydrogen in the core is exhausted, temperatures and pressures increase enough to initiate helium burning. At this point, the star migrates off the main sequence and swells to become a red giant. Hydrogen burning continues in a shell surrounding the helium-burning core, as illustrated in Figure 4.4. Subsequently, the core burns carbon, neon, oxygen, and silicon, as the ashes of one stage become the fuel for the next. Silicon burning finally produces a core of iron nuclei. The fusion of lighter elements into heavier ones results in the conversion of a small amount of mass (M) into energy (E), as described by Einstein's famous $E = MC^2$, where C is the speed of light; this sustains the star's luminosity and explains how the Sun shines.

The production of elements heavier than iron requires more energy than is released by adding particles to the nuclei. No more energy can be gained by fusing iron into heavier nuclei, so the star begins to collapse due to its own gravity. The intense pressures cause the iron core to be converted into a dense mass of neutrons, and a strong neutron flux develops above this interior stellar core. Nuclides beyond iron are mostly formed by capture of these neutrons, with each captured neutron increasing the mass of the nucleus by one unit. If the resulting nuclide is stable, it remains an isotope of the original element. If not, the atom undergoes β-decay, in which a neutron emits an electron to become a proton, thereby changing into an isotope of the next heavier element. Two neutron-capture processes are illustrated in Figure 4.5. The zigzag line of yellow boxes represents stable nuclides, the horizontal arrows illustrate neutron capture, and the diagonal arrows are β-decay. In the slow – or "s" – process, neutrons are captured at a rate that is slow compared to the rate of β-decay. This occurs in the pulsing outer shells of red giants and accounts for about half the elements heavier than iron. The rapid – or "r" – process is neutron capture on such a rapid timescale that multiple neutrons are captured before β-decay occurs. Some stable nuclides, such as ^{87}Rb or ^{94}Zr, can only be reached by the r-process (Figure 4.5).

Nucleosynthesis by the r-process occurs when the collapsing mass of the star bounces off the dense neutron core, producing a supernova explosion. The merger of neutron stars, first observed in 2017, also results in r-process nucleosynthesis, although the relative role of this source versus supernovae in producing heavy elements is unclear.

A supernova event, occurring when the collapsing stellar core implodes, ejects the newly synthesized elements back into interstellar space. These elements can be incorporated subsequently into new stars, so that the "metal" content increases with each new generation of stars. First-generation stars contain only hydrogen and helium. The heavy element content of the Sun requires that it formed in a later generation from elements produced in earlier stars. Since planets are formed from heavy elements, they are only associated with stars that are metal-rich.

In addition to **stellar nucleosynthesis**, there are two other ways to form new elements. Many of the nuclides formed during a supernova are unstable and undergo radioactive decay to form stable "radiogenic" nuclides over timescales of millions to billions of years. Of course, the s- and r-processes produce new nuclides from unstable nuclides as well, but on an extremely short timescale, so we distinguish that from radioactive decay that occurs outside stars. Other so-called "cosmogenic" nuclides are produced through irradiation by high-energy cosmic rays. Both of these processes are important in accounting for the chemical composition of the solar nebula and of the planets that formed within it.

4.3 Composition of the Solar System

High-resolution measurements of the spectrum of light emitted from the Sun reveal thousands of absorption lines (Fraunhofer lines, as described in Section 2.4) caused by the various elements absorbing light at specific wavelengths. The widths of these lines are proportional to the element abundances. Combined with a model of the solar atmosphere and an understanding of how the lines are formed and interact, the **solar spectrum** can be used to determine the chemical composition of the Sun.

An independent measure of the chemistry of the Sun can be gained from analysis of the solar wind – charged particles emitted from the solar atmosphere. The *Genesis* spacecraft collected solar wind during its 2.5 years in orbit outside the Earth's magnetic field (which repels charged solar particles). The solar wind sample (10^{20} atoms = 0.4 mg) was returned to Earth in 2004. A parachute unfortunately failed to open and the spacecraft crashed in the Utah desert, shattering many of the

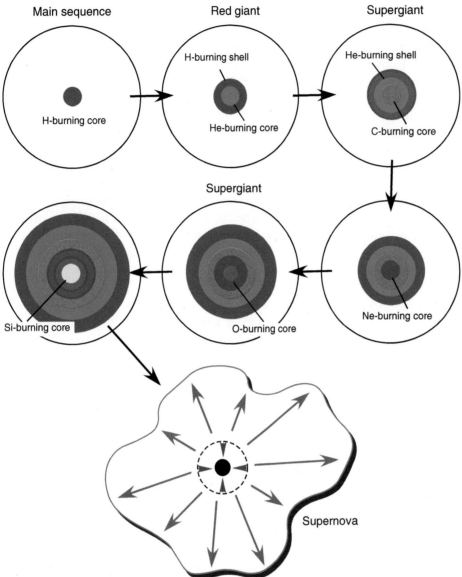

Figure 4.4 Schematic evolution of stars of greater than ten solar masses. Stars on the main sequence fuse hydrogen in their cores, giving way to helium burning as they become red giants. Cores become burning shells, as carbon, neon, oxygen, and silicon successively fuse. Once iron forms in the core, fusion ends and the star explodes as a supernova. Modified from McSween and Huss (2010).

solar wind collector plates. However, by careful cleaning and analysis, scientists have been able to salvage measurements of the solar wind composition. Corrections must be applied to the measured values, because the elements were fractionated on expulsion from the Sun, which preferentially ejects elements that are easier to ionize. This solar wind sample has been especially important in providing data on the isotopic composition of the Sun.

The solar composition is no longer the same as it was when first formed. Nuclear burning has transformed some hydrogen into helium, and in the process has modified the isotopic compositions of hydrogen, helium, carbon, oxygen, and nitrogen. Also, lithium has become severely depleted in the Sun due to nuclear reactions.

For many years, it has been recognized that certain meteorites (chondrites; see Box 4.1) have chemical compositions that mimic that of the Sun. We can think of chondrites as a sort of solar sludge; if you could cool and condense a gas of solar composition into solid, rocky matter, you would make a chondrite. Not all chondrites

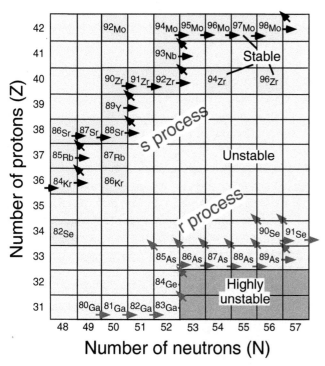

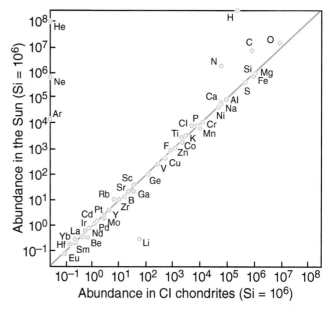

Figure 4.5 A portion of the chart of the nuclides, illustrating how the s-process (upper arrows) and r-process (lower arrows) produce new elements by neutron capture. Yellow boxes are stable nuclides, and light gray boxes are unstable nuclides that undergo β-decay that shifts them up and to the left, as shown by diagonal arrows. Neutrons are captured much faster in the r-process, until a nuclide finally reaches a highly unstable (dark gray) box and undergoes a series of β-decays until a stable nuclide is reached. Modified from McSween et al. (2003).

Figure 4.6 Comparison of element abundances in the Sun with those in CI carbonaceous chondrites. The only elements that deviate are gaseous elements that are hardest to condense into solids, and lithium which has been destroyed in stellar nucleosynthesis. After McSween et al. (2003).

are the same, and one kind in particular – CI carbonaceous chondrites – provides the best match. Figure 4.6 compares element abundances in the Sun and in CI chondrites, both normalized to the same number of silicon atoms. The main deviations are hydrogen, helium, nitrogen, carbon, and oxygen, which exist largely in the gas phase even at room temperature and thus are hard to condense, explaining their lower abundances in chondrites. Lithium has been depleted in the Sun during the time since the chondrites formed, so for this element the chondrites record the composition of the nebula better than does the present-day Sun. Because meteorites can be analyzed in the laboratory, this incredible agreement provides a way to estimate relative solar abundances for elements that have low solar abundances or do not have prominent absorption lines.

The abundances of elements in the Solar System, relative to a million silicon atoms, are illustrated in Figure 4.9. To define the abundances of some elements the solar value is used, whereas for others CI chondrite analyses provide the best determination. The Solar System

element abundances are often called **cosmic abundances**, a term coined by geochemist Victor Goldschmidt in 1937. However, stellar spectroscopy indicates that our Solar System does not have exactly the same composition as most of the galaxy, which we might call the cosmos. Relative to nearby solar-sized stars of comparable age, the Sun has a higher proportion of "metals," reflected in its 50 percent higher Fe/H ratio. Because this book is about planetary geoscience, we will focus on Solar System abundances.

Let's take a more careful look at Figure 4.9. First, note that the abundances are plotted on a logarithmic scale, to allow for the fact that they vary over many orders of magnitude. Hydrogen and helium, originally produced in the so-called Big Bang that formed the Universe, make up more than 98 percent of the mass of the Solar System. Moving to the right in the diagram, we see that lithium, beryllium, and boron have very low abundances. For the most part, these elements are not made in stars and were formed in only small amounts during the Big Bang or through irradiation by cosmic rays. Carbon, oxygen, neon, and silicon are relatively abundant, reflecting their roles as products of fusion in massive stars (Figure 4.4). The peak in the abundance of iron occurs because it is a very stable nuclide and represents the end point of normal fusion reactions. For the elements heavier than iron that are formed during supernovae, abundances decrease with increasing mass because progressively more

BOX 4.1 **A CRASH COURSE IN CHONDRITE PETROLOGY AND CLASSIFICATION**

Chondrites take their name from chondrules (Greek for "grain" or "seed"), millimeter-sized droplets of melted silicates. The chondrules were incorporated, along with irregularly shaped inclusions and metal grains, into a fine-grained assortment of minerals that we refer to as matrix. Chondrites have ultramafic compositions, being composed mostly of olivine and pyroxene. A photomicrograph of a chondrite thin section is shown in Figure 4.7. This image is similar to what was first seen by Henry Sorby, inventor of the petrographic microscope, when he first described chondrules as "drops of a fiery rain."

Chondrites exhibit compositional and textural differences that allow their classification into chemical groups (vertical axis of Figure 4.8) that are generally interpreted to indicate their formation in separate parent bodies. The "carbonaceous" chondrites are separated into the CI, CM, CR, CB, CH, CV, CO, and CK groups, with "C" indicating carbonaceous and the second letter standing for the name of a type meteorite specimen for the group. The "ordinary" chondrites, so called because they are the most common meteorite type to fall to Earth, are divided into the H, L, and LL groups that differ in oxidation state, which controls the amounts of iron partitioned into metal versus silicates; "H" and "L" stand for high and low contents of iron metal, and "LL" denotes a group with even lower iron metal content that was recognized after the H and L groups had already become established in the literature. The Rumaruti (R) group is similar to ordinary chondrites but

Figure 4.7 Thin-section photomicrograph of the Tiechitz H3 chondrite that fell in what is now the Czech Republic in 1878. Numerous millimeter-sized chondrules composed of olivine, pyroxene, and glass are contained in a fine-grained matrix.

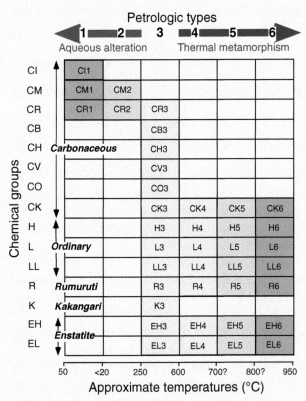

Figure 4.8 The classification scheme for chondritic meteorites. Chondrites are divided into chemical groups (shown on the vertical axis) and characterized by petrologic type, reflecting the degree of thermal metamorphism or aqueous alteration (top of the figure). Only the filled boxes (for example, CM2 or L5) are known chondrite types. Modified from McSween and Huss (2010).

more highly oxidized, and the Kakangari (K) group is somewhat similar to carbonaceous chondrites. "Enstatite" chondrites, identified by "E," are extremely reduced, resulting in a high abundance of Mg–pyroxene and the occurrence of some unusual minerals not found on our planet. They are divided into high-iron and low-iron groups ("EH" and "EL," respectively).

Regardless of chemical group, all chondrites are basically cosmic sediments composed of agglomerations of solid grains that once existed in the solar nebula. Although the chondrules were melts in the nebula, the host chondritic rocks have never been melted or chemically differentiated. This is the reason they have retained nearly solar compositions. That is not to say, however, that they have not been modified at all. Most chondrites have either experienced thermal metamorphism or aqueous alteration within their parent bodies, depending on whether the body initially formed from rock alone or as combinations of rock and ice which subsequently melted to form fluids. These changes are described as "petrologic types" (numbers at the top of Figure 4.8). Type 3 chondrites are nearly unaltered, whereas types 4, 5, and 6 indicate increasing intensities of thermal metamorphism. Dry chondrite metamorphism resulted in recrystallization that blurred the chondrule outlines, as well as the erasure of zoning in mineral grains and the crystallization of chondrule glass into feldspar. Petrologic types 2 and 1 reflect increasing aqueous alteration. Melting of ice produced fluids that altered olivine and pyroxene to form phyllosilicates like serpentine, and formed other secondary minerals. Combining the chemical group and petrologic type, a chondrite might be classified as CM2 or L5, as illustrated in the labeled boxes of Figure 4.8. Boxes with interior labels in this figure indicate chondrite types that are found in the world's collections. This quirky but serviceable taxonomic system of petrologic types is the result of becoming widely adopted before it was recognized that two different processes were involved.

Even after a century of study, the chondrites remain enigmatic objects. We still do not know exactly how chondrules formed, although ideas abound. Most enigmatic are the CI chondrites which, unlike all the other groups, contain no chondrules at all. They are essentially all matrix material that has been heavily altered by fluids to form serpentine, magnetite, carbonates, and other secondary minerals. How what might be considered a congealed mud puddle has so faithfully retained its solar composition is a mystery.

energy is required to force more neutrons into the nuclei of heavier atoms. Finally, the peculiar sawtooth pattern in Figure 4.9 results from the higher abundances of elements with even rather than odd atomic numbers. The even-numbered elements are more stable because their nuclei are bound together more strongly. To remove the sawtooth structure from diagrams that plot element abundance versus atomic number, such as rare earth element diagrams, geochemists typically normalize the abundances to chondrites.

The chemical composition of the Solar System can thus be understood as a mixture of primordial hydrogen and helium with "metals" that are the products of nucleo-synthesis in earlier generations of stars whose matter has been recycled. These earlier stars provided not only the heavy elements needed for the formation of rocky planets like the Earth, but also the biogenic elements used by our planet's life forms. As Carl Sagan noted, "we are all made from star stuff."

4.4 Minerals, Ices, and Organic Matter

4.4.1 Condensation of Minerals

Models of the physical and dynamic conditions in the solar nebula suggest that the interior near the early Sun became very hot, with temperature decreasing outward. The solid dust grains in the interior would have been vaporized and then recondensed as the nebula cooled.

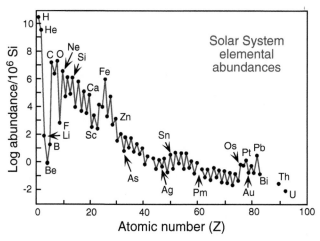

Figure 4.9 Solar System abundance of the elements, normalized to a million silicon atoms.

Nebular pressures (estimated at only 10^{-3} to 10^{-5} atm) were too low to allow liquids to be stable, so nebular vapor would have condensed directly into solid minerals.

The **condensation** temperature of an element is the temperature at which the most refractory solid phase containing that element first becomes stable in a cooling gas of solar composition. Equilibrium condensation can be modeled using thermodynamic data for all potential gas and mineral species containing each element in the nebula. At any temperature, the gas or mineral species that has the lowest Gibbs free energy is the stable phase. Condensation calculations were done long before digital computers arrived, but modern calculations using computers allow more complex computations involving 23 elements with the highest Solar System abundances.

So what does the condensation sequence look like? The curves labeled with element symbols in Figure 4.10 show the fraction of each element condensed from the nebular gas as temperature decreases. However, most elements condense as minerals rather than in native form. The condensation temperatures of various solid phases are illustrated by the positions of italicized mineral names in this diagram. Not every mineral in the sequence actually condenses directly from the vapor; instead, some early-condensed minerals react with the cooling gas to produce new minerals or to modify the compositions of solid solution minerals. This situation is analogous to Bowen's reaction series, which describes the discontinuous crystallization or continuous reactions of minerals in a cooling magma.

Corundum (Al_2O_3) is the first condensate that contains a major element, although rare elements like osmium and zirconium may condense as nuggets at even higher temperatures. The next mineral to condense is perovskite ($CaTiO_3$). Corundum then reacts with the vapor to form spinel ($MgAl_2O_4$) and melilite ($Ca_2[MgSi, Al_2]SiO_7$), and the latter reacts in turn to form diopside ($CaMgSi_2O_6$) at lower temperature. This is followed by the condensation of iron–nickel metal, anorthite ($CaAl_2Si_2O_8$), and forsterite (Mg_2SiO_4); the latter subsequently reacts with the gas to form enstatite ($Mg_2Si_2O_6$). All of these minerals appear above 1250 K.

Below this temperature, solid solutions come into play. Anorthite begins to incorporate sodium and potassium to form other feldspars, and olivine and pyroxene become increasingly iron-rich. Metallic iron reacts with sulfur in the vapor to produce troilite (FeS), and then oxidizes to form magnetite (Fe_3O_4). Finally, hydrated silicates like serpentine can form by reaction of previously condensed silicates with water vapor.

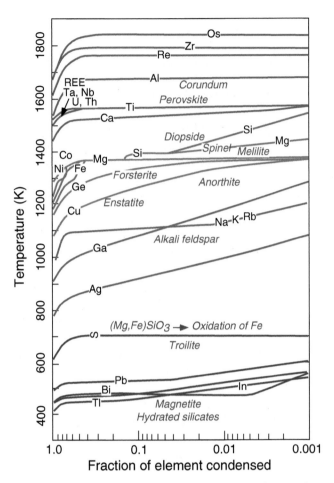

Figure 4.10 The calculated condensation sequence for a cooling gas of solar composition at 10^{-4} atm pressure. Condensation points for minerals are shown in italics, and curves show the fraction of each element condensed as a function of temperature. Modified from Grossman and Larimer (1974).

The chondrites contain direct evidence of nebular condensation. The mineral assemblage of some bizarre refractory inclusions (Figure 4.11) found in chondrites is very similar to that predicted by the condensation sequence. Also called **calcium–aluminum inclusions** (CAIs) because they are dominated by those elements, the CAIs consist of corundum, melilite, spinel, diopside, anorthite, and perovskite – all minerals predicted to condense at high temperatures. The CAIs sometimes contain nuggets of platinum-group elements like osmium and iridium as well. The mineral hibonite ($CaO·9Al_2O_3$), which was not included in the original calculations because thermodynamic data were not available, takes the place of corundum in some CAIs. Another type of inclusion, called amoeboid olivine aggregates (AOAs), consists of forsterite, with anorthite, spinel, diopside, and metal, representing the minerals predicted to condense at somewhat lower temperatures.

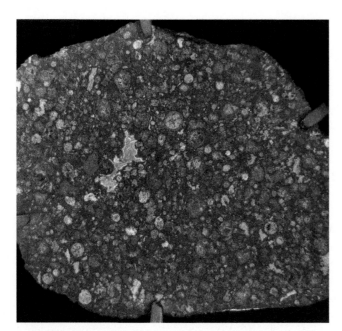

Figure 4.11 Cut slab of the Allende (Mexico) carbonaceous chondrite, showing a large white calcium–aluminum inclusion and numerous chondrules.

We should also note that evaporation is the opposite of condensation, and an alternative explanation for these inclusions could be that they are residues remaining after partial evaporation. In either case it is clear that the solar nebula experienced temperatures high enough to vaporize (or distill) nebular solids, before cooling to low temperatures.

4.4.2 Making Organic Molecules

Although the carbon atoms in our bodies were originally produced by nucleosynthesis in stars, these atoms now occur as parts of complex **organic molecules.** (Note: Although the term "organic" resembles "organism," it does not necessarily connote a connection to life; organic compounds can be abiotic combinations of carbon with hydrogen, and sometimes oxygen, nitrogen, and sulfur.) Chondrites contain organic matter too. How and where did these organic molecules form?

Using telescopes fitted with infrared and microwave spectrometers, astronomers have identified many organic compounds in molecular clouds. Diagnostic absorption bands occur when the bonds in organic molecules bend and stretch as they absorb certain wavelengths of energy. Molecular clouds can contain organic matter because they are relatively opaque to ultraviolet radiation, which destroys organic molecules in interstellar space. Within a dense cloud, simple carbon compounds like CO and CH_4 predominate, although somewhat more complex compounds also occur. In the frigid conditions within molecular clouds, deuterium (^2H) is utilized preferentially over hydrogen in chemical reactions, so the organic hydrocarbons formed there are distinctly deuterium-rich.

When the solar nebula formed, organic compounds inherited from the parent molecular cloud were further heated to form even more complex molecules that were stable at higher temperatures. Such molecules are composed of carbon chains (aliphatic compounds) and rings (aromatic compounds) with side branches containing other elements. The nebular-processed organic matter was then incorporated into planetesimals, the parent bodies of chondrites, where further reactions occurred.

Carbonaceous chondrites and the comet grains returned by the *Stardust* spacecraft provide direct sampling of extraterrestrial organic compounds for analysis using the panoply of methods available in the laboratory. Some molecules are easy to characterize, because they are soluble in various solvents and can be easily extracted from the meteorites. These include amino acids, the building blocks of life, as well as other compounds built of hydrocarbon chains or connected multiple rings with as many as 30 carbon atoms. Most organic matter in chondrites, however, consists of large, insoluble molecules (kerogen) that can only be separated by dissolving away the minerals of the host rock. Characterizing this gooey material poses a daunting challenge, and requires harsh degradation techniques that break the macromolecules into smaller fragments. The average composition of this material, expressed relative to 100 carbon atoms, is $C_{100}H_{70}N_3O_{12}S_2$ in one well-studied carbonaceous chondrite. The insoluble material is intimately mixed with flakes, tubes, and spheres of graphite.

The deuterium enrichment in extraterrestrial organic matter indicates a connection with molecular clouds. However, most of that material was likely processed in the nebula by heating and cosmic-ray irradiation or within meteorite parent bodies during aqueous alteration. The organic matter in comets appears to be more primitive and may be related more closely to molecules inherited from interstellar space.

4.4.3 Condensation of Ices: The Only Stuff Left

We now turn to the simple gaseous molecules, H_2O, CH_4, and NH_3, that were abundant in the nebula but not used completely in the formation of complex organic molecules. What happened to these gases? In the condensation sequence (Figure 4.10), we saw that some water can be incorporated in hydrous silicates at very low

BOX 4.2 STARDUST IN CHONDRITES

Vaporization of all the solid dust would have formed a homogeneous gas, thereby erasing any isotopic heterogeneity in the solar nebula. However, some variations in the amounts of s- and r-process isotopes are found in chondrites, indicating that not everything in the solar nebula was vaporized. Can the solid carriers of these isotopes be separated from normal Solar System material comprising the bulk of the chondrites? Presolar grains formed around other stars or expelled from supernova explosions should be recognizable by their exotic isotopic compositions. Surviving presolar grains are referred to more prosaically as stardust.

In 1987, cosmochemist Edward Anders and his coworkers first isolated stardust grains from a chondrite. Their method employed stepwise dissolution of a carbonaceous chondrite using harsh acid, while monitoring the presence of exotic nuclides (in this case, some s- and r-process isotopes of neon and xenon) in the residue at each step. After numerous steps, less than one-thousandth of the original mass remained, and the purified residue suddenly turned from black to white. Anders' prize was a tiny amount of carbon dust that proved to be miniature (nanophase) diamonds (Figure 4.12). These grains likely condensed from vapor as a carbon-rich red giant sloughed off its exterior gas; the grains were subsequently tagged with distinctive neon and xenon isotopes from the star's interior when it exploded and propelled the newly synthesized nuclides outward.

Buoyed by this success, the team soon discovered grains of another stardust mineral, moissanite (SiC). Measurement of the isotopic compositions of silicon and carbon identified this phase as presolar. Since then, other kinds of stardust, including graphite as well as unusual oxides and nitrides that occur as minute inclusions inside graphite and moissanite, have been recognized. Even though presolar silicate grains are more abundant than the carbon-rich stardust, they were not recognized until later because they are destroyed by the harsh chemical treatments used to isolate the carbon-rich grains. The identification of presolar olivine and pyroxene had to wait until they could be picked out in isotopic maps of chondrite thin sections made using the ion microprobe.

The s- and r-process isotopes used to identify presolar grains provide ground truth for the theoretical astrophysical models of nucleosynthesis in stars, as well as identify new nucleosynthesis pathways not yet considered. Who could have dreamed that the inner workings of stars could be better understood from tiny motes of dust in meteorites?

Figure 4.12 Tiny (nanophase) presolar diamonds, greatly magnified under an electron microscope. The yellow color results from trapped nitrogen and other gases. Image courtesy of E. Anders.

temperatures, but there is no obvious place to put methane or ammonia. At temperatures below about 200 K these gases would condense as **ices**.

Our understanding of ices in the Solar System is rudimentary, because they have been studied only by remote sensing. Ices that condensed at very low temperatures may have originally been amorphous, but they would crystallize when warmed. Ices were important in the formation of the giant planets and their satellites, which consist of rocky cores with mantles of ice or their high-pressure equivalents. At even greater distances from the Sun – in the Kuiper belt and Oort cloud – condensed ices occur as major constituents of comets. The preponderance of ice at greater heliocentric distance has led to the concept of a nebular "snow line" beyond which water ice was stable. The snow line must have been located within the present asteroid belt, since the most distant asteroids originally contained

Solar nebula composition

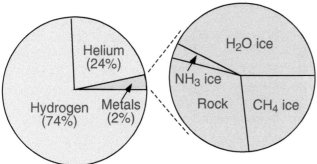

Figure 4.13 Abundance (in wt%) of the major components of the solar nebula. The diagram on the right expands the "metals" field (astronomical jargon) into condensed ices and rock.

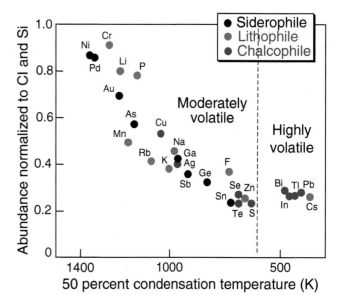

Figure 4.14 Element abundances in CV carbonaceous chondrites as a function of volatility, expressed as the temperature at which half of the element has condensed from vapor to solid. In the nebula, volatility ruled, regardless of the element's geochemical affinity. After Palme and Boynton (1993).

ice that melted to cause the aqueous alteration observed in some chondrites.

The proportion of ices to rock by mass in the solar nebula was appreciable. Figure 4.13 illustrates the nebular composition in terms of hydrogen, helium, and metals. The right-hand diagram expands the metal portion into condensed H_2O, CH_4, and NH_3 ices and rock. Jupiter and Saturn formed mostly from nebular hydrogen and helium, Uranus and Neptune primarily from ices, and the terrestrial planets from rock.

4.5 Chemical Fractionations in the Solar Nebula

Condensation, described in the previous section, offers opportunities to separate elements and isotopes. These fractionations can have profound effects on the compositions of planets.

4.5.1 Element Fractionations

The geochemical behavior of elements can be distinguished as follows: **lithophile** (rock-loving) elements tend to occur as silicates or oxides; **siderophile** (iron-loving) elements combine with iron into metal alloys; **chalcophile** (sulfur-loving) elements react with sulfur to form sulfides; and **atmophile** elements tend to form gases and reside in the atmosphere. These geochemical affinities are not absolute, as many elements can exhibit multiple preferences. Elements in planets are partitioned into core, mantle/crust, and atmosphere depending on these geochemical affinities. Even within the crust, elements can be concentrated into ore deposits based on their geochemical preferences.

In the solar nebula, though, element distributions were mostly governed by volatility. Elements that condense from the nebular vapor at high temperatures are said to be **refractory**, and those that condense at low temperatures are **volatile**. The importance of volatility (easier to say that refractorility) is illustrated in Figure 4.14, a plot of the abundance of elements in carbonaceous chondrites, normalized to CI chondrite (i.e., solar values), versus the temperature at which 50 percent of the element has condensed (this is a commonly used measuring stick for element volatility in the nebula). The element abundances decrease systematically with volatility, regardless of whether they are lithophile, siderophile, or chalcophile. This diagram shows that moderately and highly volatile elements did not fully condense in most chondrites (CI chondrites are the exception).

Documenting the separation, or fractionation, of elements in the nebula is important for understanding the compositional differences among planets. The various groups of chondrites, which are surviving nebular samples, show at least three types of chemical fractionations, relative to the Solar System composition:

1. enrichment or depletion of refractory elements (for example, calcium, aluminum, and titanium);
2. depletion of moderately volatile elements (for example, manganese, potassium, and germanium) and highly

volatile elements (for example, bismuth, indium, and lead), as already shown in Figure 4.14;

3. enrichment or depletion of siderophile elements (for example, iron, nickel, iridium, and gold).

For the first two types of fractionation, the common property is volatility, and this can be accomplished by condensation (or evaporation). The differentiated planets, as well as igneous meteorites (achondrites) from differentiated planetesimals, are even more depleted than the chondrites in volatile elements. It is not known whether the building blocks of planets were already depleted by condensation before planet formation, or evaporative loss of volatiles occurred as the planets were melted and differentiated. Either way, fractionation of elements by volatility was a significant process in the early Solar System.

Physical sorting of solid grains is another process that leads to chemical fractionation. Concentration or depletion of siderophile elements resulted when metal grains were separated from silicate material. The high density of metal relative to silicate could have promoted size sorting or preferential settling of metal grains toward the nebula midplane. Similarly, CAIs (and the refractory elements they contain) are much more abundant in carbonaceous chondrites than in other chondrite classes, suggesting that some process concentrated these inclusions in the nebula region where carbonaceous chondrites formed.

4.5.2 Isotope Fractionations

During nebular processes such as condensation or evaporation, isotopes can also be fractionated by different amounts, depending on temperature. Isotope fractionations can also occur during geologic processes like melting and crystallization.

To illustrate how isotopes are fractionated, let's consider oxygen, which has three stable isotopes: ^{16}O, ^{17}O, and ^{18}O. Oxygen isotopes in terrestrial rocks have been fractionated according to their masses, as illustrated in Figure 4.15. In this diagram, $\delta^{17}O$ and $\delta^{18}O$ refer to the ratios of those isotopes to ^{16}O, relative to a standard. The difference between ^{18}O and ^{16}O is two mass units, twice the difference between ^{17}O and ^{16}O. Thus, any mass-dependent process that can modify $^{17}O/^{16}O$ will be twice as effective in modifying $^{18}O/^{16}O$. As a consequence, geologic processes spread out the oxygen isotopic compositions of terrestrial rocks along a mass-fractionation line with slope = ½ in Figure 4.15. Various classes of meteorites each define their own mass-fractionation lines, parallel to the terrestrial line, above or below it depending on their starting

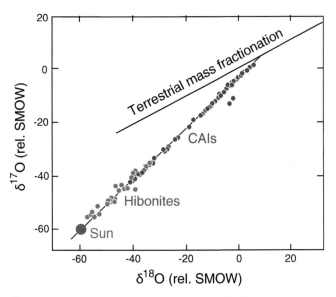

Figure 4.15 Oxygen isotopic compositions of refractory inclusions (CAIs) and individual hibonite grains from CAIs in chondrites. The δ notation refers to the ratio of the oxygen isotope to ^{16}O (per mil), relative to standard mean ocean water (SMOW). The composition of the Sun, as measured from solar wind, lies on an extension of the CAI–hibonite line. Modified from McKeegan et al. (2011).

compositions. Most isotopic fractionations in planets and planetesimals were mass-dependent.

The CAIs in chondrites, however, define a line with slope = 1 in Figure 4.15. Hibonites, the earliest condensing phase in CAIs, show even more extreme oxygen isotopic compositions. When oxygen isotopes in CAIs were first measured in 1973, Robert Clayton and his coworkers interpreted this steeper line to reflect mixing of two end members: nebular gas plotting on or above the terrestrial mass-fractionation line, and presolar grains containing pure ^{16}O (plotting on an extension of the CAI line). Subsequently, this trend has been explained as resulting from mass-independent fractionation in the solar nebula. Carbon has a high Solar System abundance, so much of the nebular oxygen was originally tied up in carbon monoxide (rather than carbon dioxide) gas. Because ^{16}O is much more abundant than the other oxygen isotopes, there were large differences in the amounts of $C^{16}O$, on the one hand, and $C^{17}O$ and $C^{18}O$ on the other. One idea is that this difference caused the less abundant $C^{17}O$ and $C^{18}O$ molecules to be preferentially dissociated into elements by irradiation. Another idea is that the oxygen isotopes became fractionated as vapor condensed into solids. In either case, the heavy oxygen isotopes then reacted with nebular hydrogen to form water vapor, which in turn reacted with condensates to form the compositions of the Earth and other bodies. Measurement of the oxygen isotopic composition of the Sun (Figure 4.15) from solar wind collected by the *Genesis* spacecraft has confirmed that it is ^{16}O-rich. Thus, it is not the CAIs that are isotopic oddballs – it is the planets and planetesimals whose oxygen isotopic compositions were determined by fractionation in the solar nebula.

Summary

The abundance of heavy elements in the Sun indicates that the Solar System formed from elements synthesized in earlier generations of stars. Condensation of a cooling nebular gas of solar composition produced a sequence of minerals that comprise the chondrites, which are surviving samples of nebular solids. Condensation also formed ices at greater heliocentric distances. Organic matter inherited from interstellar space was processed in the nebula and parent bodies to form more complex molecules. Fractionation by element volatility or physical sorting of solid grains resulted in the compositional variations now seen in Solar System bodies.

Having examined the raw materials available for Solar System formation, we will next consider the transformation of a swarm of gas molecules, mineral dust, and organic molecules around the infant Sun into a planetary system.

Review Questions

1. Why does the Sun shine?
2. How do stars evolve on the H-R diagram? What nucleosynthesis processes occur at different stages? How do elements get recycled?
3. What do we mean by Solar System element abundances, and how are they determined?
4. What are chondrites, and why are they important?
5. What is the condensation sequence?
6. How do the geochemical and cosmochemical behaviors of elements differ?

SUGGESTIONS FOR FURTHER READING

Ebel, D. S. (2006) Condensation of rocky material in astrophysical environments. In *Meteorites and the Early Solar System II*, eds. Lauretta, D. S., and McSween, H. Y. Tucson, AZ: University of Arizona Press, pp. 253–277. An excellent summary of modern condensation calculations.

Lodders, K. (2003) Solar System abundances and condensation temperatures of the elements. *Astrophysical Journal*, **591**, 1220–1247. A comprehensive discussion of solar system abundances of the elements from the perspective of a cosmochemist.

McSween, H. Y., and Huss, G. R. (2010) *Cosmochemistry*. Cambridge: Cambridge University Press. A more thorough discussion of the origin of the elements, their Solar System abundances, and fractionations.

Tolstikhin, I., and Kramers, J. (2008) *The Evolution of Matter*. Cambridge: Cambridge University Press. A more comprehensive treatment of Solar System abundances, stellar evolution, and nucleosynthesis than in this book.

REFERENCES

Grossman, L., and Larimer, J. W. (1974) Early chemical history of the solar system. *Review of Geophysics and Space Physics*, **12**, 71–101.

McKeegan, K. D., Kallio, A. P. A., Heber, V. S., et al. (2011) The oxygen isotopic composition of the Sun inferred from captured solar wind. *Science*, **332**, 1528–1532.

McSween, H. Y., and Huss, G. R. (2010) *Cosmochemistry*. Cambridge: Cambridge University Press.

McSween, H. Y., Richardson, S. M., and Uhle, M. E. (2003) *Geochemistry: Pathways and Processes*, 2nd edition. New York: Columbia University Press.

Palme, H., and Boynton, W. V. (1993) Meteoritic constraints on conditions in the solar nebula. In *Protostars and Planets III*, eds. Levy, E. H., and Lunine, J. I. Tucson, TX: University of Arizona Press, pp. 979–1004.

5

Assembling Planetesimals and Planets

We discuss the formation of a dusty accretion disk around an infant star and, from that, the planets. Telescopic observations of young stars suggest planet formation required only a few tens of millions of years, in agreement with a Solar System timescale based on measurements of radioactive isotopes in meteorites. The age of the Solar System, 4.567 billion years, is determined from calcium–aluminum inclusions (CAIs), the first-formed solids. Numerical simulations of planetary accretion further support this timescale and constrain the widths of feeding zones. The compositions of the terrestrial planets are broadly chondritic, but depletions in volatile elements suggest their assembly from already differentiated planetesimals. The ice giants have rocky cores that directly accreted nebular ices, and the even more massive gas giants have ice giant-like cores that swept up nebular gas. Leftover planetary building blocks – asteroids and comets – provide more detailed insights into planet formation processes. We complete this story by discussing the origin of the Moon by a giant impact, and the related topic of orbital perturbations possibly caused by migrations of the giant planets.

5.1 Dust to Disk

The origin of the Sun's planetary system has captured scientists' attention for centuries. Our modern understanding comes in part from constraints on processes, compositions, and chronology provided by studies of nebular materials in our own Solar System. Complementing this approach are astronomical observations of other solar systems at various evolutionary stages, as well as numerical models of the assembly of planets from gas and dust. These different approaches provide views of the solar nebula from inside and out. In this chapter, we will consider all these ways of studying planet formation.

Two actual images of swirling dust around young stars are illustrated in Figure 5.1. A new star will consume almost all of the nebular mass, and the small fraction that remains around the star necessarily acquires most of the system's angular momentum. In these images, the orbiting clouds of dust have become flattened into rotating **accretion disks** as dust settles to the nebular midplane. In contrast to these fuzzy images, an artist's conception of the solar nebula (Figure 5.2) suggests a rather turbulent dust disk. Our Sun's retinue of planets somehow emerged from such a debris-laden nebula. Let's see how this happened.

5.2 Stages of Accretion

Planets form around young, rapidly evolving stellar objects, and the stages of planet formation can be related to the evolution of their parent stars.

5.2.1 Evolution of Stellar Objects
The process of star formation can be inferred from observations of stellar objects at various evolutionary stages.

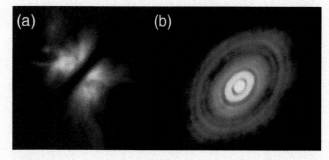

Figure 5.1 Two images of real accretion disks around young stars. (a) Star HH-30, viewed edge-on, is an image from the Hubble Space Telescope. (b) HL Tauri, from the Atacama Large Millimeter Array, shows gaps that were probably opened by unseen planets.

Figure 5.2 Artist's rendition of a turbulent solar nebula.

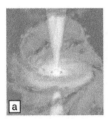

Figure 5.3 Sketches of the evolution of stellar objects.
(a) FU-Orionis stage, in which the protostar is enclosed in an optically thick nebula. (b) Classic T-Tauri stage, in which the gas and dust have formed an accretion disk. (c) Naked T-Tauri stage, in which accretion onto the star has ceased and the formation of planetesimals has cleared the gas and dust. After McSween and Huss (2010).

In the first stage, a young protostar rapidly accretes matter from the enclosing nebula. At this stage, we cannot actually see the star because it is embedded within an optically thick swarm of dust and gas (Figure 5.3a), so the star's energy is irradiated from the nebula itself. Objects at this stage periodically eject matter from their poles; the violent ejections are sometimes called "FU-Orionis outbursts" after the star in which they were first observed. Orbiting particles rotate faster than the gas, so they effectively encounter a "headwind" that slows them down and causes them to drift inward toward hotter regions of the nebula. The formation of water containing isotopically heavy oxygen and the condensation (or evaporation) of solids to form CAIs, both described in Section 4.5, likely occurred in the dusty nebula when the infant Sun was at this stage.

The next stage, called a "classical T-Tauri star" after its type sample, has an optically thick nebula that has been flattened into an accretion disk (Figure 5.3b). Accretion of matter to the star continues, but at a slower rate. This

protostar may have bipolar outflows and strong stellar winds. Grains in the disk began to coagulate, and chondrules may have formed at this stage. Clumps of accreted matter grow into **planetesimal** size, and the earliest-formed bodies are heated to the point of melting by the rapid decay of short-lived radionuclides.

A stellar object at the final stage has an optically thin disk (Figure 5.3c), leading to its description as a "naked T-Tauri star." Nebular gas and dust are mostly gone, and the lack of stellar outbursts suggests that the infall of matter into the star has ceased. The formation of larger planetesimals accelerated when the Sun reached this stage, with continued growth forming **planetary embryos** of 100–1000 km in size, which eventually were accreted to form planets. Collectively, this stage and the previous ones took only about ten million years. At this point, the Sun joined the main sequence.

5.2.2 Planet Formation

As small planetesimals accrete to form larger objects, gravitational interactions become increasingly important. When the relative velocities of the planetesimals are larger than the escape velocity, accretion occurs with orderly growth. As accretion continues and bodies grow even larger, the relative velocities become less than the escape velocity, favoring runaway growth of the most massive objects as they gobble up the smaller ones (Morbidelli et al., 2012).

The formation of planets from embryos has been studied using numerical simulations, starting with a few dozen objects in initially circular orbits. Close encounters between bodies cause them to be gravitationally perturbed into eccentric orbits that cross each other, allowing the bodies to collide and accrete to form a few planets. The timescale for the formation of planets in stable orbits in these models varies from a few tens of million to 100 million years. The succession of the orderly growth of planetesimals, runaway growth of planetary embryos, and planet formation is called the standard model of planetary accretion.

An example of the output of an accretionary model is illustrated in Figure 5.4a (O'Brien et al., 2006). The system initially contains ~1000 planetesimals with orbits characterized by semi-major axis (distance from the Sun) and eccentricity of their orbits. The simulation shows the progressive coagulation of bodies into a few larger planets with relative diameters indicated by the size of the open circles. Three planets have formed by 10 Ma and continue to grow larger for the rest of the simulation. Such models are stochastic and produce different results each time. Figure 5.4b shows the different planetary systems formed in four simulations; the planets in the model shown in

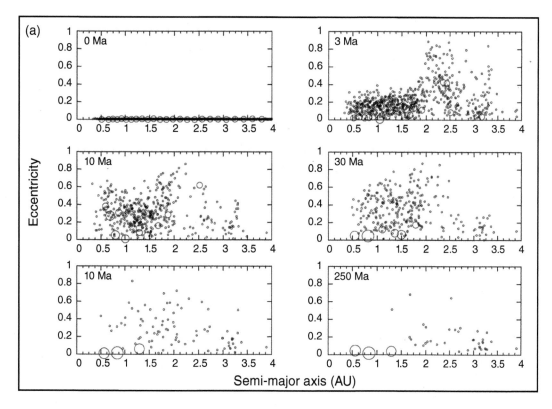

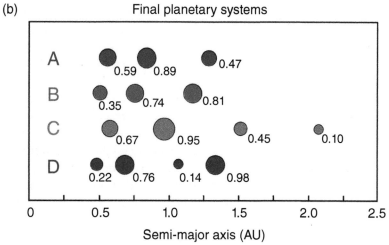

Figure 5.4 (a) Snapshots of an accretion model for planetesimals that coagulate to form planets (open circles). Orbital semi-major axis (distance from the Sun) and eccentricity for each body are shown in the time progression. (b) Final planetary systems produced by four stochastic simulations; A is the result of the accretion model in (a). The numbers by each planet indicate its mass relative to Earth. Courtesy of D. O'Brien.

Figure 5.4a are system A, and the numbers beside each planet indicate its mass relative to Earth. Such roll-of-the-dice models are sometimes called "Monte Carlo" simulations, after the famous gambling capital. By tracking the fate of each planetesimal, accretion models can also estimate the width of the nebular region from which each growing planet accretes, the so-called "feeding zone."

The accumulation of planetary embryos into planets entailed many collisions between bodies of comparable size. Giant impacts between massive bodies may have stripped off some of Mercury's mantle and formed the Earth's Moon, as described in Section 5.6.

Unlike the terrestrial planets, the gas giants Jupiter and Saturn contain large amounts of hydrogen and helium and must have formed early, while gas was a significant component of the nebula. The favored mechanism for their formation is the core accretion model. The cores initially resembled the rocky terrestrial planets, augmented by mantles of ices, but when they became massive enough they attracted gas directly from the surrounding nebular disk. A competing, and less popular, model is the gravitational instability hypothesis, which posits the formation of a giant gaseous protoplanet by gravitational collapse of a clump of nebula. The ice giants Uranus and Neptune apparently came late to the party. They contain less hydrogen and helium than Jupiter and Saturn, suggesting that they may have accreted more slowly and thus never quite reached the conditions for runaway gas accretion before the gas was dissipated. Instead, they accreted large amounts of ices, both containing ~65 percent water by mass. The ice giants are analogous to the cores of the gas giants.

5.3 Solar System Chronology, by the Numbers

We use both long-lived and short-lived radioactive isotopes, introduced in Section 3.1, to determine the age of the Solar System and the timing of major events in its evolution. Most of these events occurred about 4.5 billion years ago, but the precision of most measurements of long-lived isotopes is not sufficient to distinguish them. To do that, we use short-lived radionuclides to unravel the chronological details.

Because **CAIs** are the first condensates (or evaporative residues) formed in the solar nebula (Section 4.4), it is expected that they should have the oldest ages of any Solar System materials. The ages for CAIs determined from the most precise isotopic chronometer available – the Pb–Pb system, based on decay of ^{235}U and ^{238}U – are between 4567 and 4568 Ma (Figure 5.5). (The sequential numbers "4567" are easy to remember.) This is taken as the age of the Solar System. To be fair, we should acknowledge an added uncertainty of about a few million years in that age because we do not precisely know the decay rates of uranium isotopes.

Measurements of the decay product of the short-lived radioisotope ^{26}Al (half-life = 0.7 Ma) provide further support for the idea of using CAIs to date the Solar System. We know that the ^{26}Al was live when the CAIs formed, because its daughter isotope, ^{26}Mg, is now sited in aluminum-bearing minerals like spinel and hibonite. This nuclide was more abundant in CAIs than in any other Solar System materials, consistent with the early formation of these refractory solids. The former presence of ^{26}Al has also been documented in chondrules and in igneous meteorites (achondrites), but always in lower abundances than in CAIs. An example in Figure 5.6 compares the ^{26}Al content, expressed as its ratio relative to stable, nonradiogenic

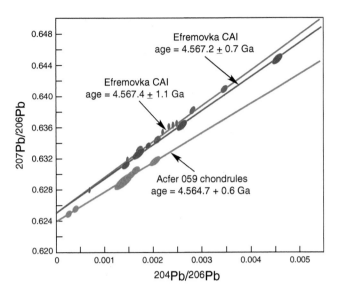

Figure 5.5 Pb–Pb diagram for two CAIs from the Efremovka chondrite and chondrules from the Acfer 059 chondrite. Slopes of CAI isochrons indicate an age of 4.567 Ga, and chondrules are younger. Modified from Amelin et al. (2002).

^{27}Al, on the left vertical axis, and the corresponding ages relative to CAIs on the right axis, for chondrules in several classes of chondrites. This diagram demonstrates that chondrules formed a few million years after CAIs. These are relative ages, but they can be converted to absolute ages using the Pb–Pb age of CAIs as an anchor.

Because the solar nebula gas would have been magnetized, the measured magnetism of ancient meteorites can constrain the lifetime of the nebula. The age of the oldest igneous meteorites is 4563 Ma, and these samples formed in a near-zero magnetic field (Wang et al., 2017). This is taken as evidence that the solar nebula gas had been lost within approximately four million years of Solar System formation.

Calcium–aluminum inclusions and chondrules are components of chondrites. Can we date the time of formation of the chondrites themselves, that is, the accretion ages of chondritic planetesimals? This is tough, because accretion does not reset isotopic chronometers. Accretion must have begun after the chondrules formed, and must have ended by the time of metamorphism or aqueous alteration in parent bodies. Using ^{26}Al and other short-lived radioisotopes, the times of chondrite parent body accretion have been bracketed at approximately 4–7 million years after CAIs.

Until recently, planetary scientists believed that the chondrites, composed of nebular materials, accreted early. Melted and differentiated meteorites (achondrites and irons) were thought to have formed later, when

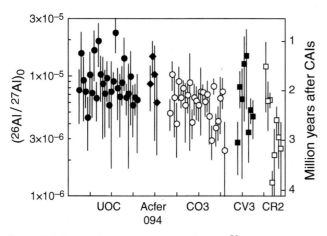

Figure 5.6 Ion probe measurements of initial ^{26}Al contents and ages of chondrules in several meteorite groups. Reprinted by permission from Cambridge University Press: Cosmochemistry, Harry Y. McSween Jr. and Gary R. Huss, Copyright (2010).

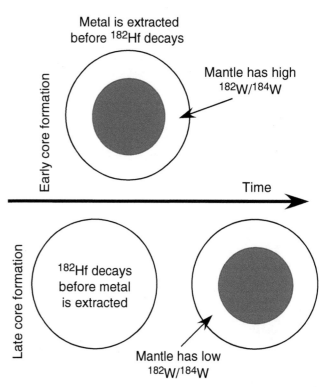

Figure 5.7 Timing of core formation is revealed by the decay of ^{182}Hf (lithophile) to form ^{182}W (siderophile). If core formation was early (rapid), the silicate mantle would now contain excess ^{182}W, expressed as a ratio with nonradiogenic ^{184}W. Slower (late) core formation would allow ^{182}W to partition into the metallic core, leaving the mantle depleted.

chondritic precursors melted. However, the advent of high-precision measurements of short-lived radionuclides in achondrites and irons has turned this notion upside down. The formation of iron meteorites, the cores of differentiated planetesimals, occurred about one million years after CAIs (Kruijer et al., 2014), earlier than the formation of the chondrites. The isotope ^{182}Hf decays rapidly to ^{182}W (half-life = 9 Ma), and this short-lived radionuclide has been especially useful in this determination. Hafnium is lithophile and tungsten is siderophile, so ^{182}Hf will partition into silicates, but its decay product will be concentrated in metal. The timing of core differentiation can be determined, based on whether hafnium was retained in and subsequently decayed within the silicate mantle, or rapidly decayed into tungsten and was subsequently extracted into the core (Figure 5.7).

Determining the age of the Earth is not straightforward, because there is no preserved geologic record of its earliest history. Only the Pb–Pb and ^{182}Hf–^{182}W isotopic systems provide useful constraints, and both of these systems basically define the time of planetary differentiation. That may not pose a problem though, because of the consensus view that accretion and core formation on the Earth were concurrent. The Pb–Pb age of the Earth, first measured by Claire Patterson in 1956 (recalculated to 4.48 Ga using a new uranium decay constant) postdates the beginning of the Solar System by nearly 100 million years. Subsequently measured Pb–Pb ages are a few hundred million years older than Patterson's date (Tera, 1980). Ages of the Earth determined using the hafnium–tungsten isotopic system are also slightly older, ranging from 11–15 million years after CAIs

(Halliday, 2004). Exactly what the "age" of the Earth really means can be complicated, because it likely continued to grow by accretion over some time interval, all the while undergoing differentiation. The age of the Moon is uncertain. The Pb–Pb data for the Moon (Barboni et al., 2017) suggest it formed 60 million years after CAIs. However, a review of the reliability of various radiometric ages for lunar rocks thought to have crystallized soon after the Moon formed indicates a preponderance of younger ages from 4.34 to 4.37 Ga (Borg et al., 2015), which has been used to argue that the Moon formed at least 200 million years after CAIs. An age for Mars determined from ^{182}Hf–^{182}W in martian meteorites is approximately ten million years after CAIs (Kruijer et al., 2017). The ages of the other terrestrial planets have not been determined, because we have no samples for analysis.

Because Jupiter and Saturn are made mostly of hydrogen and helium, they had to have formed when nebular gas was present. Thus these giant planets accreted rapidly during the first few million years. Uranus and Neptune must have formed more slowly or later.

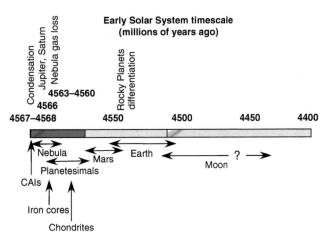

Figure 5.8 Solar System timeline, constructed from Pb–Pb and short-lived radionuclide isochrons for meteorites and planetary samples.

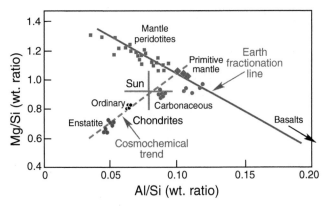

Figure 5.9 Mg/Si versus Al/Si for chondrites and terrestrial mantle rocks. Intersection of the cosmochemical and fractionation trends serves as an estimate of the primitive mantle composition. Modified from Righter et al. (2006).

An early Solar System timeline, summarizing all these ages, is shown in Figure 5.8. All Solar System bodies have subsequently suffered large impacts that can sometimes be dated because some isotopic systems are reset by shock. Earlier in Section 5.2, we attempted to identify specific processes like condensation and planetesimal accretion with the astronomical stages of the Sun. That connection, rough at best, utilized the relative ages portrayed in this timeline and took into account the thermal and dynamic characteristics of the Sun's stages.

5.4 Recipes for Planets

In order to understand the materials that accreted to form a planet, we must determine its bulk composition. However, a differentiated planet's bulk composition must be estimated indirectly because there are no samples anywhere on or within it that represent the entire body.

5.4.1 The Terrestrial Planets

Let's focus initially on the Earth's composition, since it is the composition that is best constrained. The Earth is commonly assumed to be broadly chondritic in composition, based on the ratios of refractory elements that have similar geochemical characteristics and thus should not be fractionated from each other during planetary processes.

The Earth is mostly oxygen, magnesium, silicon, and iron, like chondrites. Nearly 80 percent of our planet's iron is in the core, so it must have been accreted as metal, and metal is abundant in some chondrite groups. The

Mg/Si and Al/Si ratios of chondritic meteorites and the Sun are illustrated in Figure 5.9. These compositions define a cosmochemical trend. The compositions of peridotites from the Earth's mantle define a terrestrial fractionation trend on this diagram. The intersection of these two trends is taken to represent the Earth's bulk composition. Three published estimates of the primitive (unfractionated) mantle are also shown, and all plot near this intersection. The Earth is most similar to carbonaceous chondrites in terms of these three elements, but its oxygen isotopic composition is like enstatite chondrites and its osmium isotopic composition is like ordinary chondrites. If the Earth is chondritic, it probably formed from some mixture of chondrites like those in our present meteorite collections.

The ratios of refractory and volatile elements having similar geochemical behavior reveals that the Earth is depleted in volatile elements, relative to chondrites. Figure 5.10 illustrates two such ratios of volatile to refractory elements, K–U and Rb–Sr. Each pair of elements travel together during igneous processes, so their ratio remains nearly constant although their absolute amounts may change. From this figure we can see that volatile potassium and rubidium are depleted in the Earth, and in Mars, the Moon, and achondrites (HEDs, angrites) as well, relative to refractory uranium and strontium. This is a general characteristic of all differentiated bodies. Figure 5.11 shows that the depletion of volatiles in the Earth extends to other elements besides potassium and rubidium; volatility in this diagram is indicated by the temperatures at which half of the element condenses from a gas of solar composition. In contrast, refractory elements occur in roughly

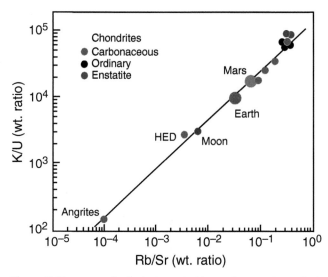

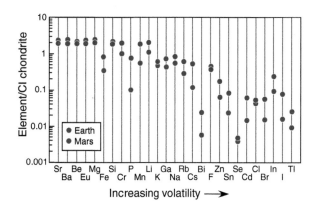

Figure 5.12 Comparison of the bulk compositions of Mars and Earth, both normalized to CI chondrites. After Taylor (2013).

Figure 5.10 Ratios of volatile (K and Rb) to refractory (U and Sr) elements in meteorites and planets. Planets and differentiated planetesimals (HED and angrite meteorites) are depleted in volatile elements, relative to chondrites. Modified from Tolstikhin and Kramers (2008).

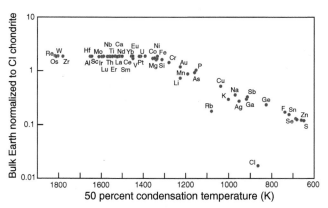

Figure 5.11 Elemental composition of the bulk Earth, normalized to CI chondrites. The refractory elements, which condense at high temperatures, have nearly chondritic abundances (raised slightly above 1 by the loss of volatiles). Volatile elements are systematically depleted, according to their volatility. After Albarede (2009).

chondritic proportions. The Earth's volatile element depletion could be explained if it accreted in large part from already differentiated planetesimals, rather than from chondrites. In Section 5.3 we learned that the earliest-formed planetesimals melted and differentiated to form cores, and chondritic bodies formed later. These differentiated bodies may have been abundant in the inner Solar System, because accretion is thought to have begun closer to the Sun where the density of

dust was higher. Thus, differentiated bodies would have been likely building blocks for the terrestrial planets. Alternatively, the Earth and other differentiated bodies could have formed from chondritic material, and somehow lost volatiles during differentiation. In either case, though, the bulk compositions of these bodies would still be chondritic, except for depletions in volatile elements.

Even with unlimited samples to analyze, determining the Earth's bulk composition, as illustrated in Figure 5.11, requires considerable effort. Estimating the composition of other planets is even more challenging. Let's take the case of Mars. Three different approaches have been utilized, with varying success. The first approach mixes chondrite classes (components of the nebula) to produce a planet having element ratios like those in martian meteorites. The second also mixes chondrite classes, but in this case to reproduce the distinctive oxygen isotopic composition of martian meteorites. The third uses geophysical data to estimate how mineralogy varies with depth, and the bulk composition is calculated from the mineralogy. All three approaches give roughly similar results, mainly differing in the degree of volatile element depletion. A recent reassessment of the Mars bulk composition (Taylor, 2013) is compared with that of the Earth in Figure 5.12. The compositions are rather similar, except that Mars has higher concentrations of moderately volatile elements (in the middle of the diagram).

We might expect planetesimals, and the planets made from them, to vary in some regular way with distance from the Sun. If the nebula temperatures decreased outward, planets should show decreasing refractory element abundances and increasing volatile element abundances

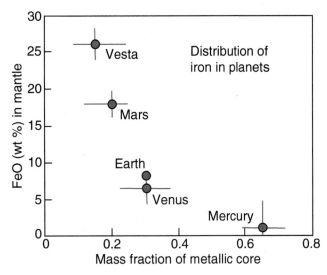

Figure 5.13 The mass fractions of iron as metal in cores and as FeO in mantles vary systematically in the terrestrial planets and asteroid Vesta. After Righter et al. (2006).

as we move outward from Mercury, to Venus, Earth, and Mars. However, this pattern is not observed. We do observe, however, systematic changes in the oxidation state of these planets. We can see this most readily by comparing the amounts of iron occurring as metal in cores and as FeO in mantles. (We will explain how core masses in other planets are determined in the following chapter.) Mercury has a massive iron core with very little FeO in its mantle, Earth and Venus have intermediate-sized cores, and Mars has a small core and an FeO-rich mantle (Figure 5.13).

5.4.2 The Giant Planets

Jupiter and Saturn, the gas giants, are made mostly of hydrogen and helium, indicating that they formed early from the surrounding nebular gas. However, their densities suggest that heavier elements are enriched (roughly equivalent to ~20 Earth masses) over solar abundances. Thus Jupiter and Saturn likely contain rocky cores. The ice giants, Uranus and Neptune, are composed mostly of rock and ices, so they likely formed after the gaseous nebula was lost or condensed into solid form.

5.5 The Leftovers: Asteroids and Comets

Billions of small bodies – asteroids and comets – represent the detritus from planetary accretion. They were not used as planetary building blocks, but carry important information about planet formation.

5.5.1 Asteroids

Many thousands of unused planetary building blocks litter the asteroid belt and regions around and outside the orbit of Jupiter (Burbine, 2017). These objects still exist because massive Jupiter's gravity field prevented them from being swept up into a planet. The Main asteroid belt has gaps caused by orbital resonances with Jupiter. Asteroids in these gaps feel Jupiter's tugs more often and can become perturbed into Earth-crossing orbits, where they can eventually fall to Earth as meteorites. Figure 5.14 shows the distribution of asteroid mass with heliocentric distance and inclination from the ecliptic.

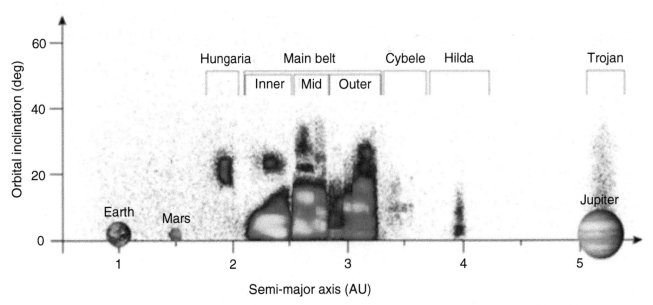

Figure 5.14 The distribution of mass within the asteroid belt, showing the major divisions and gaps produced by orbital resonances with Jupiter. Modified from DeMeo and Carry (2014).

The small asteroids imaged by visiting spacecraft (shown previously in Figure 1.10) show irregular shapes, as appropriate for collisional fragments. The fragmentations, too, are the result of Jupiter pumping up their orbital eccentricities and thus increasing the opportunities for collision. Phobos and Deimos, the tiny moons of Mars, are probably captured asteroids and are included in that figure. Intact Ceres and Vesta (see Box 5.1), the two most-massive asteroids (Ceres is now classified as a dwarf planet, and Vesta might have been before a giant collision removed a significant part of its mass), are spherical or nearly so. Many small asteroids have densities, estimated from their gravitational effects on nearby spacecraft, considerably less than solid rock. Increased porosity, resulting from impact disruption and gravitational re-accretion of the resulting fragments, is thought to be the likely explanation, although asteroids containing ice as well as rock will also have lower densities. Figure 5.15 illustrates a metamorphosed chondritic asteroid, with internal shells containing different petrologic types, and a rubble pile asteroid consisting of jumbled fragments gravitationally reassembled after collisional disruption. A differentiated body, consisting of core, mantle, and crust, might also form a rubble pile asteroid after a collision.

Asteroid spectra provide a means of compositionally classifying asteroids (Box 2.2), and comparisons of their spectra with those of meteorites suggest some connections. The distribution, by mass, of asteroid types is shown in Figure 5.16. The locations and classifications of the largest asteroids, which comprise most of the mass, are also shown. A rough regularity can be discerned: E-, S-, and V-class asteroids are mostly concentrated closest to the Sun, C- and B-class asteroids occur mostly in the middle and outer Main belt, and P- and D-class asteroids are most abundant at greater heliocentric distance. This distribution may reflect, in part, increased heating nearer the Sun. The innermost bodies were thermally metamorphosed or melted. The E-class asteroids are spectrally similar to enstatite chondrites and their melted equivalents, the aubrites; many S-class asteroids appear to be ordinary chondrites and their partly melted equivalents, the acapulcoites; and V-class asteroids are parent bodies of the igneous eucrites and diogenites. Also occurring in this region but not illustrated in Figure 5.16 are lesser amounts of M-class asteroids, interpreted to be fragments of metallic cores that have been stripped of the overlying silicate mantles. The C- and B-class bodies in the middle belt, spectrally similar to carbonaceous chondrites, once contained ice, which on heating produced aqueous fluids that altered them. Melting the ice used thermal energy, which moderated temperatures in these asteroids. At greater

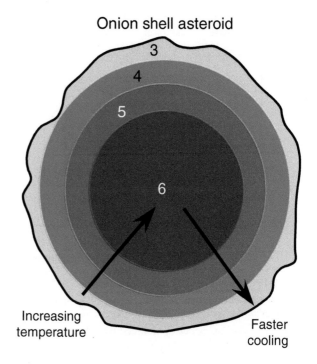

Onion shell asteroid

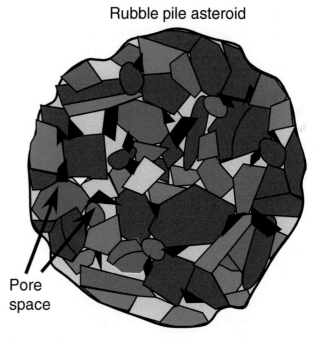

Rubble pile asteroid

Figure 5.15 Comparison of structures for chondritic asteroids. The onion shell model has concentric metamorphic zones, with type 6 chondrites in the interior and type 3 on the outside. A rubble pile structure is produced by collisional disruption and reassembly of an onion shell body, jumbling the fragments and adding porosity to the asteroid.

distances ice never melted, so the P- and D-class bodies are relatively unaltered. Meteorites from these bodies have not been recognized, although they may have been sampled as some interplanetary dust particles. The

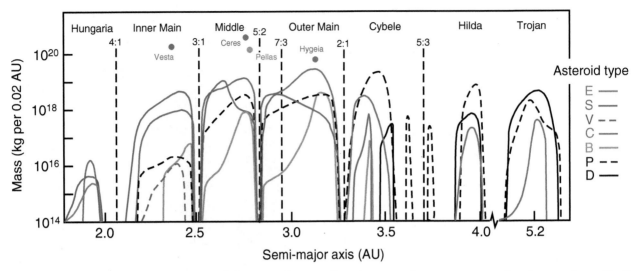

Figure 5.16 Asteroid spectral classes versus heliocentric distance. Abundances are shown as masses rather than numbers of bodies. A general distribution pattern, with E-, S-, and V-class asteroids occurring closer to the Sun, C- and B-class asteroids concentrated in the middle and outer belt, and P- and D-class asteroids occurring farthest from the Sun, is observed, although the orbits of many asteroids have been scrambled. Modified from DeMeo and Carry (2014).

distributions of asteroid types do not show clean separations, suggesting that their orbits have been scrambled to some degree. A possible explanation is discussed in Section 5.6.

5.5.2 Comets

The radial distance at which ice condensed in the nebula is referred to as the **snow line;** determining its actual location is complicated, and its exact position probably varied with time. The giant planets, as well as comet reservoirs, are located beyond the snow line. The sources of comets are the Kuiper belt, located beyond the orbit of Neptune and extending from 30 to 50 AU, and the even more distant Oort cloud. Pluto and its moon Charon are actually the innermost Kuiper belt objects. Current models suggest that most of the icy bodies now in the Kuiper belt and Oort cloud originally formed among the giant planets, but were ejected into these nether regions.

The ices in comets sublimate as they are warmed when perturbed into orbits that bring them closer to the Sun. Their bright appearance, caused by escaping gas and dust, is in stark contrast to the comet nuclei themselves, which are dark as black velvet. Images of comet nuclei visited by spacecraft were shown in Figure 1.11. The dark crusts are apparently lag deposits of dust, and the underlying ice-rich materials only show in outbursts. Comets are weakly consolidated objects with high porosity, and tidal forces during a close encounter with a planet can cause them to split into fragments, as occurred when comet Shoemaker–Levy passed by Jupiter in 1992.

Samples of comet Wild 2 collected and returned to Earth by the *Stardust* spacecraft (Figure 5.17) contain the same minerals found in chondrites (Brownlee, 2014). Surprisingly, some of these minerals occur in CAIs and chondrules, thought to have formed by condensation and melting close to the Sun. The expectation, before this mission, was that comets consisted of ices and tiny interstellar grains with organic mantles. Although this is just one comet, its mineralogy is similar to IDPs that are thought to sample many comets. From this evidence it appears that comets are not as primitive as previously believed, and that nebular materials must have been transported outward over vast distances.

5.5.3 A Hole in the Solar Nebula?

Stable isotopic analyses of meteorites and larger bodies for which we have samples (Earth, Moon, Mars, and Vesta) reveal a compositional gap (Warren, 2011; Scott et al., 2018) (Figure 5.20a). The axes of this figure require a little explanation. In Section 4.5, we learned that oxygen has three isotopes, and $\Delta^{17}O$ is the deviation in $\delta^{17}O$ from the terrestrial mass fractionation line. The horizontal axis, $\varepsilon^{54}Cr$, is $^{54}Cr/^{52}Cr$ relative to the ratio in chondrites. Both isotopes are hypothesized to identify materials from different regions of the solar nebula. A number of other isotope ratios, such as $^{50}Ti/^{47}Ti$ and $^{95}Mo/^{94}Mo$, also show a similar

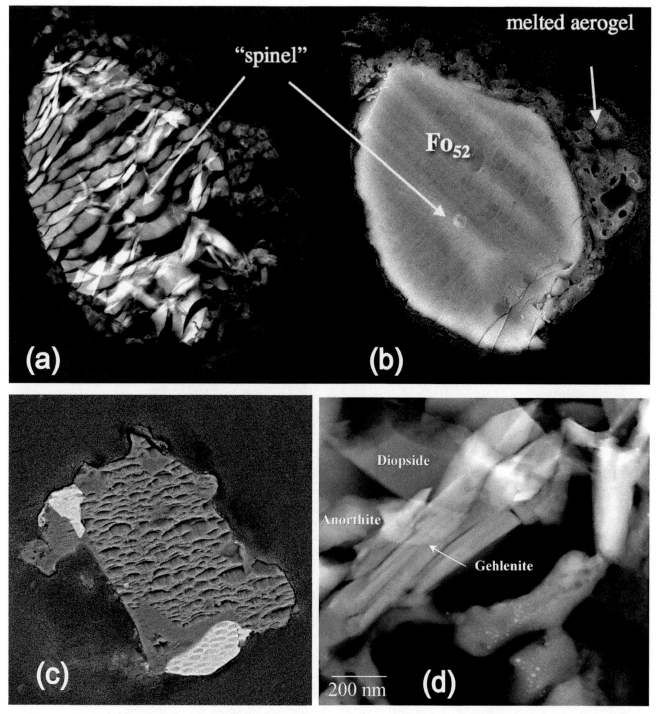

Figure 5.17 Examples of comet particles collected by the *Stardust* spacecraft. (a,b) Plane-polarized light and SEM (scanning electron microscope) images, respectively of an olivine grain containing spinel, FOV = 4 μm. (c) Particle composed of pyroxene and sulfide, FOV = 1 μm. (d) Particle composed of minerals in CAIs. Images by D. Brownlee; reprinted by permission from Cambridge University Press: Cosmochemistry, Harry Y. McSween Jr. and Gary R. Huss, Copyright (2010).

compositional gap when plotted against $\Delta^{17}O$. This isotopic gap has been interpreted to represent a physical gap in the early nebula, separating regions with distinct compositions.

In this model, the ordinary and enstatite chondrites, the Earth, Moon, Mars, and Vesta, as well as numerous achondrites and irons, formed in the inner Solar System. The carbonaceous chondrites, along with a few types of

BOX 5.1 **FIRE AND ICE: VESTA AND CERES**

Vesta (Russell et al., 2015) and Ceres (Russell et al., 2016) are the most massive representatives of only a handful of remaining intact planetesimals. Both asteroids have been explored by the *Dawn* spacecraft, which provided views of the geologic complexity of protoplanetary bodies. Vesta is a differentiated body of rock and metal. In contrast, Ceres has a rocky interior with a large amount, ~25 percent, of ice. These bodies may represent compositional extremes in the asteroid belt, but both kinds of planetesimals must have been building blocks for the Earth.

A **shape model** (Figure 5.18a) illustrates the rugged topography of Vesta's surface. A geologic map of Vesta, compiled using stratigraphic procedures described in Section 3.2, is shown in Figure 5.18b. Cratered highlands in the northern hemisphere represent the oldest crust. The excavation of two giant impact basins, first Veneneia and then Rheasilvia, near the south pole scattered ejecta over the southern hemisphere and produced equatorial girdles of ridges and troughs (the Saurnalia Fossae and Divalia Fossae, respectively). Marcia is among the most recent impact craters on Vesta.

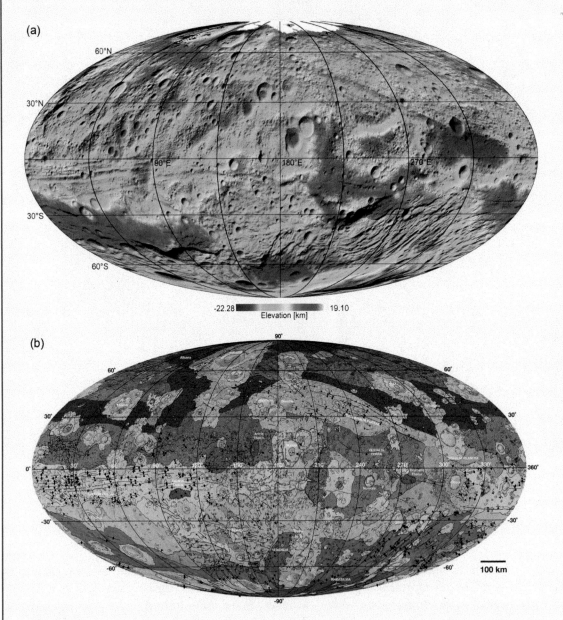

Figure 5.18 (a) Vesta shape model, showing the rugged topography of the asteroid's surface. (b) Geologic map of Vesta. See Williams et al. (2015) for key to mapped units.

Ceres has an icy mantle with a dark coating of phyllosilicates on the surface. Bright spots are the sites of recent activity that concentrated carbonate minerals. A shape model and geologic map of Ceres are shown in Figure 5.19a,b. It has less topographic relief than Vesta, but the lack of relaxation of its small impact craters requires that its crust contains no more than ~40 percent ice.

Although these rocky and ice-bearing asteroids are distinct, the compositional boundary between asteroids and comets is actually rather fuzzy. A few distant, ice-bearing asteroids show periodic cometary outbursts, and the aqueous alteration experienced by carbonaceous chondrites indicates that many C-class asteroids once contained ice. In the early Solar System, asteroids beyond the snow line may have resembled comets before their ice melted.

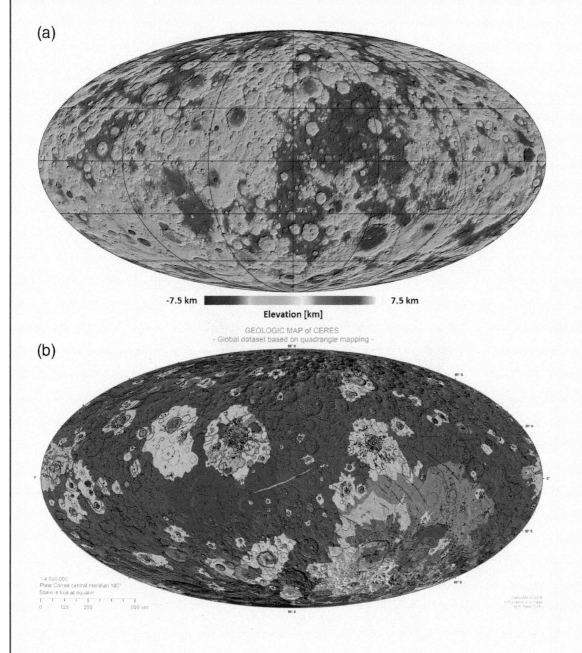

Figure 5.19 (a) Ceres shape model. (b) Geologic map of Ceres. Adapted from Williams et al. (2018), see that reference for key to mapped units.

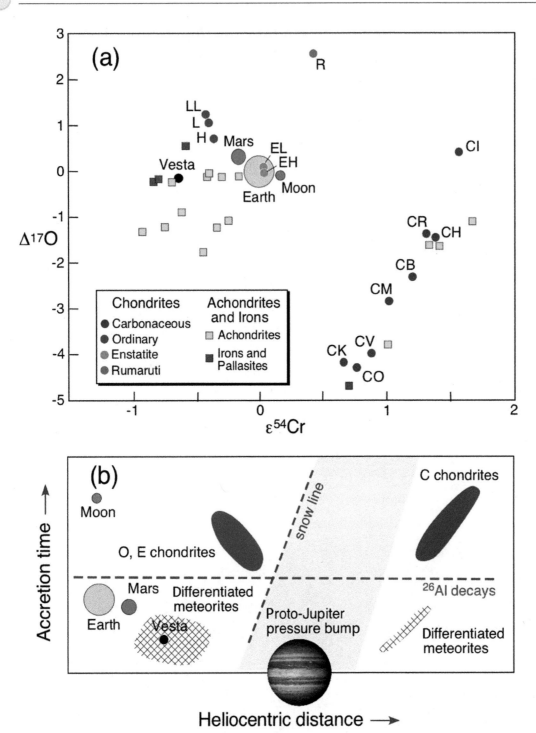

Figure 5.20 (a) Distinct stable isotopic compositions may point to a gap in the solar nebula, with most chondrites, achondrites, irons, and terrestrial planets forming in the inner Solar System, and carbonaceous chondrites and some achondrites and irons forming in the outer reaches beyond the orbit of Jupiter. The diagram shows the average compositions of meteorite groups and of other bodies for which we have samples. Modified from Scott et al. (2018). (b) Interpretation of the gap, caused by early accretion of proto-Jupiter, which produces a bump in pressure that prevents objects farther out from entering the inner Main belt and terrestrial planet region. Bodies outboard of the snow line contain ices. Bodies that accreted early, before too much ^{26}Al decayed, experience melting and differentiation, whereas those that accreted later experienced thermal metamorphism or aqueous alteration.

Figure 5.21 Artist's depiction of the Moon-forming impact.

oxidized achondrites and irons, are interpreted to have formed in the outer Solar System, past the snow line and beyond the orbit of Jupiter (Figure 5.20b). The early accretion of proto-Jupiter may have kept these compositionally distinct regions from mixing. Once Jupiter's nucleus became large enough, it would have accreted the nebular gas around it, so that gas pressure was locally lowered. This pressure bump would have effectively isolated the regions inside and outside of Jupiter's orbit. As discussed in Section 5.6.2, later migrations of the giant planets would then have implanted the carbonaceous chondrite bodies into the asteroid belt.

5.6 Whence Earth's Moon?

The Earth's Moon is unique, in that it is so large compared to its neighboring planet. The origin of the Moon has captivated scientists' imagination, and its explanation prompts novel ideas about the scale and importance of impact processes in the early Solar System.

5.6.1 Origin of the Moon

It is generally thought that a glancing blow to the Earth by a Mars-sized impactor (Figure 5.21) produced fragments and vaporized material that went into orbit around the Earth and subsequently accreted to form the Moon. The impactor presumably had already differentiated, and

numerical simulations suggest that its core would have combined with the Earth's core, whereas parts of its mantle would have been blown away to eventually form the Moon.

The Moon's strong depletion in volatile elements provides geochemical evidence for the large impact hypothesis. For example, the lunar K–U and Rb–Sr ratios are a factor of approximately four lower than in the Earth. Not all of the orbiting vapor would have condensed, accounting for the scarcity of water, potassium, rubidium, and other volatile components. The oxygen isotopic composition of lunar samples deviates only very slightly from the Earth's mass fractionation line, suggesting that the impactor and the Earth originated in the same nebular neighborhood, or were efficiently mixed during the collision.

5.6.2 Orbital Scrambling

The Moon-forming impact was gigantic, but impacts of other massive bodies to create large basins occurred later on the Moon itself, as well as on Mercury, Venus, and Mars (and on Earth, although the record has been erased by geologic processes). Dating of these large impacts is based mostly on lunar rocks returned by *Apollo* astronauts. This has led to the concept of a **Late Heavy Bombardment** that appears to have ended at about 3.9 Ga. However, there has been considerable disagreement about whether the bombardment was a Solar System-wide punctuated event or represents a gradually declining, protracted accretion of ever-larger objects. In any case, the radial mixing of asteroid spectral classes (Figure 5.16) and the scattering of cometary bodies originally formed within the giant planet region into the Kuiper belt also point to a dynamic, rather than static, Solar System.

The need for a way to perturb large asteroids into the inner Solar System and eject comets outward into the Kuiper belt has prompted models that explore the possible orbital migration of the giant planets. Two possible models are illustrated in Figure 5.22. The Grand Tack model (Walsh et al., 2011) posits the inward and then outward migration of Jupiter and Saturn during the nebula phase; the name alludes to a sailing maneuver for changing direction. Gravitational interactions with inward-drifting Jupiter would have scrambled the orbits of smaller objects in the accretion disk, and ejected many of them from the Main belt. This model accounts for the otherwise puzzling small size of Mars and the asteroid belt, by depleting these regions of planetary

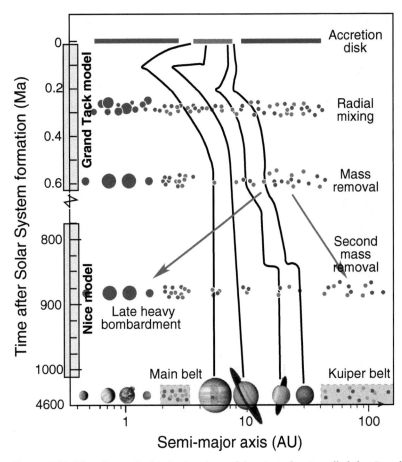

Figure 5.22 The effects of orbital migrations of the giant planets, called the Grand Tack and Nice models, on planetesimals in the early accretion disk and later on asteroids in the Main belt. Modified from DeMeo and Carry (2014).

building materials. The Nice model (based on a trio of papers published in 2005) describes later wanderings of the giant planets, reshuffling asteroid orbits and sending them careening into the inner Solar System, where they could impact the terrestrial planets (perhaps accounting for the Late Heavy Bombardment). Icy bodies formed in the more collision-rich giant planet region could also have been scattered outward to the Kuiper belt by migrations of Uranus and Neptune. Planetary dynamic models are providing a new paradigm for Solar System evolution that has implications for how planetary geology is interpreted.

Summary

The planetary system formed from coagulation of gas and dust into planetesimals and then planets within an accretion disk. The timescale for the Solar System has been determined from measuring radiogenic isotopes in meteorites; its age, based on the earliest solids (CAIs), is ~4.567 billion years, and planet formation was complete in a few tens of millions of years. The terrestrial planets are broadly chondritic in composition, but are depleted in volatile elements. This characteristic may have been inherited if they accreted from already differentiated protoplanets. The giant planets have much greater proportions of volatiles; Jupiter and Saturn accreted hydrogen and helium directly from the nebula, and Uranus and Neptune accumulated large quantities of ices. Asteroids and comets represent leftover planetary building blocks. A very large impact into the Earth is thought to have formed the Moon. Asteroid impacts onto other planets, as well as scattering of icy bodies into the outer Solar System, may have resulted from orbital scrambling by migrations of the giant planets.

Review Questions

1. What is an accretion disk, and how does it evolve into a planetary system?
2. What do observations of young stars and numerical simulations of accretion reveal about planet formation?
3. Why do we use refractory inclusions to date the age of the Solar System, and what is its age?
4. How are short-lived and long-lived radionuclides used in combination to determine the timing of early solar system processes?
5. What evidence suggests that the Solar System has experienced a dynamic orbital evolution?

SUGGESTIONS FOR FURTHER READING

Burbine, T. H. (2017) *Asteroids.* Cambridge: Cambridge University Press. A modern summary of asteroid astronomy, composition, physical properties, and impact threat.

DeMeo, F. E., and Carry, B. (2014) Solar System evolution from compositional mapping of the asteroid belt. *Nature,* **505**, 629–634. A very accessible review of models predicting the scrambling of asteroid orbits and the Late Heavy Bombardment.

Halliday, A. N. (2004) The origin and earliest history of the Earth. In *Treatise on Geochemistry,* Vol. 1, *Meteorites, Comets, and Planets,* ed. Davis, A. M. Oxford: Elsevier, pp. 509–557. An excellent summary of constraints on the age of our planet.

Libourel, G., and Corrigan, C. M. (2014) Asteroids: new challenges, new targets. *Elements,* **10**, 11–17. This paper and the following papers in the same issue are short and easy-to-digest introductions to properties, spectra, and composition of asteroids.

Morbidelli, A., Lunine, J. I., O'Brien, D. P., et al. (2012) Building terrestrial planets. *Annual Review of Earth and Planetary Sciences,* **40**, 251–275. A nice description of our current understanding of planetary accretion models.

Taylor, G. J. (2013) The bulk composition of Mars. *Chemie der Erde,* **73**, 401–420. A wonderful review of the methods by which planetary bulk compositions are determined.

Tolstikhin, I., and Kramers, J. (2008) *The Evolution of Matter,* Cambridge: Cambridge University Press. This book provides a rigorous explanation of planetary accretion and compositions.

REFERENCES

Albarede, F. (2009) *Geochemistry,* 2nd edition. Cambridge: Cambridge University Press.

Amelin, Y., Krot, A. N., Hutcheon, I. D., et al. (2002) Lead isotopic ages of chondrules and calcium-aluminum-rich inclusions. *Science,* **297**, 1678–1683.

Barboni, M., Boehnke, P., Keller, B., et al. (2017) Early formation of the Moon 4.51 billion years ago. *Science Advances,* **3**, e1602365.

Borg, L. E., Gaffney, A. M., and Shearer, C. K. (2015) A review of lunar chronology revealing a preponderance of 4.34–4.37 Ga ages. *Meteoritics & Planetary Science,* **50**, 715–732.

Brownlee, D. (2014) The Stardust mission: analyzing samples from the edge of the Solar System. *Annual Review of Earth and Planetary Sciences,* **42**, 179–205.

Burbine, T. H. (2017) *Asteroids.* Cambridge: Cambridge University Press.

DeMeo, F. E., and Carry, B. (2014) Solar System evolution from compositional mapping of the asteroid belt. *Nature,* **505**, 629–634.

Halliday, A. N. (2004) The origin and earliest history of the Earth. In *Treatise on Geochemistry,* Vol. 1, *Meteorites, Comets, and Planets,* ed. Davis, A. M. Oxford: Elsevier, pp. 509–557.

Kruijer, T. S., Touboul, M., Fischer-Godde, M., et al. (2014) Protracted core formation and rapid accretion of protoplanets. *Science,* **344**, 1150–1154.

Kruijer, T. S., Kleine, T., Borg, L. E., et al. (2017) The early differentiation of Mars inferred from Hf–W chronometry. *Earth & Planetary Science Letters,* **474**, 345–354.

McSween, H. Y., and Huss, G. R. (2010) *Cosmochemistry.* Cambridge: Cambridge University Press.

Morbidelli, A., Lunine, J. I., O'Brien, D. P., et al. (2012) Building terrestrial planets. *Annual Review of Earth and Planetary Sciences,* **40**, 251–275.

O'Brien, D. P., Morbidelli, A., and Levinson, H. F. (2006) Terrestrial planet formation with strong dynamical friction. *Icarus,* **184**, 39–58.

Righter, K., Drake, M. J., and Scott, E. R. D. (2006) Compositional relationships between meteorites and terrestrial planets. In *Meteorites and the Early Solar System II,* eds. Lauretta, D. S., and McSween, H. Y. Tucson, AZ: University of Arizona Press, pp. 803–828.

Russell, C. T., McSween, H. Y., Jaumann, R., et al. (2015) The Dawn mission to Vesta and Ceres. In *Asteroids IV*, eds. Michel, P., DeMeo, F. E., and Bottke, W. F. Tucson, AZ: University of Arizona Press, pp. 419–432.

Russell, C. T., Raymond, C. A., Ammannito, E., et al. (2016) Dawn arrives at Ceres: exploration of a small, volatile-rich world. *Science*, **353**, 1008–1010.

Scott, E. R. D., Krot, A. N., and Sanders, I. S. (2018) Isotopic dichotomy among meteorites and its bearing on the protoplanetary disk. *The Astrophysical Journal*, **854**, 164–176.

Taylor, G. J. (2013) The bulk composition of Mars. *Chemie der Erde*, **73**, 401–420.

Tera, F. (1980) Reassessment of the "Age of the Earth". *Carnegie Institution of Washington Year Book*, **79**, 524–531.

Tolstikhin, I., and Kramers, J. (2008) *The Evolution of Matter*, Cambridge: Cambridge University Press.

Walsh, K. J., Morbidelli, A., Raymond, S. N., et al. (2011) Sculpting the inner Solar System by gas-driven orbital migration of Jupiter. *Nature*, **475**, 206–209.

Wang, H., Weiss, B. P., Bai, X-N., et al. (2017) Lifetime of the solar nebula constrained by meteorite paleomagnetism. *Science*, **355**, 623–627.

Warren, P. H. (2011) Stable-isotopic anomalies and the accretion assemblage of the Earth and Mars: A subordinate role for carbonaceous chondrites. *Earth and Planetary Science Letters*, **311**, 93–100.

Williams, D. A., Blewett, D. T., Buczkowski, D. L., et al. (2015) Complete global geologic map of Vesta from Dawn and mapping plans for Ceres. *Lunar and Planetary Science Conference*, **46**, 1126.

Williams, D. A., Buczkowski, D. L., Crown, D. A., et al (2018) High-resolution global geologic map of Ceres from NASA Dawn mission. In *Planetary Geologic Mappers Annual Meeting*. Houston, TX: Lunar and Planetary Science Institute, abstract 7001.

6

Planetary Heating and Differentiation

We explain how heat is produced by radioactive decay, segregation and exothermic crystallization of metallic cores, impacts, and tidal forces. Planetesimals in the early Solar System were most affected by the decay of short-lived radionuclides. Larger, rocky planets were heated primarily by large impacts and core segregation. Because rocks are poor conductors, heat retention in rocky bodies is a function of planet size. Large-scale melting to produce magma oceans was likely a common process facilitating differentiation to form cores, mantles, and crusts. Metallic liquids are probably necessary for core segregation. Primary crusts, formed during planetary differentiation, are rarely preserved. Mantles are residues from the extraction of silicate crustal melts and core materials. Differentiation of the giant planets was driven by density variations in high-pressure forms of gases, ices, and rock more than by heating and melting. The importance of the various planetary heat sources changes over time; in modern planets the effective heat sources are decay of long-lived radioisotopes and, for the Earth, exothermal crystallization of the liquid outer core.

6.1 Too Hot to Handle

The geologic evolution of planetesimals and planets is fueled by heat from various sources. Rocks are notoriously poor conductors of heat, so heat generated by processes in the deep interiors of large rocky planets can be retained over geologic timescales. Besides conduction, other mechanisms of heat loss such as convection can cause molten bodies to cool more rapidly. On the other hand, small bodies lose heat more rapidly than large planets, because their smaller rock masses allow heat to be conducted outward more efficiently. Once the heat reaches the surface, it is radiated into space, and the greater ratio of surface area to volume

for small bodies promotes more rapid cooling. This is illustrated in Figure 6.1, a diagram showing the apparent duration of igneous activity on various rocky bodies as a function of body size. For small bodies like asteroid Vesta, igneous activity lasted for only a few million years. Igneous activity ceased on the Moon after several billion years, and still persists today on large planets like the Earth.

The giant planets, Jupiter, Saturn, Uranus, and Neptune, are warmer than can be explained by solar radiation alone, so they too must have internal heat sources. These sources are roughly comparable in magnitude to the energy they receive from the Sun.

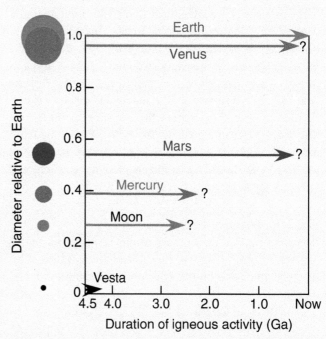

Figure 6.1 Relationship between planetary size and the duration of igneous activity, demonstrating the effect of cooling in smaller bodies.

6.2 Heat Sources

Heat is generated by varying combinations of processes, some endogenic and some exogenic. Regardless of the source, the added energy is converted to atomic motions and thence to heat.

6.2.1 Accretion and Impacts

Impacts provide a planet with energy as well as mass, and can be a potent heat source. This is a particularly important heat source during planetary accretion, but later impacts also cause heating. The energy (E) comes from the encounter velocity (V) and the gravitational energy gained as the body falls to the planet's surface at roughly the **escape velocity** (GM/R):

$$E = V^2/2 + GM/R \qquad (6.1)$$

where G is the gravitational constant, and M and R refer to the target planet's mass and radius. Note that impact velocity is squared, so this is a particularly important factor. This energy can be converted into a temperature change using the expression:

$$\Delta T = E/(MC_p) \qquad (6.2)$$

where C_p is the heat capacity. Accretion of small bodies involves modest velocities (as well as modest masses) and does not generate nearly as much energy as large impacts. Also, impacts provide only limited heat to planetesimals, because the GM/R term refers to the small target bodies.

Approximately 70 percent of the kinetic energy from large impacts is available for heating, with the remainder carried off with escaping ejecta. In large impacts, much of the energy is buried deep within the target. Because the waning stages of planetary accretion involve impacts between large bodies of comparable size, impact heating was especially important at that stage. Planetary erosion is the opposite of accretion; the crossover point, where the mass of ejecta exceeds the mass of the impactor, depends on planet size. The crossover occurs at velocities of 20 km/s for the Moon (which is presently eroding) and 45 km/s for Mars (which is presently accreting).

6.2.2 Radioactive Decay

Decay of radioactive isotopes results in mass loss that is converted into energy, mostly in the form of heat. This energy can be substantial, typically a few thousand joules per mole.

The decay of long-lived ^{40}K, ^{235}U, ^{238}U, and ^{232}Th currently accounts for radioactive heat production in planets. The heat generated depends on the decay rates of the radioisotopes and their abundances. As an illustration, ^{40}K and ^{87}Rb have comparable half-lives, but ^{40}K accounts

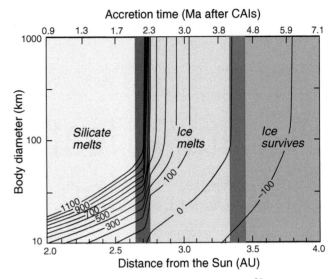

Figure 6.2 Model for asteroids heated by decay of ^{26}Al, assuming accretion starts closer to the Sun and progresses outboard. Contours show the maximum temperatures reached within bodies of various sizes, and the vertical red and blue bars separate regions where rock and ice melt. After Grimm and McSween (1993).

for 15 percent of the heat generated in the Earth's crust, whereas the contribution from ^{87}Rb is insignificant because of its relatively low abundance. Long-lived radioisotopes are not particularly important in heating small planetesimals, because the heat can diffuse out faster than it is generated by radioactive decay.

Despite its low abundance, the short-lived radioisotope ^{26}Al decayed very fast and thus caused significant heating within the first few million years of Solar System history. One estimate of the heating effect of ^{26}Al decay in planetesimals is shown in Figure 6.2. In this model, accretion is assumed to have occurred first nearer the Sun, where the density of gas and dust was greater, and to have swept outward to greater heliocentric distance with time. This accretion model is certainly a simplification, but it may explain the pattern of heating inferred from asteroid spectroscopy (discussed in Section 5.5). Bodies that accreted earlier would have had a greater amount of live ^{26}Al, which continued to decay as the accretion front moved outward. Thus, bodies of 100 km diameter nearer the Sun, falling within the "silicate melts" field of this diagram, were melted. Bodies that were smaller or accreted later were still heated and thermally metamorphosed, but their interiors never got hot enough to melt. In bodies accreted even later in the middle of the asteroid belt and beyond the snow line, temperatures were high enough to melt ice and cause aqueous alteration. At even greater heliocentric distance, bodies accreted after ^{26}Al had mostly decayed, so even ice survived.

6.2.3 Core Segregation and Core Crystallization

In the early Earth, **core segregation** caused heating, as gravitational potential energy (that is, massive metal distributed away from the planet's center of mass) was converted to thermal energy as metal fell to the planet's center.

The energy necessary to melt a solid includes that required to raise its temperature to the melting point, plus more energy to convert it from solid to liquid (the latter is called the latent heat of melting). Conversely, crystallization of a melt releases latent heat, so crystallization is exothermic. As the Earth has cooled over billions of years, its liquid outer core has been crystallizing, adding to the volume of the solid inner core. This crystallization releases heat that drives convection in the molten outer core and heats the mantle from below.

6.2.4 Tidal Forces

Tides extend beyond distorting water – the familiar ocean tides. **Tidal forces** can also distort solid bodies. Distortion occurs because the gravitational attraction on a revolving body is stronger on one side than on the other. If the revolving body is close enough to the more massive object, the revolving body responds by deforming into a football shape, analogous to the way the oceans deform on Earth.

If the body's orbit is eccentric (to maintain eccentricity over time, a third body in orbital resonance is needed), the gravitational field changes during the orbit, and this produces flexing, as the gravity differential changes along the orbit. This flexing produces heating by shear friction. Tidal heating is the cause of volcanic eruptions on Jupiter's moon Io, where the present-day heat flow is ~200 times that expected for radioactive decay. Internal heating by tidal forces also accounts for thermal anomalies and erupting ice fountains on Saturn's moon Enceladus.

6.3 Magma Oceanography

When planetesimals formed in the early Solar System, the decay of short-lived radionuclides was the most important heat source. Calculations suggest that small bodies that accreted within the first million years after CAIs should have melted completely.

For larger planets like the early Earth, the kinetic energy delivered by large impacts was the most potent heat source, and the decay of short-lived radionuclides played a significant but lesser role because of the time lapse before planet formation. The terrestrial planets may also have experienced large-scale melting to form **magma oceans** (Elkins-Tanton, 2012). Because these collisions

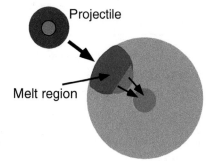

Projectile

Melt region

Impact melting and core segregation

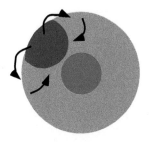

Isostatic adjustment   Global magma ocean

Figure 6.3 Origin and evolution of a magma ocean formed as a consequence of a giant impact. A hemispherical magma ocean forms, and the iron core of the projectile merges with the target body core. Isostatic adjustment causes the magma ocean to form a globally distributed layer.

occurred over some time interval, the planets may actually have had several, transient magma oceans. Thus, the Moon-forming impact (Section 5.6) may have been only the last major collision affecting the early Earth. If a colliding body already had a core, then the gravitational potential energy of merging the cores could have added additional heat. The possible formation and evolution of a magma ocean is illustrated in Figure 6.3.

Bulk melting of a rocky planet or planetesimal would have produced a magma ocean having an ultramafic composition. Experimental data suggest that the viscosity of such a magma ocean would be very low, allowing vigorous convection and rapid heat loss. The lifetime of the Earth's convecting magma ocean may have been only a few thousand years. However, if an insulating crustal lid developed, its lifetime could have been longer, on the order of 100 million years.

The observational evidence for magma oceans on the terrestrial planets is sparse, but the theoretical argument is compelling. The best evidence for a magma ocean comes from the Moon, as described in Box 6.1. Although lunar rocks reveal the structure of the solidified magma ocean, they are not applicable to understanding the crystallization of magma oceans on the terrestrial planets. The low pressures within the Moon allowed feldspar to

BOX 6.1 **A LUNAR MAGMA OCEAN**

The ~50 km thick crust of the Moon is composed predominantly of plagioclase, forming anorthosites, with minor amounts of pyroxene and olivine. The feldspathic crust is thought to have formed by plagioclase flotation in a magma ocean; olivine and pyroxene either crystallized at the bottom or sank to form the mantle. The late-stage residual liquid, rich in incompatible elements that were excluded from fractionating minerals, was trapped between the crust and mantle and crystallized to form a selvage of KREEP rocks (the acronym stands for potassium, rare earth elements (REE), and phosphorus – all incompatible and thus concentrated in late-stage melts).

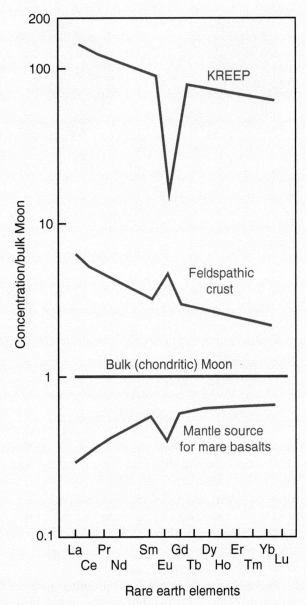

Rare earth element patterns of lunar rocks (Figure 6.4) provide geochemical support for this model. These elements are generally similar in size and charge (trivalent), and thus travel together during melting and crystallization. An exception is europium (Eu), which is divalent in the highly reduced Moon, and substitutes readily for calcium in plagioclase. The feldspathic crust has a positive Eu anomaly, reflecting the accumulation of plagioclase in the crust. The mantle has a complementary negative Eu anomaly, which was inherited by subsequent basaltic melts of the mantle that have erupted to form the maria. KREEP rocks, representing the last dregs of the magma ocean, are extremely enriched in REEs.

Crystallization models for a completely melted and half-melted Moon are contrasted in Figure 6.5. The depth of the magma ocean is constrained by the amount of plagioclase that can be formed, which

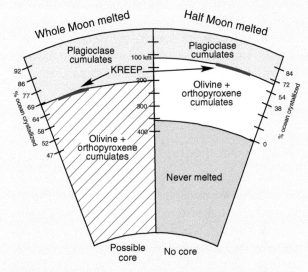

Figure 6.4 Rare earth element patterns, normalized to the bulk Moon composition, provide evidence of a magma ocean. The europium (Eu) anomalies indicate accumulation or depletion of plagioclase, which floated to form the lunar crust. KREEP, the residual liquid from the magma ocean, is extremely enriched in incompatible REEs.

Figure 6.5 Two models for crystallization of the lunar magma ocean, involving different amounts of melting. In both scenarios, cumulus olivine and pyroxene sink and cumulus plagioclase floats. KREEP, the last gasp of melt, is trapped between the crust and mantle cumulates. Modified from Ryder (1991).

in turn depends on the extent to which the Moon is enriched in refractory elements (including aluminum). In both models, olivine and orthopyroxene crystallized early and formed a cumulate mantle. After about 70 percent fractional crystallization, which increased the Al_2O_3 content of the remaining magma, plagioclase began forming and floated upward to form the crust. The amount of crystallization before the onset of plagioclase is shown as tick marks. The late-stage residual melt (KREEP) was sandwiched between the crust and mantle. Later mixing (not illustrated) may have resulted from sinking of the dense KREEP layer and rising of plumes of mantle cumulates.

crystallize and accumulate to form its crust. The corresponding high-pressure aluminous mineral is garnet, which would have remained in the mantle in larger planets with deeper magma oceans.

6.4 Differentiation of Rocky Planets and Planetesimals

Planetary **differentiation** to form cores, mantles, and crusts is arguably the most fundamental geologic process. What we know, or infer, about differentiation is pieced together from samples, experiments, and theory.

6.4.1 Getting to the Heart of the Matter: Cores

The formation of planetary cores occurs because of the large density contrast between silicates and metal. Cores are predominantly iron, because of its high Solar System abundance and density. The separation of solid iron from solid silicate is too sluggish to have been a significant process (Stevenson, 1990). However, iron alloyed with other elements, like oxygen, sulfur or silicon, has a lower melting temperature than mantle silicates. Core formation may thus have involved the movement of molten iron through a solid silicate mantle. In this case, percolation of liquid metal depends on melt connectivity (Rubie et al., 2007). Liquid can "wet" silicate grains if the dihedral angle (θ) between solid–liquid boundaries connected at a triple junction is less than ~60° (Figure 6.6). In this case, the liquid is fully interconnected and can migrate through the solid. If $\theta > 60°$, the liquid forms isolated pockets and cannot segregate unless the melt fraction constitutes at least several volume percent. At high pressures, appropriate for the mantles of the terrestrial planets, dihedral angles exceed 60° and some metal remains stranded in the mantle. The dihedral angle is less relevant if shear occurs, which tends to favor percolation.

In the case of a magma ocean, the molten core of an impacting body would emulsify into small droplets that "rain" through the melt. Liquid iron would pond at the base of the magma ocean. From this point on, large blobs

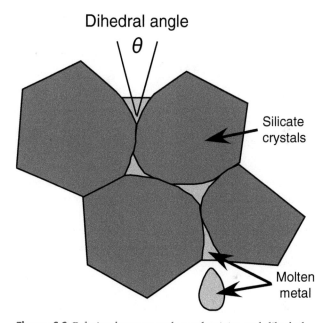

Figure 6.6 Relation between melt conductivity and dihedral angle in an aggregate of crystals containing dispersed metallic melt.

of molten metal would sink rapidly through the solid silicate mesh or intrude downward as dikes.

Small metal droplets can chemically equilibrate with the enclosing magma ocean, providing a test for this mechanism. Conversely, large descending blobs of liquid metal would probably preclude equilibration with the enclosing silicates. Siderophile element abundances in mantle rocks can be used to test for equilibration. Compared to the abundances of lithophile elements, siderophile and chalcophile elements in the Earth's mantle are strongly depleted (Figure 6.7). We have seen diagrams like this before (look back at Figures 4.14 and 5.11), where elements are plotted according to the temperature at which 50 percent of the element would have condensed. The refractory lithophile elements have constant (chondritic) relative abundances, whereas the abundances of volatile lithophile elements decrease with volatility. On the other hand, the siderophile element abundances show no

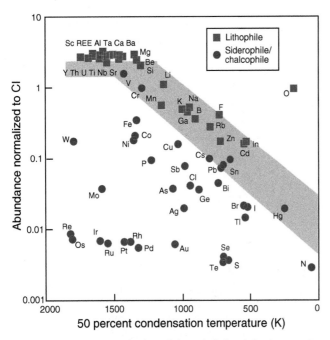

Figure 6.7 Depletions of siderophile and chalcophile elements in the Earth's mantle, compared to lithophile elements. All elements are plotted according to the temperature at which 50 percent of the element would have condensed from a nebula gas. Mantle abundances of siderophile elements are consistent with high-pressure equilibration, perhaps at the base of a magma ocean. Modified from Carlson et al. (2014).

relationship to volatility, because they have been modified by core formation.

The partitioning of siderophile elements depends on their relative affinity for metal versus silicate, which varies with pressure and temperature, as well as the oxidation state, which controls the amount of metallic Fe versus FeO. Experiments conducted under different conditions predict the partitioning of siderophile elements between mantle silicate and metal, and their results can be compared with measured siderophile element abundances in the mantle (Righter, 2011). These data indicate that the siderophile element partitioning occurred at high pressure, likely at the base of a deep magma ocean (Badro et al., 2015). However, the occurrence of the most highly siderophile elements (the elements at the bottom left of Figure 6.7) in the mantle in nearly constant (chondritic) relative proportions (Figure 6.7) suggests that the liquid metal was extracted completely into the core, and a veneer of chondritic material was accreted later. A small amount, less than 1 percent of the Earth's mass, of added chondrite would be sufficient to replace the highly siderophile elements in the mantle, but would not significantly affect other element abundances.

Another model for core formation in the Earth (Wade and Wood, 2005) posits that the core grew gradually by accretion of already differentiated bodies. Each incoming protoplanet added its core to the growing metal mass at the center of the Earth. As the planet progressively increased in size, a global magma ocean deepened, and the pressures and temperatures at which metal and silicates equilibrated increased. This model is not applicable to core formation in small bodies, but it illustrates the complexities that are likely in planetary-sized bodies.

6.4.2 Going Up: Crusts

Crusts form from melts that ascend buoyantly and erupt on the surface or are emplaced as plutons. A critical amount of melting, usually a few percent (Maaloe, 2003), must occur before magmas can segregate from their source regions. Although melts can rise upward along grain boundaries, the formation of veins along fractures can drain melts more efficiently. Heat loss is also important in controlling magma ascent. The magma cooling rate is proportional to its surface area times the temperature difference with the surrounding rock. Calculations indicate that, to reach the surface, magma moving in a planar fracture must move 10^4 times faster than in a sphere of equivalent volume. As a consequence, spherical ascending magmas (diapirs) are more likely, at least until the magma nears the surface. On planets, tectonic control of the locations of melting allows magmas to be injected repeatedly in the same place, leading to subsurface magma chambers and volcanic edifices (McCoy et al., 2006). Multiple heat sources within planets allow crust formation to occur over billions of years. On small bodies, crust formation occurs without tectonic influences.

Taylor and McLennan (2009) distinguish "**primary**" and "**secondary**" **crusts**. A primary crust is formed during planetary differentiation. The Moon's anorthositic crust is primary, having crystallized directly from the lunar magma ocean. Primary crusts have commonly been destroyed by cataclysmic impacts and foundering during mantle overturn, although that appears not to have been the case for the Moon. A secondary crust is produced from magmas formed by partial melting of the mantle. The lunar mare basalts are an example of secondary crust.

Almost none of the Earth's primary (Hadean-age, earlier than 3.9 Ga) crust remains, and the evidence used to surmise its composition is enigmatic and contradictory. Its only unambiguous remnants are a handful of detrital zircon crystals with ages as old as 4.37 Ga that occur in younger sedimentary rocks. These zircons apparently crystallized in felsic magmas, but suggestions for an early, widespread granitic crust do not appear to be supported by other evidence.

BOX 6.2 **METAL CORES IN ASTEROIDS**

Vesta (~500 km diameter) is the only intact asteroid known to have a metallic core, with an estimated diameter of 220 km (Russell et al., 2012). Iron meteorites provide direct samples of core materials and demonstrate that core formation was common in planetesimals.

Iron meteorites are composed mostly of iron and nickel, with accessory minerals like troilite FeS, schreibersite $(Fe,N)i_3P$, and graphite C. Slow cooling produces the Widmanstatten pattern (Figure 6.8a), an intergrowth of low-nickel kamacite and high-nickel taenite crystals. The coarseness of the intergrowth structure and the diffusional zoning of nickel at crystal boundaries can be used to estimate the rate of cooling following solidification. The estimated cooling rates correspond to asteroids having diameters of 50–200 km. Because iron metal is so thermally conductive, cores should have uniform cooling rates throughout, and most iron groups bear this out. However, the meteorites in a few iron groups show radically different cooling rates. The explanation for this is that the insulating silicate mantle was stripped off by oblique (sometimes called "hit-and-run") impacts, allowing metal at various depths within the naked cores to cool at different rates.

Pallasites, composed of olivine and metal (Figure 6.8b), were long thought to be samples of the boundaries between cores and the overlying mantles in asteroids. An alternative idea is that they are impact-generated mixtures of molten metal and mantle olivine.

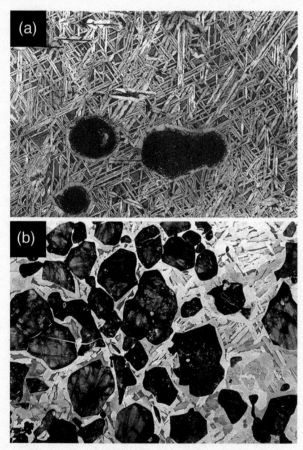

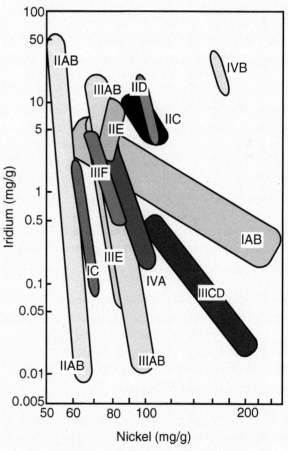

Figure 6.8 (a) Polished and etch slab of the Mount Edith (Australia) iron meteorite, exhibiting a characteristic Widmanstatten pattern formed by the intergrowth of two iron–nickel minerals, kamacite and taenite. The dark blobs are FeS (troilite). (b) Slab of the Fukang (China) pallasite, composed of olivine and metal. Smithsonian Institution images.

Figure 6.9 Iridium versus nickel abundances in iron meteorites. Compositional differences serve to classify irons. The trends are due to fractional crystallization and other processes during solidification of molten iron cores in asteroids. Modified from Scott and Wasson (1975).

Differences in the abundances of siderophile elements (iridium, gallium, germanium, nickel) are used to classify irons and pallasites. The iridium and nickel abundances of the most abundant iron meteorite groups are illustrated in Figure 6.9. There are lots of additional iron groups with only a few meteorites. Altogether, as many as 60 different asteroid cores may be represented in our meteorite collections. The taxonomy of irons reflects the tortured evolution of the classification system. Originally, irons were assigned to compositional groups I through IV, and some were later subdivided (III became IIIA and IIIB) and sometimes recombined (IIIAB) as more data became available.

The sloping chemical trends in Figure 6.9 reflect solidification processes in the molten cores. Fractional crystallization accounts for most of the observed trends, although more complex models including liquid immiscibility and trapping of melt between growing solids have also been advocated (Chabot and Haack, 2006). It is unclear whether asteroid cores solidified from the center outward (as in the Earth) or from the core–mantle boundary inward, and whether the crystals formed concentric layers or elongated dendrites.

Even though it seems likely that the Earth had an early magma ocean, it is unlikely to have formed an anorthositic crust like that of the Moon. Plagioclase is not stable below about 40 km depth, and even if crystallized it would have sunk in the wet (less dense) terrestrial magma ocean. Mantle temperatures during the Hadean were higher than at present, due to increased radiogenic heat production and large impacts, prompting suggestions for an early ultramafic crust of lavas (komatiites) formed by large degrees of melting. If an ultramafic crust once existed, it has vanished without leaving a recognizable geochemical signature. The most probable composition for the Earth's primary crust is basaltic. The rare ancient zircons, the first vestige of Earth's secondary crust, could have formed by remelting of such a basaltic precursor. The planet's earliest geologic record is composed of basalts and sodium-rich felsic rocks (tonalite, trondhjemite, granodiorite, collectively called the "TTG suite") of Archean (3.9–2.5 Ga) age. The TTG rocks formed by partial melting of foundering or subducted basalt, and are distinct from the potassium-rich granitic rocks of the post-Archean continental crust, which might be termed "tertiary." The modern crust of our planet is dominated by basaltic magmatism at spreading centers and felsic magmatism at subduction zones.

An important characteristic of the Earth's crust is its profound enrichment in **incompatible elements** (Figure 6.10) – a consequence of multiple periods of partial melting, each further concentrating these elements into the melt phase. The incompatible elements have large ionic size and/or charge, favoring their incorporation in magma over confining crystal structures. Sequestering of incompatible elements is also likely in the crusts of other planets, although probably not to the extent seen on Earth, where we even find ore deposits of these and other elements concentrated by hydrothermal fluids or in highly evolved granitic magmas.

Extrapolating what has been learned about the Earth's crust to other planets and planetesimals is not very useful, because our planet's crust is compositionally unique. This is due mainly to our planet's liquid water, which enables plate tectonics and is ultimately responsible for the siliceous continental crust. Although the crusts of Mercury, Venus, and Mars are all dominated by basaltic rocks, consistent patterns in crustal evolution are elusive (Taylor and McLennan, 2009). The compositional variations among achondrites that represent crustal rocks from asteroids are even more pronounced, and the crusts of the satellites of the giant planets are stranger still. The compositions of igneous crusts on various Solar System bodies and the processes that generate and modify them are discussed more fully in Chapter 10.

6.4.3 What's Left: Mantles

We consider mantles last, because they are what remain after the extraction of metal to form cores and of silicate partial melts to form crusts. We previously noted that the Earth's mantle is strongly depleted in siderophile and chalcophile elements (Figure 6.7); conversely, its lithophile element abundances mimic those of the bulk Earth. The mantle and crust also have complementary geochemical patterns – whatever is concentrated in one is depleted in the other.

Whether or not mantles formed by solidification of a magma ocean or as residues from incomplete melting, they are composed of ultramafic rocks – a natural consequence of having chondritic bulk compositions. Any planet's primitive upper mantle is basically peridotite, composed of olivine + orthopyroxene + clinopyroxene + an aluminous mineral (plagioclase, spinel, or garnet, depending on pressure). Here we are not considering extremely high-pressure phases that occur in the lower mantles of the large terrestrial planets; these will be

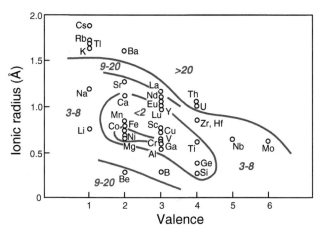

Figure 6.10 Incompatible elements have high size and charge, and are enriched in the Earth's continental crust relative to their abundances in the primitive mantle. Contours group elements with similar degrees of enrichment (italicized blue numbers). Modified from Taylor and McLennan (2009).

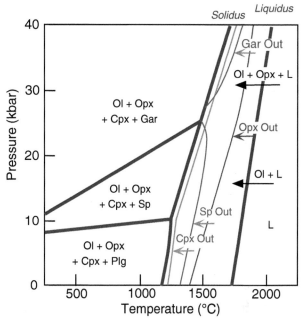

Figure 6.11 Schematic phase diagram for mantle peridotite, illustrating the effects of partial melting. With increasing pressure, the aluminous phase changes from plagioclase (Plg) to spinel (Sp) to garnet (Gar). With increasing temperature, melting begins at the solidus and first clinopyroxene (Cpx) and then garnet or spinel are exhausted from the residue, leaving olivine (Ol) + orthopyroxene (Opx). Continued melting exhausts orthopyroxene, leaving only olivine, until complete melting at the liquidus. After McSween et al. (2003).

introduced in Chapter 7. Partial melting of peridotite yields basaltic magma. To see the effects of extracting basaltic magma on the Earth's mantle, let's examine a pressure–temperature diagram showing the stability of minerals in mantle residues (Figure 6.11). The mineralogy of the mantle at different pressures is shown on the left side of the diagram. With increasing temperature, solid rock crosses the solidus and begins to melt. The formation of basaltic magma first exhausts clinopyroxene and spinel or garnet, yielding a residue of olivine + orthopyroxene (this rock is called harzburgite). Further melting exhausts orthopyroxene, leaving only olivine (dunite). Basaltic magmas on Earth sometimes carry xenoliths of harzburgite or dunite, representing the mantle residues in their source regions. Although the various terrestrial planets may have somewhat different mantle compositions and phase relations, this example can in general explain how their mantle mineralogies vary when partial melts are extracted to form crusts. Interior pressures within planetesimals are lower, precluding the formation of garnet and spinel, but the basic melting relationships are similar except in mantles with very low oxidation states.

6.4.4 Another View: Partial Differentiation

Although planets have fully differentiated, the fate of planetesimals is less clear. Melted (achondritic) and unmelted (chondritic) meteorites are conventionally thought to have formed on different parent bodies. However, this picture has been muddied somewhat by evidence that may suggest some planetesimals experienced incomplete melting, resulting in differentiated interiors and outer, chondritic crusts (Weiss and Elkins-Tanton, 2013). The so-called primitive achondrites are residues from small degrees of

partial melting, and it seems likely that the near-surface crusts of such bodies must be chondritic. The IIE iron meteorites appear to be mixtures of molten iron with a variety of silicate rocks, including basalts and chondrites, perhaps resulting from impact disruption of a partly differentiated planetesimal. Paleomagnetic measurements of some chondrites suggest that their magnetic fields originated within the parent bodies themselves, most easily explained if they had molten cores. In fact, there could have been all gradations between fully melted and differentiated planetesimals and unmelted chondritic bodies.

6.5 Differentiation of the Giant Planets

It is probable that Jupiter and Saturn each contain rocky cores. The main source of internal heat in these planets is gravitational potential energy gained during the bodies' accretion and, in the case of Saturn, by sinking of dense material. High-pressure experiments indicate that both gas giants exhibit a continuous transition from molecular hydrogen to metallic (ultradense, electrically conducting) hydrogen. In this case, the atmosphere can be considered a differentiated layer. Helium may be immiscible in metallic hydrogen, allowing it to sink into the deep interior, at least in Saturn.

The interior structures of Uranus and Neptune are compositionally layered. The relative proportions of gaseous hydrogen and helium, ice, and rock are similar in both planets, and rock and ice constitute most of their masses. Some of the rock may be concentrated in cores. Like Jupiter and Saturn, the ice giants are mostly warmed by the heat left over from accretion.

The giant planets, so different in composition from the terrestrial planets, are clearly differentiated, but the processes that caused differentiation are profoundly different. Stratification depends on density contrasts among rock, ices, and (in the case of Jupiter and Neptune) high-pressure polymorphs of gaseous species, and is not driven by heating and melting.

6.6 Hot, and Then It's Not

The heat sources that caused planetary differentiation have changed with time. The primary heat source in early planetesimals was the decay of ^{26}Al. The short half-life of this radionuclide means that it was an effective heat source for only a few million years. As we saw in Figure 6.11, the aluminous phase melts early so that ^{26}Al is concentrated into partial melts that segregate into the crust, thereby further depriving the interior of the ability for more

sustained melting. This short-lived radionuclide played a lesser role in large planets because they took longer to form.

Similarly, radiogenic heating from long-lived radioisotopes has decreased over time, at different rates for different isotopes. For example, heat production in the Hadean Earth was nearly four times the present value. About 90 percent of the current radiogenic heat production in the Earth comes from ^{238}U and ^{232}Th, but before about 2.5 Ga the heat production from ^{40}K would have contributed more than 30 percent of the total (Taylor and McLennan, 2009). Potassium, uranium, and thorium are incompatible elements that tend to be fractionated into melts that form the crust. Even though heat from the crust can escape faster than heat from the mantle, these are still very important heat sources for planets.

Currently, the Earth's heat flow is 42 terawatts (42×10^{12} W). The decay of long-lived radionuclides is estimated to produce ~18 TW. The remainder is probably attributable to exothermic crystallization of the liquid outer core, although some portion is likely to be fossil accretional heat.

Heating of planets by impacts was much more significant in the early Solar System, especially during the waning stages of accretion, when large bodies of comparable size collided. Impact heating is no longer a significant contributor.

Summary

The sources for heating planetesimals and planets are:

- decay of long-lived (^{40}K, ^{235}U, ^{238}U, ^{232}Th) and short-lived (especially ^{26}Al) radionuclides, resulting in mass loss that is transformed into energy;
- segregation of metal cores, which transforms gravitational potential energy into thermal energy, and exothermic crystallization of molten core metal;
- large impacts that convert kinetic energy of the impactor into heat; and
- gravitational tides, where frictional heating occurs as a body flexes along its orbital path.

Short-lived radionuclide decay was particularly important in heating planetesimals in the early Solar System; long-lived radionuclides are not very important for small bodies because heat escapes faster than it can build up. Planets early in their histories were primarily heated by large impacts and core segregation; planetary heating in recent times is dominated by long-lived radionuclide decay and exothermic crystallization in bodies with partly liquid cores. Tidal heating affects some satellites of the giant planets.

Large-scale melting to produce magma oceans is thought to have been common for terrestrial planets in the early Solar System, although its evidence remains elusive (except for the Moon). Pervasive melting would have aided differentiation. Metal drops could sink through, and equilibrate with, a magma ocean. Metallic liquids that could "wet" silicates were probably required to separate core materials from a solidified mantle, and once accumulated into large blobs could displace but probably not equilibrate with mantle silicates. Primary crusts and mantle cumulates could form from magma oceans; secondary crusts formed by later remelting of the mantle.

Differentiation has also occurred in the giant planets. In Jupiter and Saturn, molecular hydrogen gives way to metallic hydrogen at very high pressures, and both may contain highly compressed rocky cores. Uranus and Neptune contain rocky cores with ice mantles. Although still retaining residual accretional heat, stratification in the giant planets results from differences in density, rather than melting.

Having seen how and why planetary differentiation occurred, in the next chapter we will explore the nature of the deep interiors of planets.

Review Questions

1. Why does radioactive decay release heat, and which long-lived radioisotopes are responsible for planetary heating?
2. Which heating mechanisms were responsible for the differentiation of the Earth? Which heating mechanisms are important now?
3. What evidence suggests that the Moon's differentiation involved a magma ocean?
4. What controls the separation from the mantle of core materials and of magmas that become the crust?
5. How is differentiation in the giant planets different from the terrestrial planets?

SUGGESTIONS FOR FURTHER READING

Badro, J., and Walter, M. (2015) *The Early Earth: Accretion and Differentiation*. New York: John Wiley & Sons. Everything you need to know about our planet's formation and differentiation, all in one place.

Carlson, R. W., Garnero, E., Harrison, T. M., et al. (2014) How did early Earth become our modern world? *Annual Reviews of Earth and Planetary Sciences*, **42**, 151–178. This up-to-date review summarizes the Earth's geochemical inheritance from protoplanets, the crystallization of its early magma ocean, and the formation of its core, mantle, and crust.

Elkins-Tanton, L. T. (2012) Magma oceans in the inner Solar System. *Annual Review of Earth and Planetary Sciences*, **40**, 113–139. A thoughtful assessment of the formation and solidification of magma oceans on planetesimals and planets.

Elkins-Tanton, L. T., and Weiss, B. P., eds. (2017) *Planetesimals: Early Differentiation and Consequences for Planets*. Cambridge: Cambridge University Press. This book describes the compositional diversity of planetesimals and their magmatic differentiation.

Rubie, D. C., Nimmo, F., and Melosh, H. J. (2007) Formation of Earth's core. In *Treatise in Geochemistry*, Vol. 9, eds. Holland, H. D., and Turekian, K. K. Oxford:

Elsevier, pp. 51–90. This chapter provides a comprehensive survey of the physics of core formation, and its compositional and temporal constraints.

Taylor, S. R., and McLennan, S. M. (2009) *Planetary Crusts: Their Composition, Origin and Evolution*. Cambridge: Cambridge University Press. Here's a whole book on crusts, providing a superb overview of the nature and origin of crusts of different planets.

REFERENCES

Badro, J., Brodholt, J. P., Piet, H., et al. (2015) Core formation and core composition form coupled geochemical and geophysical constraints. *Proceedings of the National Academy of Sciences*. DOI: 10.1073/pnas.1505672112.

Carlson, R. W., Garnero, E., Harrison, T. M., et al. (2014) How did early Earth become our modern world? *Annual Reviews of Earth and Planetary Sciences*, **42**, 151–178.

Chabot, N. L., and Haack, H. (2006) Evolution of asteroidal cores. In *Meteorites and the Early Solar System II*, eds. Lauretta, D. S., and McSween, H. Y. Tucson, AZ: University of Arizona Press, pp. 747–771.

Elkins-Tanton, L. T. (2012) Magma oceans in the inner Solar System. *Annual Review of Earth and Planetary Sciences*, **40**, 113–139.

Grimm, R. E., and McSween, H. Y. (1993) Heliocentric zoning of the asteroid belt by [26]Al heating. *Science*, **259**, 653–655.

Maaloe, S. (2003) Melt dynamics of a partially molten mantle with randomly oriented veins. *Journal of Petrology*, **44**, 1193–1210.

McCoy, T. J., Mittlefehldt, D. W., and Wilson, L. (2006) Asteroid differentiation. In *Meteorites and the Early Solar System II*, eds. Lauretta, D. S., and McSween, H. Y. Tucson, AZ: University of Arizona Press, pp. 733–745.

McSween, H. Y., Richardson, S. M., and Uhle, M. E. (2003) *Geochemistry: Pathways and Processes*, 2nd edition. New York: Columbia Press.

Righter, K. (2011) Prediction of metal–silicate partition coefficients for siderphile elements: an update and assessment of PT conditions for metal–silicate equilibrium during accretion of the Earth. *Earth and Planetary Science Letters*, **304**, 158–167.

Rubie, D. C., Nimmo, F., and Melosh, H. J. (2007) Formation of Earth's core. In *Treatise in Geochemistry*, Vol. 9, eds. Holland, H. D., and Turekian, K. K. Oxford: Elsevier, pp. 51–90.

Russell, C. T., Raymond, C. A., Coridini, A., et al. (2012) Dawn at Vesta: testing the protoplanetary paradigm. *Science*, **336**, 684–686.

Ryder, G. (1991) Lunar ferroan anorthosites and mare basalt sources: the mixed connection. *Journal of Geophysical Research*, **118**, 2065–2068.

Scott, E. R. D., and Wasson, J. T. (1975) Classification and properties of iron meteorites. *Reviews of Geophysics and Space Physics*, **13**, 527–546.

Stevenson, D. J. (1990) Fluid dynamics of core formation. In *Origin of the Earth*, eds. Newsom, H. E., and Jones, J. H. New York: Oxford University Press, pp. 231–250.

Taylor, S. R., and McLennan, S. M. (2009) *Planetary Crusts: Their Composition, Origin and Evolution*. Cambridge: Cambridge University Press.

Wade, J., and Wood, B. J. (2005) Core formation and the oxidation state of the Earth. *Earth and Planetary Science Letters*, **236**, 78–95.

Weiss, B. P., and Elkins-Tanton, L. T. (2013) Differentiated planetesimals and the parent bodies of chondrites. *Annual Review of Earth and Planetary Sciences*, **41**, 529–560.

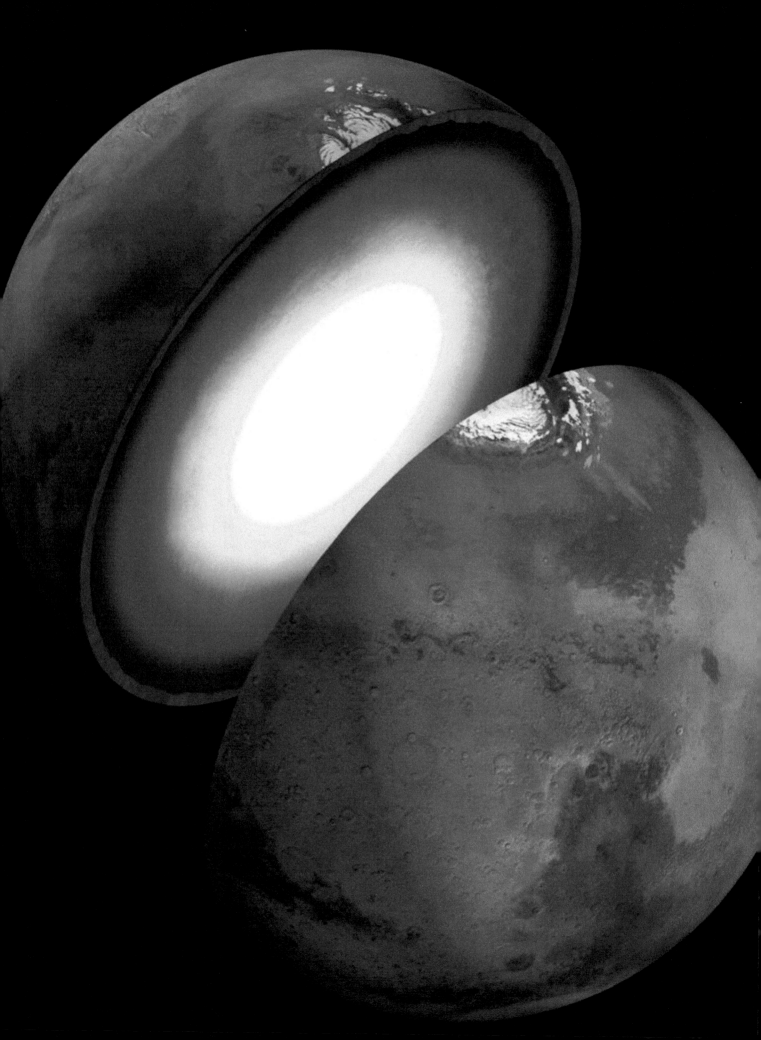

7

Unseen Planetary Interiors

The interior structures of the Earth and Moon are determined from seismic data. The existence and sizes of cores in other planets are inferred from observations of planetary sizes, masses, and shapes, which constrain their uncompressed mean densities and moment of inertia factors. Mantle and crust thicknesses can also be estimated from gravity data obtained by orbiting spacecraft. Successful models of planetary interiors constructed from compositional data must be consistent with observed densities and moments of inertia. High-pressure laboratory experiments can constrain the mineralogy of mantles and cores and the partitioning of elements between silicate and metal in the terrestrial planets. The interiors of the giant planets are not well understood, because of uncertainties in their compositions and internal temperatures and pressures. The states of hydrogen and helium in the interiors of Jupiter and Saturn, and the crystalline forms of ices in Uranus, Neptune, and icy satellites, are inferred from experimentally determined or calculated phase diagrams. The giant planets may have small rocky cores, with successive layers of either metallic hydrogen (Jupiter and Saturn) or ices (Uranus and Neptune), and molecular hydrogen. Planetary mantles and cores evolve over geologic time, through cooling and extraction (or reintroduction, in the case of Earth) of crustal components.

7.1 Hardened Hearts

Seismic data for the Earth and Moon provide information on their interior structures, but such data are not yet available for other planets. Instead, we use other indirect means to model their interiors. Critical observations of planets are their masses, volumes, shapes, and gravity fields. When combined with compositions and phase diagrams from high-pressure experiments or calculations, we can construct and test models of planetary interiors.

In Section 5.4, we learned how the bulk composition of a planet can be estimated from element ratios in rocks, and in Section 6.4 we explored how differentiation processes transform a compositionally homogeneous planet into one with a dense metal core, an ultramafic silicate mantle, and a crust of more easily melted silicates. Now let's be more specific in characterizing the unseen planetary interiors.

7.2 Inside the Planet We Know Best

Study of the Earth's interior illustrates techniques that might be brought to bear on other planets. These hard-won discoveries about the Earth illustrate the complexities that can occur at high pressures and temperatures.

7.2.1 Seismology

Earthquakes create vibrations that rumble through our planet along curved paths and eventually emerge back on the surface. Making sense of this cacophony of sound can be a challenge; one geophysicist has suggested that reaching conclusions from the Earth's interior from seismic echoes is like trying to reconstruct the inside of a piano from the sounds it makes while crashing down stairs. Despite the challenges of interpreted seismic data, we have learned much about the hidden interior of Earth using this indispensable tool.

Seismic energy travels as waves that vibrate parallel (compressional, or P waves) or perpendicular (shear, or S waves) to the direction of propagation. Let's review a few basic rules that explain how seismology works. First, the velocity of a seismic wave depends on the ratio of rigidity to density of the rocks through which it travels: Rigidity increases faster than density inside the Earth, so velocity increases downward. Regardless of rock density, P waves travel faster than S waves. Second, seismic waves are refracted (bent) or reflected by gradients or

Velocity (km/s) or density (Mg/m³)

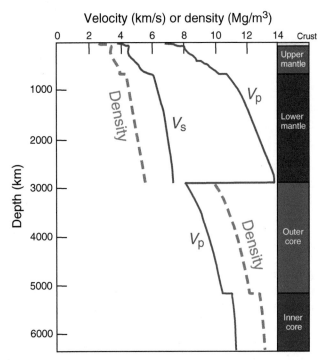

Figure 7.1 Changes in the velocities of P and S waves with depth in the Earth, and corresponding density variations.

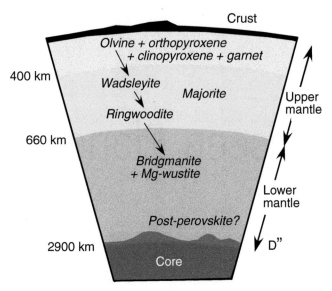

Figure 7.2 Mineralogic interpretation of the Earth's mantle, based on densities determined from seismic data.

discontinuities in material properties, chiefly density but also the ability to resist change in volume (compressibility) and shape (rigidity). Third, liquids cannot transmit S waves, so the seismic properties of solid and molten materials are different.

Figure 7.1 is a seismic cross-section of the Earth, showing measured velocities (V) of P and S waves as a function of depth. The change in density (ρ) with depth or planetary radius (r) can be calculated from those seismic velocities using the Adams–Williamson equation:

$$\mathrm{d}\rho/\mathrm{d}r = -(\rho_r g_r)/\phi_r \qquad (7.1)$$

where ρ_r is the density at radius r, g_r is the gravitational acceleration at radius r, and ϕ_r is a seismic parameter that incorporates the seismic velocities (V) for P and S waves at radius r: $\phi = V_P^2 - 4/3\, V_S^2$. The calculated densities are also shown in Figure 7.1.

Starting at the top of Figure 7.1, the slowest seismic velocities correspond to the crust. There is an abrupt jump in velocity that defines the crust–mantle boundary, called the Mohorovicic discontinuity after its discoverer; this boundary is commonly referred to as the "Moho" for reasons that should be obvious. Velocities within the mantle increase progressively, with sudden jumps at depths of ~400 km and ~660 km. Another discontinuity at a depth of 2900 km marks the boundary between mantle and the outer core. Here the P wave velocity suddenly plummets and the S wave disappears altogether.

One final discontinuity at 5150 km depth corresponds to the top of the inner core.

Figure 7.2 gives a geologic interpretation of the bulk density changes within the mantle portion of Figure 7.1. The more-or-less continuous increase in density with depth reflects compression due to increasing pressure. The sudden discontinuities at the crust–mantle and mantle–core boundaries indicate changes in composition. Crustal rocks of intermediate to mafic composition give way in the mantle to ultramafic rocks composed of denser silicates like olivine and pyroxene, and mantle silicates give way to molten iron metal in the core. The other discontinuities reflect phase changes in minerals. The uppermost mantle consists of olivine + orthopyroxene + clinopyroxene + spinel or garnet. Between 400 and 660 km depth (pressures of 13.5–23 GPa), olivine transforms to high-pressure polymorphs having approximately the same composition – first wadsleyite (having the β-spinel structure), and then ringwoodite (having the γ-spinel structure). Another phase that occurs at these depths is majorite, a garnet-group mineral with complex solid solution. Not illustrated in the upper mantle of Figure 7.1 is the asthenosphere, a lubricating layer between the elastic lithosphere above and the viscous mantle below. At 660 km depth, ringwoodite transforms to bridgemanite ($Mg,Fe)SiO_3$. This mineral has the composition of olivine with the structure of perovskite. (In literature before 2014, this phase is called Mg-perovskite). This transformation defines the lower mantle. Since bridgemanite is stable throughout much of the mantle, it is actually the most abundant silicate mineral on Earth. Bridgmanite coexists in the lower mantle with magnesiowüstite ($Mg,Fe)O$. Just above the mantle–core boundary is the D″ ("D double-prime") layer, variously interpreted as a phase transformation to some undefined post-perovskite structure or to melt. This layer has also been ascribed to chemical interactions between the core and mantle, or described as a graveyard for subducted slabs.

The explanation for the curious seismic behavior at 2900 km depth is not to be found in additional changes in atomic packing of minerals that comprise the mantle. The core must have a fundamentally different composition. Iron is the only plausible element, because of its high Solar System abundance coupled with its depletion in the crust and mantle. An alloy of iron and nickel provides a better match for core seismic velocity than does iron alone. The absence of S waves in the outer core indicates that it is substantially molten, and its P wave velocity suggests a few percent of some light element (probably silicon and oxygen, perhaps some sulfur or hydrogen). The density of the inner core is consistent with solid iron–nickel metal.

Earthquake ray paths are actually complex, because velocities do not simply increase with depth. Instead rays bend toward the vertical as they travel out of the Earth (Figure 7.3). P waves are not recorded by seismometers at angular distances between 104° and 140° because of refraction at the mantle–core boundary. S waves do not reappear on the surface at all beyond 104° because, as noted earlier, they are not transmitted through the liquid outer core. These "shadow zones" are useful for determining the size of the Earth's core; however, shadow zones likely will have no application to other planets unless we can have global seismometer coverage.

P wave velocities in the inner core are anisotropic, suggesting that the inner core may itself have a complex structure. One suggestion is that iron–nickel metal crystals in its outermost regions are aligned in a north–south direction, whereas crystals near the center of the core point roughly east–west. Grain orientations may have been established during core crystallization or later deformation.

Large earthquakes can also generate **free oscillations**, stationary standing waves that vibrate in different modes. Free oscillations can persist for days or months, and cause the Earth to ring like a bell. The dispersion of free oscillations provides further information on the velocity structure of the crust and mantle, as well as confirmation of a liquid outer core.

7.2.2 Samples from the Mantle

Samples of upper mantle rocks are sometimes brought up as xenoliths in erupting magmas, or tectonically exposed in mountain belts. These mantle samples are spinel or

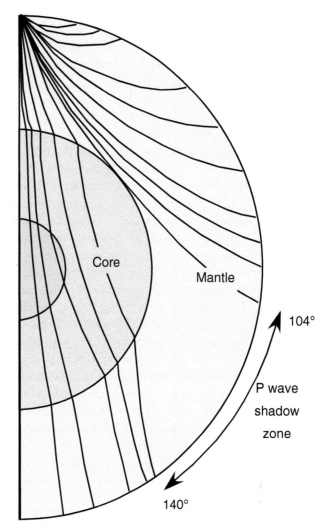

Figure 7.3 Refracted P wave paths through the Earth's core produce a shadow zone where seismic signals are not received on the surface.

garnet peridotite or residues from partial melting (harzburgite or dunite, as described in Section 6.4). Although rocks from deeper parts of the mantle that contain high-pressure phases like ringwoodite and bridgmanite are not directly sampled, the ultramafic compositions of upper mantle rocks provide useful comparisons.

7.2.3 High-Pressure Experiments

A multi-anvil cell consists of a compressed cavity containing steel anvils that converge on a small sample, so that pressure is amplified by reducing the area over which force is applied. These cells can apply pressures up to 25 GPa at temperatures as high as 3000 K, appropriate to upper mantle conditions. The high-pressure silicate minerals of the upper mantle have now been produced experimentally (although they don't last at surface pressures). Even bridgmanite, the primary constituent of the lower mantle, has been synthesized.

BOX 7.1 **HIGH-PRESSURE NAMESAKES FOR HIGH-PRESSURE MINERALS**

In the late 1950s and 1960s, experiments were not capable of producing the high-pressure silicate minerals that occur in the mantle. A. E. Ringwood, an experimentalist at the Australian National University, recognized that germanates (minerals in which germanium substitutes for silicon) could serve as low-pressure analogs for high-pressure silicates. He conducted experiments on the germanate equivalents of olivine and pyroxene, observing phase changes into denser polymorphs having spinel structures. With that insight, Ringwood predicted that olivine would transform to β- and γ-spinel phases in the Earth's mantle. A few years later, Ringwood succeeded in synthesizing the spinel form of silicate olivine. At that time, no naturally occurring high-pressure polymorphs of olivine had been discovered; natural occurrence is a precondition for naming a new mineral. In 1969, a naturally occurring sample of γ-spinel structured olivine was identified in a highly shocked meteorite, and was named ringwoodite in honor of Ringwood's seminal work in mantle mineralogy.

Percy Bridgman, a physics professor at Harvard, won the 1946 Nobel Prize for his work on the properties of materials at high pressure. A malfunction in machinery led him to modify his pressure apparatus, resulting in a huge increase in experimentally achievable pressures to more than 10 GPa. His career included studies of the compressibility, electrical and thermal conductivity, and viscosity of metals and crystals. Although Bridgman did not work directly on mantle minerals, his pressure apparatus was instrumental in developing later experiments. The existence in the mantle of $(Mg,Fe)SiO_3$ with the perovskite structure was first suggested in 1962, and synthesized in high-pressure experiments in the next decade. By the late 1970s it was hypothesized to be the mineral that characterized the lower mantle below 660 km depth. However, silicate perovskite, the most abundant mineral in the Earth (comprising up to 38 percent of its volume), had never been found in nature, precluding its naming. In 2014, this phase was finally discovered in another shocked meteorite, and named bridgmanite in honor of Bridgman's pioneering work.

Majorite and wadsleyite derive their names from other mineralogists, and were also first discovered in highly shocked meteorites. Meteor impacts provide the only natural occurrences of these mantle minerals because they can achieve the extreme pressures necessary to form them on the Earth's surface.

Extreme conditions, even approaching those in the core (pressure = 140–360 GPa, temperature = 4000–6200 K), can be reached in experiments using another apparatus. A diamond anvil cell consists of two opposing diamonds that compress a small sample between their tips. The sample can be analyzed through the diamonds by X-ray diffraction and fluorescence. Dynamic shock experiments utilize gas guns to compress target materials to similarly high pressures, but for only fractions of a second, requiring very fast characterization of the high-pressure minerals.

The core is made of dense **polymorphs** of iron, having either hexagonal closest-packed or face-centered cubic structures. The uncertainty in structure arises from the effects of other included elements like nickel, silicon, and oxygen. Experiments so far have focused on the phase relations and properties of iron at core pressures and the melting point of iron in the outer core (~6200 K, hotter than the Sun's surface).

Experiments also constrain the geochemical partitioning of elements between the Earth's mantle and core (Righter and Drake, 1997), as described in Section 6.4.

The results of experiments conducted at different pressures can be compared to mantle siderophile element abundances to estimate the depth at which core metal and mantle silicate equilibrated. Estimates by Badro et al. (2015) suggest core formation occurred in a magma ocean not exceeding 1800 km depth.

7.2.4 Seismic Tomography and Convection

Planetary heat loss by convection is much more efficient than by conduction. Convection in the outer core accounts for the Earth's magnetic field, and in the Earth's mantle drives plate tectonics. Seismologists wring information about convection from earthquake waves by using **seismic tomography** – basically three-dimensional imaging of the Earth's velocity structure compiled from many travel-time measurements of P and S waves. We have already seen that seismic velocity depends on density, but it also varies with elasticity, which is a function of temperature. A tomographic image shows where within the Earth seismic waves pass through hot or cold material, which slows them down or speeds them up, respectively.

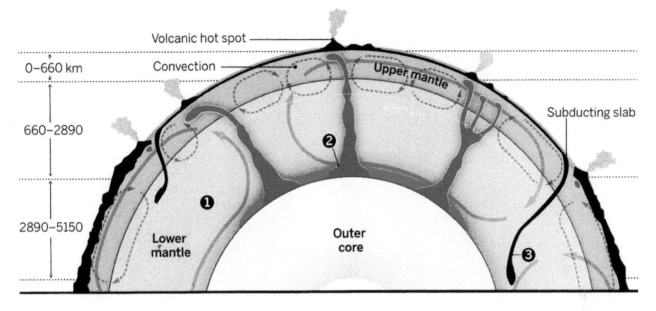

Figure 7.4 Sketch of upward convection (1) causing plumes rising from ultra-low velocity zones (2), and downward convection carrying subducted slabs (3). Adapted from Hand (2015).

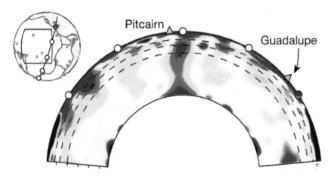

Figure 7.5 Tomographic cross-section (location shown in inset map) of the mantle, showing a plume extending from the core–mantle boundary; its surface expression is a volcano (Pitcairn). Broken lines indicate depths of 400, 660, and 1000 km. Adapted from French and Romanowicz (2015).

A long-standing argument in geophysics has been whether convection occurs throughout the whole mantle, or whether the upper and lower mantles convect separately (Figure 7.4). In other words, do subducted slabs stop at the 660 km discontinuity or penetrate into the lower mantle, and do rising plumes occur only within the upper mantle or extend from the core–mantle boundary to the surface? Tomography has now imaged cold subducted slabs, some stalling within the upper mantle and some extending downward well into the lower mantle. More recently, tomography images (Figure 7.5) have revealed hot plumes rising from the base of the mantle (French and Romanowicz, 2015).

Another revelation is the existence of ultra-low velocity zones at the core–mantle boundary. These zones constitute the D″ layer (Figure 7.2). An emerging model suggests that descending slabs herd detritus at the bottom of the mantle into large piles of unknown composition and state. Mantle plumes rise from these ultra-low-velocity piles, so they must somehow channel heat from the crystallizing outer core, causing convection in the overlying mantle.

The Earth's magnetic field also arises from electric currents resulting from convection in the core. Paleomagnetic measurements of ancient rocks indicate that the Earth has had a magnetic field for at least 3.5 Ga. A transition in geomagnetic strength between 1.5 and 1.0 Ga could be explained by nucleation of the solid inner core during this period (Biggin et al., 2015).

7.3 Inside Other Rocky Planets

Seismic data provide the most accurate picture of the Earth's interior, but seismometers have not generally been placed on other Solar System bodies, so in most cases we must use more indirect means of assessing their interiors.

7.3.1 Seismology

Moonquakes occur because of tidal forces between the Moon and the Earth, rather than from tectonic activity. Seismometers were placed on the Moon by astronauts on the *Apollo 12, 14, 15,* and *16* missions (Figure 7.6).

Figure 7.6 The first seismometer placed on the Moon by *Apollo 12* astronauts. NASA image.

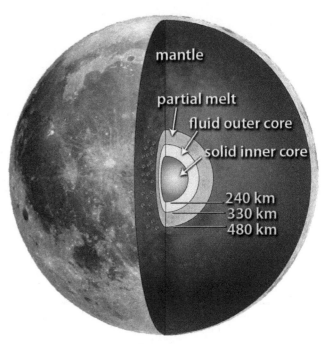

Figure 7.7 Model of the Moon's interior, showing a small core. Red dots are moonquake locations.
From Weber et al. (2011), with permission.

They remained functional until they were switched off in 1977. Early attempts to model lunar seismic data were frustrated by noisy signals reverberating within the pulverized crust. However, an improved array-processing technique has allowed researchers to use old *Apollo* data to trace the path of each signal as it traveled through the interior. The result (Weber et al., 2011) was the discovery of a small core, 330 ± 20 km in radius (Figure 7.7). A solid inner core is surrounded by a partly molten outer core, implying that 40 percent by volume of the core has solidified. Density estimates are consistent with an inner core that is pure iron, whereas the outer core may contain ~6 wt% sulfur. Nests of moonquakes about 1000 km deep are activated by tides raised by the Earth, and the deepest mantle is attenuating and may be partly molten. Seismic data have also been used to measure the average thickness of the lunar crust at ~50 km (Wieczorek et al., 2006).

The *Viking* landers carried seismic recorders to Mars, but no useful information was obtained. The mechanism for unlocking the caged sensor on *Viking 1* failed to operate, and gusty winds jiggled the instrument on *Viking 2* and hampered detection of any tremors. The *InSight* mission is devoted to studying the interior of Mars. The lander carries a seismometer, comparable to the best terrestrial instruments. The analysis of marsquakes should provide new information on the size and structure of that planet's core, mantle, and crust.

7.3.2 Mean Density

Calculating a planet's **mean density** requires measurement of its mass and volume. Mass is determined from its gravitational effect on a nearby satellite or spacecraft. A planet's diameter can be measured by occultation of stars or of orbiting spacecraft; volume can be calculated from the diameter geometrically by assuming a spherical shape. Because planets are not really spheres, however, a more detailed shape model is needed for precise determination of its volume. A shape model can be made from a digital terrain model obtained using radar altimeter data on an orbiting spacecraft or from stereoscopic topographic maps based on estimated heights from spacecraft imagery.

In comparing planets, we use **uncompressed mean density**, which corrects for gravitational compression within planets of different sizes. The observed mean densities and the calculated uncompressed mean densities of the terrestrial planets and the Moon are given in Table 7.1. To first order, the density differences among rocky planets reflect variations in the proportions of rock (with densities of ~3.0–3.5 g/cm^3) and iron metal (with much higher densities of ~7–8 g/cm^3). The relative proportions of rocky mantles and metal cores estimated from the uncompressed mean densities of the terrestrial planets and the Moon are illustrated in Figure 7.8.

Table 7.1 **Mean densities and moment of inertia factors for planets (de Pater and Lissauer, 2010; Lissauer and de Pater, 2013)**

Planet	Density/uncompressed density	Moment of inertia factor
Mercury	5.430/5.3 g/cm^3	0.353
Venus	5.204/4.3 g/cm^3	0.33
Earth	5.515/4.4 g/cm^3	0.33
Moon	3.344/3.3 g/cm^3	0.393
Mars	3.933/3.74 g/cm^3	0.365
Jupiter	1.326/ g/cm^3	0.254
Saturn	0.687/ g/cm^3	0.210
Uranus	1.318/ g/cm^3	0.23
Neptune	1.638/ g/cm^3	0.23

Uncompressed mean densities

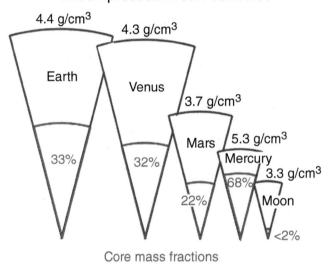

Core mass fractions

Figure 7.8 Comparison of the relative sizes of iron metal cores and silicate mantles for the terrestrial planets and the Moon, as constrained by uncompressed mean densities. Red numbers indicate core mass percentages.

7.3.3 Moment of Inertia

The **moment of inertia factor** is a dimensionless quantity that characterizes the internal distribution of mass within a planetary body. It is calculated as I/MR^2, where I is the largest principal moment of inertia (normally in the polar direction), M is the mass of the body, and R is the mean radius. For bodies in hydrostatic equilibrium (when gravity is balanced by a pressure gradient force), the moment of inertia factor can be estimated from the geometric flattening of the body as it rotates. The shapes of rotating planets are actually oblate spheroids, that is, they have wider equatorial girths. This represents the equilibrium shape of a planet under the combined influence of gravity and centrifugal forces.

If density is uniform throughout the body, $I/MR^2 = 0.4$. If density increases with depth (as is normally the case),

then I/MR^2 is less than 0.4. The low values for the terrestrial planets (Table 7.1) indicate pronounced increases in density toward the center, consistent with metal cores. For planets with slow rotation rates like Mercury and Venus, non-hydrostatic effects come into play, complicating the interpretation of I/MR^2; because of this, the moment of inertia of Venus is little more than an educated guess.

A tight constraint on the moment of inertia of Mars was made by Doppler tracking of the *Mars Pathfinder* lander, which allowed its position to be precisely determined (Folkner et al., 1997). Comparison with a previous measurement of a *Viking* lander's position 20 years earlier revealed that in the interim the planet's rotational pole had precessed – that is, the tilt of the planet had changed, much like the wobble of a spinning top. The rate of precession is governed by the moment of inertia, so this analysis provided an independent measurement of the internal distribution of mass within Mars.

7.3.4 Gravity and Tides

NASA's *GRAIL* mission used maps of the Moon's gravitational field to constrain its interior structure. Two small orbiters, named *Ebb* and *Flow*, exchanged telemetry so that any changes in the distance between them were precisely measured. That distance varied, depending on the gravity below the spacecraft. The resulting gravity map provided details on the structure of the lunar crust.

In the absence of other data, gravity can sometimes be modeled to estimate core mass. An example is the determination of the size of the core in asteroid Vesta, by using precise tracking of the orbit of the *Dawn* spacecraft. Like the moment of inertia, the gravitational flattening of the body as it rotates, called J_2, can be used to constrain the distribution of mass. Russell et al. (2012) constructed models of Vesta's mantle and core that were consistent with the mean density and J_2 measured by *Dawn*, deriving a core radius of ~110 km and core mass fraction of ~0.18.

Tides can sometimes provide another useful constraint on internal structures. The magnitude of the tidal response can be used to infer that Mars and the Moon have liquid cores and that Titan has a subsurface ocean.

7.3.5 Models of Planetary Interiors

Constructing models of the terrestrial planets involves the following steps. To a first approximation, we can ignore the crust and consider the planet to be composed of silicate mantle and iron metal core.

- Assume a bulk chemical composition. In the absence of other information, this is normally a chondritic composition. If samples are available, we may be able to define a more specific composition for the planet, as described in Section 5.4.

- Calculate internal pressures as functions of depth, using the mass and radius of the planet. A temperature profile must also be estimated or assumed.
- Calculate the mass and radius of the core, and subtract that much iron from the bulk composition.
- Using the bulk silicate composition, estimate the minerals present and their relative abundances as a function of depth, using mineral stabilities from experiments or thermodynamic calculations.
- Test the validity of the model by calculating mean density and moment of inertia factor and comparing them with measured values for the planet.
- Change the parameters of the model and iterate until its mean density and moment of inertia best match the measured values. The model can also be refined by adding additional layers, such as a crust or multiple mantle layers.

Such models have provided insights into the interiors of other planetary bodies:

Moon: The Moon has a low mean density and its moment of inertia factor is only slightly less than the value for a homogeneous sphere, implying a small core of 300–400 km radius. A core of 330 + 20 km radius has been confirmed by seismic data. A lunar model based on mean density indicates an ultramafic mantle and a feldspathic crust with an average thickness of ~50 km, although *GRAIL* measurements indicate a 34–43 km thick crust (Wieczorek et al., 2013). Mantle seismic velocities are consistent with olivine and pyroxene. The attenuation of S waves in the deep mantle may suggest partial melting.

Mercury: The uncompressed mean density of Mercury indicates that ~60 percent of its mass is metal, comprising a core that extends out to 75 percent of the planet's radius (refer to Figure 3.12). This massive core has spawned the hypothesis that some of Mercury's mantle has been stripped off by a massive impact. Variations in the planet's rotation rate suggest that the core and mantle are decoupled, implying a liquid outer iron core containing substantial sulfur and an inner core of iron–nickel metal. Modeling suggests that a layer of solid FeS might be sandwiched between the mantle and core, but this remains uncertain.

Venus: The uncompressed mean density and moment of inertia factor for Venus and Earth are similar, suggesting that the venusian interior may resemble that of Earth. The absence of a magnetic field implies that Venus' core is not convecting, perhaps because it is all solid or, if liquid, it is cooling too slowly. Another difference is the absence of plate tectonics, possibly attributable to lack of water on Venus.

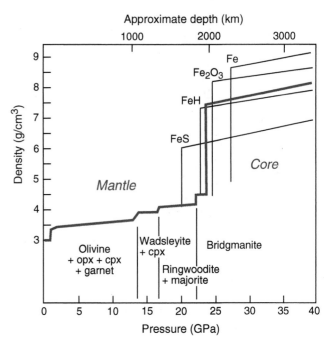

Figure 7.9 Density profile for the Mars mantle and core (heavy red line) and stability fields for mantle minerals, based on experiments using a widely accepted compositional model for Mars. A range of core compositions are shown; the density profile is for a mixture of Fe and FeS. Modified from Bertka and Fei (1998).

Mars: Data acquired by Mars spacecraft, coupled with compositional constraints from martian meteorites, allow its interior to be modeled with greater fidelity than for other planets. An example is illustrated in Figure 7.9. This model adopts a commonly accepted bulk chemical composition, derived from element ratios in martian meteorites. Experiments using this composition allowed Bertka and Fei (1998) to estimate the mineralogy of the mantle, which is similar to that of the Earth's mantle, except that the minerals are more iron-rich and phase transitions occur at greater depths in Mars owing to its smaller size and internal pressures. They also considered a variety of core compositions with different densities, which would have correspondingly different radii to meet the bulk density constraint. Their preferred density profile, assuming that the core is a mixture of Fe and FeS, is shown as a heavy red line in Figure 7.9. By assuming a 50 km thick basaltic crust, they then calculated a mean density and moment of inertia factor for this model. More recent measurements have established that the martian core has a radius of 1680 ± 160 km and is likely composed of Fe and FeS. For a core of that size, it is not clear that any bridgmanite occurs in the lower mantle.

7.3.6 Timing of Planetary Differentiation

Planetary differentiation is most directly dated using the ^{182}Hf–^{182}W isotopic system. As described in Section 5.3, this chronometer dates the fractionation of metal from silicate, and thus Hf–W ages indicate the timing of core formation (Halliday and Kleine, 2006). Tungsten isotopic measurements of martian and HED meteorites indicate rapid differentiation on Mars and asteroid Vesta during the first 10 and ~30 million years, respectively, of Solar System history. However, uncertainties in the initial Hf–W ratio allow even earlier core formation (<10 million years) for Mars. The Earth's differentiation was more protracted, with Hf–W ages in the range of 40–50 million years after CAIs. The differentiation of the Moon is interpreted to have taken place after 60 Ma.

7.4 Interiors of the Giant Planets and Icy Moons

As a first approximation, only hydrogen and helium have significant abundances in the giant planets. The amounts of nebular gases that were incorporated into the giant planets, relative to accreted solids (ices and rock), decrease monotonically from Jupiter outward to Neptune. Models of the interiors of these planets depend on temperature–pressure–density relationships that are unfortunately not well constrained.

7.4.1 Jupiter and Saturn

To model the interiors of the gas giants, we must understand the behavior of hydrogen and helium at high pressures and temperatures. A phase diagram for hydrogen, based on experiments and theoretical calculations, is illustrated in Figure 7.10. Let's focus on the high-temperature part of this diagram, applicable to the giant planets whose geothermal gradients are illustrated by heavy lines. The outer regions of Jupiter and Saturn consist of molecular H_2 in fluid form. At temperatures of 3000–4000 K and pressures >1.4 Mbar, the hydrogen molecules dissociate and form a metallic fluid. By "metallic" we mean that this phase is so densely packed that the electron clouds overlap, making it electrically conductive (like a metal).

The other important component of the gas giants is helium. Liquid helium and hydrogen are not miscible once the stability field for metallic hydrogen is reached, so they are expected to separate. Helium is denser and thus rains down into the deeper interiors, explaining the apparent depletion of helium relative to the solar H–He ratio in the atmospheres of Jupiter and Saturn. Other components detected in the atmospheres of these planets include methane (CH_4), ammonia (NH_3), phosphine (PH_3), hydrogen sulfide (H_2S), and noble gases, and these may be minor components in the interior as well.

The measured mean densities of Jupiter and Saturn (Table 7.1) are averages of molecular and metallic

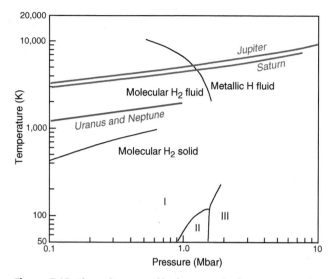

Figure 7.10 Phase diagram of hydrogen at high pressures and temperatures, showing the transition from molecular fluid H_2 to a metallic hydrogen fluid. The pressure–temperature conditions inside the giant planets are shown by heavy lines. Solid forms of molecular and metallic hydrogen occur at low temperatures, but these are not realistic for the giant planets. Modified from de Pater and Lissauer (2010).

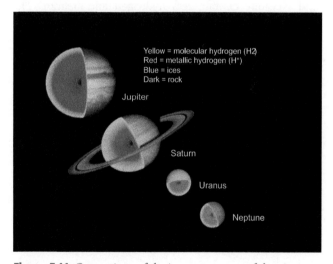

Figure 7.11 Comparison of the inner structures of the giant planets, based on uncompressed mean densities.

hydrogen plus an estimated 15–30 Earth masses of higher atomic-weight material. Small, dense rocky cores are possible but not required by current data; the denser materials could also be partly distributed in a surrounding envelope. The uncompressed mean density values of these planets are difficult to calculate, and so are not included in Table 7.1. The giant planets have very low I/MR^2 values (Table 7.1), because their interiors are so much denser than their gaseous envelopes. The hypothesized internal structures of Jupiter and Saturn are illustrated in Figure 7.11.

7.4.2 Uranus and Neptune

The geothermal gradients within the ice giants produce molecular H_2 fluids but never reach the conditions for metallic hydrogen (Figure 7.10). The lower pressures allow hydrogen and helium to be miscible, so they do not separate and the H–He ratio in their atmospheres is nearly solar. Their atmospheres contain more methane and ammonia than Jupiter and Saturn, but information on other components is lacking.

Unlike the larger gas giants, the interiors of Uranus and Neptune contain ices, likely made of water, methane, ammonia, and hydrogen sulfide. A phase diagram for H_2O is illustrated in Figure 7.12. At low temperatures water adopts a variety of crystalline forms, indicated by roman numerals; "I" is the familiar hexagonal form of ice that occurs on Earth. The geothermal gradients of Uranus and Neptune pass through this field and then into the field of supercritical fluid, with properties very different from liquid water. Phase diagrams for other ice compositions are not well determined at high pressures, but might be as complex as water.

The total mass of higher atomic-weight materials is roughly the same in all the giant planets, but Uranus and Neptune contain less hydrogen and helium. The mean densities for Uranus and Neptune (Table 7.1) represent averages of rock (possibly forming cores), ice mantles, and molecular hydrogen envelopes, as illustrated in Figure 7.10. Their moment of inertia factors (Table 7.1) reflect the concentration of mass in their interiors.

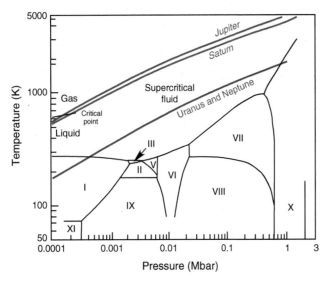

Figure 7.12 Phase diagram for H_2O at high pressures. Crystalline forms of ice are noted by Roman numerals. The geothermal gradients for the giant planets are shown by heavy lines. Modified from de Pater and Lissauer (2010).

7.4.3 Icy Moons

The icy satellites of the giant planets, and of Pluto and its moons, contain crystalline forms of water and lesser amounts of other components composed of nitrogen, carbon, and sulfur. The H_2O-ice crystalline forms in Figure 7.12 likely come into play in the interiors of these bodies. Interior pressures in the largest icy satellites are not high enough to cause phase changes in silicates, but high-pressure forms of ices can occur.

The mean densities of the icy satellites, with the exception of Europa (2.99 g/cm³), all fall in the range 1.00–1.94 g/cm³ (Johnson, 2004). These densities are interpreted to reflect mixtures of rock and ice. The icy moons can be separated into two categories: those large enough that self-compression should be considered, versus those small enough that porosity becomes important. Bodies in these categories can be roughly distinguished by spherical versus irregular shapes. Estimated porosities in the smaller bodies can range up to 40 percent. The few available measurements of moment of inertia factors (de Pater and Lissauer, 2010) are comparable to values for the terrestrial planets and indicate rocky cores.

The magnetometer on board the *Galileo* spacecraft found magnetic disturbances that are interpreted to indicate the presence of subsurface oceans on three moons of Jupiter – Europa, Ganymede, and Callisto. A slight wobble in the orbit of Enceladus, a moon of Saturn, also requires that part of its icy shell be liquid. Sprays of water vapor emanating from fractures on Enceladus are likely fed from this source. On all these satellites, liquid water is likely maintained by tidal heating, and the melting point of ice is likely decreased by dissolved salts.

7.5 Evolution of Planetary Interiors

Earth's interior has cooled over geologic time, with concomitant changes in the compositions of mantle-derived magmas from ultramafic (komatiites) to mafic (basalts). GRS-measured abundances of iron, silicon, and thorium on Mars have likewise been used to assess magmatic changes in its volcanic terrains over time. Baratoux et al. (2011) attributed these compositional variations to a hotter mantle in ancient Mars. Although the mechanisms by which Earth and Mars have lost heat differ, the end result is similar – an evolution in the compositions of extracted magmas. In both cases, incompatible heat-producing elements have also been extracted from the mantle over time. Temporal changes in thermal structure may have affected the depths of some phase transitions.

The formation of the continental crust has depleted the Earth's upper mantle in silica, incompatible elements, and some noble gases, relative to the lower mantle. Although the Earth's siliceous continents have no parallel on other planets, the extraction of crusts of whatever composition on other planets has changed the mantle source regions. In the absence of plate tectonics, mantle evolution is one-directional. On Earth, however, subduction recycles crust back into the mantle, reintroducing incompatible elements and producing complex patterns of enrichment and depletion. A plausible date for the onset of plate tectonics is ~3 Ga, based on change in the composition of the crust from mafic to felsic (Tang et al., 2016).

Planetary cooling also causes molten cores to crystallize. The Earth's solid inner core is growing progressively larger over time, and the cores of other terrestrial planets may have been similarly affected. Core crystallization concentrates less-compatible minor elements in the liquid phase, accounting for light element concentration in the Earth's outer core. It can also affect magnetic dynamos that require liquid convection.

The timescale for separation of helium from metallic hydrogen in Jupiter and Saturn is unclear, but this change represents an evolution in mantle structure. Thermal evolution in icy satellites may be reflected in the relative depths of subsurface oceans and thicknesses of icy crusts.

Summary

The vertical distributions and changing mineralogy of regions in the Earth's interior are inferred from the measured velocities and paths of propagating seismic waves and the results of high-pressure experiments. To model the interiors of other planets, we use measurements of their sizes, masses, shapes, and gravity fields to determine mean densities and moment of inertia factors. The terrestrial planets and the Moon consist of iron metal cores, ultramafic mantles, and crusts constructed from partial melts of the mantles. The giant planets consist of similar small amounts of rock, with proportions of hydrogen and helium decreasing from Jupiter outward to Neptune. The larger planets consist mostly of molecular hydrogen giving way to metallic hydrogen and helium at greater depths; the smaller planets have hydrogen mantles surrounding mantles of ices. The interiors of planets evolve in response to cooling, allowing crystallization of liquid cores and temporal changes in the compositions of magmas extracted from mantles. Interior evolution is one-directional, except on Earth where plate tectonics recycles crust back into the mantle.

Review Questions

1. Explain how seismic velocities provide information on the Earth's interior structure.
2. Draw and label a cross-section of the Earth's interior, noting each major compositional boundary and phase change.
3. Mercury's mean density is 5.43 g/cm^3. If Mercury consists entirely of rock (average density = 3.3 g/cm^3) and iron metal (density = 7.9 g/cm^3), calculate the planet's fractional abundance of iron by mass.
4. Explain how we can use the sizes, masses, and shapes of planets to model their interiors.
5. How does the high-pressure phase diagram for hydrogen help explain the interior structure of Jupiter?

SUGGESTIONS FOR FURTHER READING

de Pater, I. and Lissauer, J. J. (2010) *Planetary Sciences*, 2nd edition. Cambridge: Cambridge University Press. Chapter 6 of this authoritative text is a wonderful source of information on the geophysics of planetary interiors.

Hirose, K., Labrosse, S., and Hernlund, J. (2013) Composition and state of the core. *Annual Review of Earth and Planetary Sciences*, **41**, 657–691. An excellent summary of what is known about the Earth's core from high-pressure experiments and other data.

Lunine, J. I. (2004) Giant planets. In *Treatise on Geochemistry*, Vol. 1, ed. Davis, A. M. Amsterdam: Elsevier, pp. 623–636. An up-to-date, somewhat technical review of what is known about the giant planets.

REFERENCES

Badro, J., Brodholt, J. P., Piet, H., et al. (2015) Core formation and core composition from coupled geochemical and geophysical constraints. *Proceedings of the National Academy of Sciences.* DOI: 10.1073/pnas.1505672112.

Baratoux, D., Toplis, M. J., Monnereau, M., et al. (2011) Thermal history of Mars inferred from orbital geochemistry of volcanic provinces. *Nature,* **475,** 338–341.

Bertka, C. M., and Fei, Y. (1998) Implications of the Mars Pathfinder data for the accretion history of the terrestrial planets. *Science,* **281,** 1838–1840.

Biggin, A. J., Piispa, E. J., Pesonen, L. J., et al. (2015) Palaeomagnetic field intensity variations suggest Mesoproterozoic inner-core nucleation. *Nature,* **526,** 245–248.

de Pater, I., and Lissauer, J. J. (2010) *Planetary Sciences,* 2nd edition. Cambridge: Cambridge University Press.

Folkner, W. M., Yoder, C. F., Yuan, D. N., et al. (1997) Interior structure and seasonal mass redistribution of Mars from radio tracking of Mars Pathfinder. *Science,* **278,** 1749–1752.

French, S. W., and Romanowicz, B. (2015) Broad plumes rooted at the base of the Earth's mantle beneath major hot spots. *Nature,* **525,** 95–99.

Halliday, A. N., and Kleine, T. (2006) Meteorites and the timing, mechanisms, and conditions of terrestrial planet accretion and early differentiation. In *Meteorites and the Early Solar System II,* eds. Lauretta, D. S., and McSween, H. Y. Tucson, AZ: University of Arizona Press, pp. 775–801.

Hand, E. (2015) Mantle plumes seen rising from Earth's core. *Science,* **349,** 1032–1033.

Johnson, T. V. (2004) Major satellites of the giant planets. In *Treatise on Geochemistry,* Vol. 1, ed. Davis, A. M. Amsterdam: Elsevier, pp. 637–662.

Lissauer, J. J., and de Pater, I. (2013) *Fundamental Planetary Science: Physics, Chemistry and Habitability.* Cambridge: Cambridge University Press.

Righter, K., and Drake, M. J. (1997) Metal/silicate equilibrium in a homogeneously accreting Earth: new results for Re. *Earth and Planetary Science Letters,* **146,** 541–554.

Russell, C. T., Raymond, C. A., Coradini, A., et al. (2012) Dawn at Vesta: testing the protoplanetary paradigm. *Science,* **336,** 684–686.

Tang, M., Chen, K., and Rudnick, R. L. (2016) Archean upper crust transition from mafic to felsic marks the onset of plate tectonics. *Science,* **351,** 372–375.

Weber, R. C., Lin, P-Y., Garnero, E. J., et al. (2011) Seismic detection of the lunar core. *Science,* **331,** 309–312.

Wieczorek, M. A., Jolliff, B. L., Khan, A., et al. (2006) The constitution and structure of the lunar interior. *New Views of the Moon: Reviews in Mineralogy and Geochemistry,* **60,** 221–264.

Wieczorek, M. A., Neumann, G. A., Nimmo, F., et al. (2013) The crust of the Moon as seen by GRAIL. *Science,* **339,** 671–675.

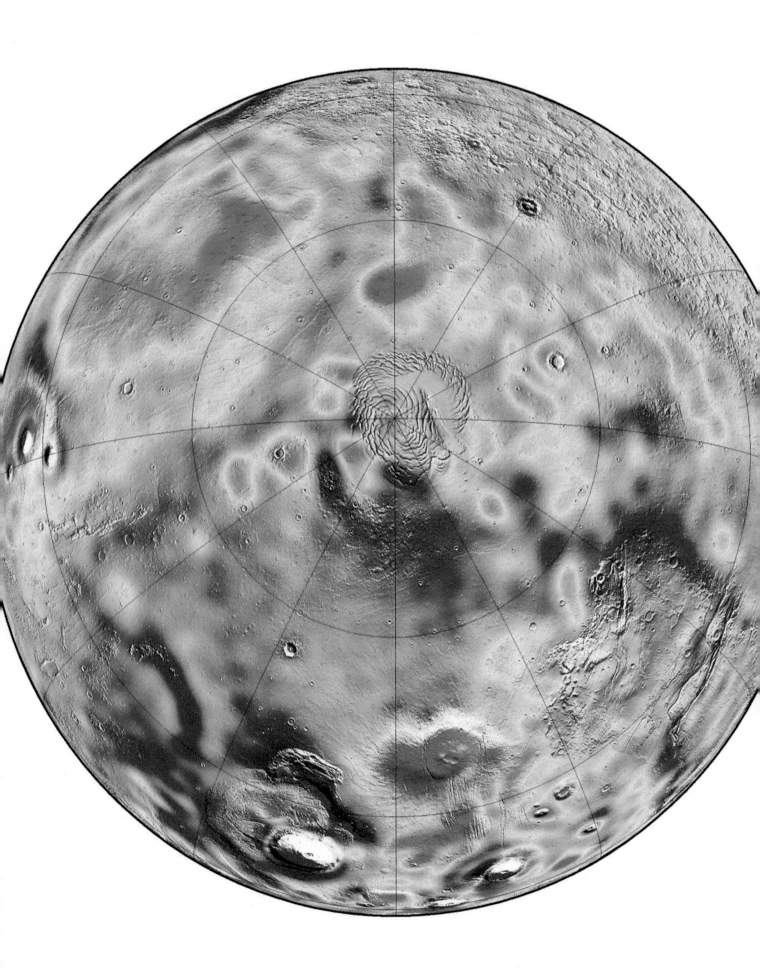

8

Planetary Geodynamics

Planetary bodies dynamically respond to applied stresses. Heat transfer out of the interior commonly leads to stresses that affect the surface. For quantitative analysis of geodynamics, numerical techniques are generally required and are applied looking at the material as a continuum. Rocks and ice in planetary bodies ultimately want to be in equilibrium with applied stresses. Equilibrium can be assessed by computing whether the stress gradients balance the applied force. The material response to stress is strain, which can be calculated from displacement gradients throughout the material. Stress and strain in a solid are related through intrinsic material properties (e.g., Young's modulus and Poisson's ratio). The material properties of rock and ice are similar enough that the icy lithospheres of the moons of the outer planets undergo the same basic processes as the rocky lithospheres of the terrestrial planets. Large lithospheric blocks are supported isostatically, floating in the asthenosphere. Topography can also be supported by the strength of the lithosphere, in which case some amount of flexure occurs as a result of the load on the surface. The distribution of mass in the subsurface can be inferred from measurements of the gravity field. From such measurements, it is possible to discern if a feature such as a mountain or volcano has a large root, or if a large mass lies beneath a surface with no topography (e.g., lunar mascons). Surface temperature is controlled for most planetary surfaces by solar heating, the effect of which generally only penetrates a few meters into the surface. Heat flows through the brittle lithosphere by conduction, but the deeper asthenosphere transfers heat through convection. The asthenosphere behaves like a fluid on geologic timescales, and its response to stress must be investigated in terms of fluid mechanics. The exact response to stress, or the rheology, depends on many factors, including temperature, composition, grain size, and the magnitude of stress. The ductile behavior of the interior is coupled to the surface, enabling geodynamicists to use observations of the surface to infer properties of the interior.

8.1 Motions in Planetary Interiors

Planetary surfaces are, in many ways, shaped by properties of their interiors and motions below their surfaces. Those properties and motions are strongly controlled and driven by planetary heating and heat trying to get out of the interior. Chapter 6 introduced planetary heating, and Chapter 7 introduced planetary interiors. We now want to take a look at techniques that are used to understand the dynamics below the surfaces of planets and the effects on the surfaces that spacecraft, and maybe eventually astronaut geologists, study.

Earth's surface, for example, is very geologically active, and most of that activity is explained in the framework of plate tectonics. Earth's surface plates move relative to each other (at rates of about 5 to 10 cm/yr, in general), and those motions induce a lot of stress in the lithosphere. If we look more deeply, we see that plate motions are driven by convection working to transfer heat from Earth's deep interior to its surface, to ultimately be radiated to space. Examining the rest of the Solar System, though, we do not see any clear evidence of global plate tectonics taking place on any other body. Does that mean that their interiors are cold and inactive? Certainly not.

Dramatic examples such as the pervasive volcanism on Io and the active geysers erupting from the south pole of Enceladus tell us that even small moons can have dynamic interiors. Geologic features such as large rift systems, mountains, and volcanoes on Venus, global thrust faults on Mercury, giant volcanoes and rifts on Mars, relaxed topography and fault systems on icy satellites, and apparently active convection in surface nitrogen ice on Pluto make it clear that intense forces are or have been acting on surfaces throughout the Solar System. This chapter introduces concepts, and their quantitative foundation, needed to investigate and understand motions in planetary interiors and how those affect planetary surfaces.

8.2 Geologic Stresses and Deformations

Force balance is fundamental to understanding the equilibrium state of a body and for investigating what happens in non-equilibrium conditions. Our goal is to compute deformation within a solid planetary surface for given stresses. Alternatively, we may be interested in the inverse problem – determining from the observed deformation state what stresses must be, or have been, acting within the solid portion of a planetary body. Rocks and ices under relatively small stresses generally display linear **elastic behavior**. Such behavior is the domain of classical elastostatics (for the equilibrium case) and elasto-dynamics (for non-equilibrium situations). Later on (Section 8.6), we'll look at the case of more fluid-like (**ductile**) behavior. Section 9.2 discusses these behaviors in the context of observed structures and tectonics. Elasto-statics/-dynamics consists of three components (or sets of equations): (1) stress equilibrium (i.e., force balance or momentum conservation), (2) definition of strain in terms of continuum deformations, and (3) constitutive equations that relate stress and strain in a solid. To solve these sets of equations quantitatively, particularly numerically (the common approach of modern geophysics/geodynamics), geodynamicists work in the domain of **continuum mechanics**.

8.2.1 Balancing Act: Stress Equilibrium

Stress (σ) is defined as a force divided by the area over which the force acts. Figure 8.1 shows the stress components on an infinitesimally small volume inside some solid (e.g., a planetary lithosphere). *Normal stresses* are those that act perpendicular to one of the surfaces of the volume (i.e., along the surface normal vector), and *shear stresses* are those that act parallel to the surface. The force

acting on each surface can be decomposed into one normal stress and two shear stresses that are all orthogonal to one another. Customarily, the first subscript refers to the direction of the surface normal, and the second to the direction of the force (but conventions vary in different fields, so it is always necessary to check conventions). Because there are no internal torques, $\sigma_{xy} = \sigma_{yx}$, $\sigma_{xz} = \sigma_{zx}$, and $\sigma_{yz} = \sigma_{zy}$, leaving six independent stress components of three normal stresses and three shear stresses.

Using the infinitesimal volume in Figure 8.1, it is possible to show that for the volume to be in equilibrium, any outside applied force in each dimension must balance the sum of normal and shear **stress gradients** in that dimension. This equilibrium state is expressed mathematically as the following set of equations:

$$
\begin{aligned}
f_x &= \frac{\partial \sigma_{xx}}{\partial x} + \frac{\partial \sigma_{yx}}{\partial y} + \frac{\partial \sigma_{zx}}{\partial z} \\
f_y &= \frac{\partial \sigma_{xy}}{\partial x} + \frac{\partial \sigma_{yy}}{\partial y} + \frac{\partial \sigma_{zy}}{\partial z} \\
f_z &= \frac{\partial \sigma_{xz}}{\partial x} + \frac{\partial \sigma_{yz}}{\partial y} + \frac{\partial \sigma_{zz}}{\partial z}
\end{aligned}
\tag{8.1}
$$

where f_x, f_y, and f_z are the external applied forces per unit volume in each dimension. In the common static case within a planetary surface, the primary external force is gravity. Box 8.1 includes an illustration of how to use the stress equilibrium equations to solve for a continuous expression for stress.

8.2.2 What Exactly Is Strain?

When a stress is applied to an object or mass of material, the material will deform. This deformation is called **strain** (ε). Like stress, strain has normal and shear components. In macroscopic terms, normal strain is generally defined as the change in length of a body in a given dimension to the original length in that dimension, and shear strain as the change in angles between faces of the body.

In continuum mechanics, strain is defined by the relative displacements of points within the body (Figure 8.2). If all points move the same amount, there is no relative displacement, just an overall translation or rotation of the body. The important quantities are, again, gradients – in this case, displacement gradients. Normal strains are defined as

$$
\begin{aligned}
\varepsilon_{xx} &= \frac{\partial u_x}{\partial x} \\
\varepsilon_{yy} &= \frac{\partial u_y}{\partial y} \\
\varepsilon_{zz} &= \frac{\partial u_z}{\partial z}
\end{aligned}
\tag{8.2}
$$

where u_x, u_y, and u_z are the displacements in each dimension. The shear strains are defined as

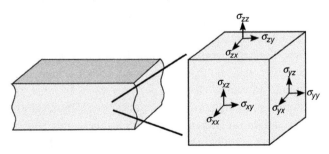

Figure 8.1 Illustration of the normal and shear components of stress acting on an infinitesimally small volume within a planetary surface.

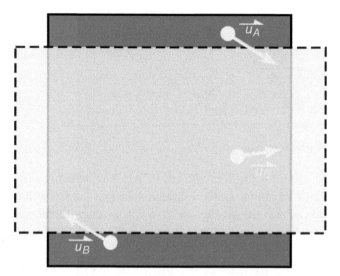

Figure 8.2 Illustration of displacement vectors within a strained volume. Dark blue (solid outline) represents the original shape. In order to deform to the light blue shape (dashed outline), points within the volume must move. Displacement gradients are found by computing differences between the displacement vectors of points within the volume.

$$\varepsilon_{xy} = \frac{1}{2}\left(\frac{\partial u_x}{\partial y} + \frac{\partial u_y}{\partial x}\right)$$

$$\varepsilon_{xz} = \frac{1}{2}\left(\frac{\partial u_x}{\partial z} + \frac{\partial u_z}{\partial x}\right) \qquad (8.3)$$

$$\varepsilon_{yz} = \frac{1}{2}\left(\frac{\partial u_y}{\partial z} + \frac{\partial u_z}{\partial y}\right)$$

Similar to the case with stresses, the complementary shear strains are equal to each other (i.e., $\varepsilon_{xy} = \varepsilon_{yx}$, $\varepsilon_{xz} = \varepsilon_{zx}$, and $\varepsilon_{zy} = \varepsilon_{yz}$).

8.2.3 Relating Stress and Strain

Now that we can relate deformation (displacements) to strains and compute stresses from applied forces, we need to be able to relate stress and strain to each other. Solid geologic materials near the surfaces of the Earth and other planetary bodies generally respond elastically to applied stresses (so long as the stresses are not too large). Rock (or ice) will deform when stress is applied, but will return to its original shape when the stress is removed. Furthermore, the response is generally linear – doubling the stress will double the strain. This behavior is sketched schematically in Figure 8.3.

Mathematically, linear elasticity is similar to stretching and compressing a spring, so Hook's law applies. The relation of normal stresses to normal strains in the same dimension can be written as $\sigma_{xx} = E\varepsilon_{xx}$, where the proportionality constant E is called *Young's modulus*. It is a property that describes how stiff a material is with respect to tension or compression. Since strain is a dimensionless quantity, E must have the same units as stress (i.e., force

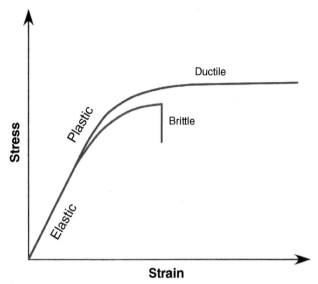

Figure 8.3 Schematic illustration of the relationship between stress and strain in most solid geologic materials. For small stress and strain, the relationship is linear and elastic. Eventually, the amount of strain is such that it cannot be recovered (plastic) and the relationship becomes nonlinear. If the stress overcomes the strength of the material plus confining pressure, fracturing occurs (brittle). Otherwise, the material will flow to accommodate applied stress (ductile).

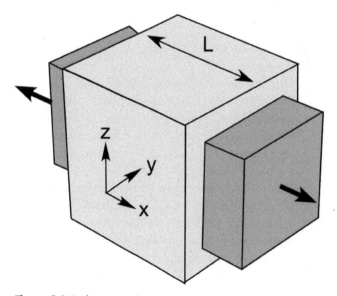

Figure 8.4 Deformation (extension or compression) in one dimension causes deformation in the orthogonal dimensions. This effect is parameterized by Poisson's ratio.

per area, or pressure). E for rocks is typically ~40–80 GPa and for ice is ~6–12 GPa.

In real materials, normal strain in one dimension leads to a normal strain in the orthogonal dimensions. This behavior is illustrated in Figure 8.2, where the original square compresses in the vertical direction and extends in the horizontal, and in Figure 8.4, where the cube extends in the x-direction and correspondingly compresses in the

y- and z-directions. *Poisson's ratio* (v) is the ratio of the normal strains in the orthogonal dimension to the strain in the dimension of primary deformation. For example, in Figure 8.4, $v = -\varepsilon_{xx}/\varepsilon_{zz} = -\varepsilon_{yy}/\varepsilon_{zz}$. The negative sign indicates that extension in the z dimension causes contraction in the x and y dimensions. v for most rocks is between 0.1 and 0.35 and for ice is typically 0.3–0.36.

Adding together the normal strains felt in each dimension from the stresses applied in all dimensions leads to the constitutive equations for normal stresses and strains:

$$\varepsilon_{xx} = \frac{1}{E}\sigma_{xx} - \frac{v}{E}\sigma_{yy} - \frac{v}{E}\sigma_{zz}$$

$$\varepsilon_{yy} = -\frac{v}{E}\sigma_{xx} + \frac{1}{E}\sigma_{yy} - \frac{v}{E}\sigma_{zz} \qquad (8.4)$$

$$\varepsilon_{zz} = -\frac{v}{E}\sigma_{xx} - \frac{v}{E}\sigma_{yy} + \frac{1}{E}\sigma_{zz}$$

Shear stresses lead to shear strains in the same directions, but not in orthogonal directions. The proportionality constant between shear stress and shear strain is called the *shear modulus* or *modulus of rigidity* (G_s), such that, for instance, $\sigma_{xy} = 2G_s\varepsilon_{xy}$. G_s can be rewritten in terms of

E and v: $2G_s = E/(1 - v)$. Therefore, the constitutive relations for shear stresses and strains can be written as:

$$\varepsilon_{xy} = \frac{1+v}{E}\sigma_{xy}$$

$$\varepsilon_{xz} = \frac{1+v}{E}\sigma_{xz} \qquad (8.5)$$

$$\varepsilon_{yz} = \frac{1+v}{E}\sigma_{yz}$$

These three fundamental sets of relations (stress equilibrium equations, strain definition equations, and constitutive equations) provide the foundation for quantitatively investigating the mechanical behavior of geologic materials.

8.3 The Weight of the World: Isostasy and Flexure

An excess (or deficit of) mass in or on the lithosphere of a planetary body induces stress. The stress can be accommodated isostatically (by buoyancy in the asthenosphere) or flexurally (by the strength of the lithosphere). We will see that these two possibilities are actually end members of a continuum geophysical description.

BOX 8.1 COMPRESSION OF ICE SPIRES ON CALLISTO

To illustrate a fairly straightforward application of the relations presented in this section, let's look at vertical erosional structures such as might occur from various processes throughout the Solar System. Figure 8.5 shows spires of ice on Callisto, one of Jupiter's large moons, resulting from long-term erosion of the surface. An idealized sketch for this problem is also shown.

We want to know how self-compression affects the dimensions of these features. If there is a significant effect on the dimensions, we would want to take the elastic response into account when studying further erosion. The two questions to answer are:

1. How much does the spire compress under its own weight?
2. How much extension ("bulging") occurs at the base from the weight?

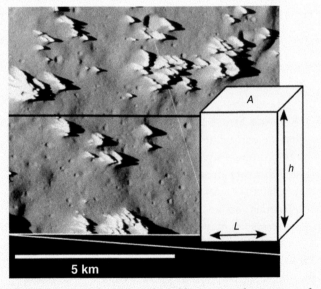

Figure 8.5 Image from NASA's *Galileo* mission shows spires of ice on Callisto. Inset shows an idealized sketch of an ice spire, showing variables used in Box 8.1.

The known dimensions and physical properties for this problem are: $\rho = 992$ kg/m^3, $E = 9$ GPa, $v = 0.33$, $h = 500$ m, $L = 100$ m, and $g = 1.24$ m/s^2.

Looking at the problem as a whole reveals a path to take to the solution: The weight of the column will impose stresses, those stresses induce strains, and we can compute overall dimension changes from strains.

The applied stress in the vertical (z) dimension is due to gravity, and there are no applied stresses in the horizontal (x and y) dimensions. From stress equilibrium

$$\frac{\partial \sigma_{zz}}{\partial z} = -\rho g \tag{8.6}$$

Solving for σ_{zz} by integration gives

$$\sigma_{zz}(z) = \int_{h}^{z} -\rho g \; dz = \rho g(h - z) \tag{8.7}$$

Since $\sigma_{xx} = \sigma_{yy} = 0$, the constitutive relations resolve to

$$\varepsilon_{xx} = -\frac{v}{E}\sigma_{zz} = -\frac{v}{E}\rho g(h - z)$$

$$\varepsilon_{yy} = -\frac{v}{E}\sigma_{zz} = -\frac{v}{E}\rho g(h - z) \tag{8.8}$$

$$\varepsilon_{zz} = \frac{1}{E}\sigma_{zz} = \frac{1}{E}\rho g(h - z).$$

The strain definition equations provide the means to convert these strains into changes in height (h) and width (L):

$$\Delta h = \int_{\Delta h}^{0} du_z = \int_{h}^{0} \varepsilon_{zz}dz = \int_{h}^{0} \frac{1}{E}\rho g(h - z)dz \tag{8.9}$$

which gives

$$\Delta h = \frac{1}{2E}\rho g h^2 \tag{8.10}$$

For the properties and dimensions given for this problem, the total compression due to the weight of overlying ice is 1.7 cm.

Since $z = 0$ at the base of the spire, the change in width at the base is given by

$$\Delta L = \int_{0}^{\Delta L} du_x = \int_{-L/2}^{L/2} \varepsilon_{xx}dx = \int_{L/2}^{L/2} -\frac{v}{E}\rho g h \; dx \tag{8.11}$$

which gives

$$\Delta L = -\frac{v}{E}\rho g h L \tag{8.12}$$

For this problem, the "bulging" of the base amounts to 1.1 mm change (extension) in width. Similar geologic features, called buttes, occur in desert regions on Earth. For a typical butte with $h = 300$ m and $L = 75$ m (and ρ, E, and v appropriate for sandstone), Δh and ΔL are 2.4 cm and 1 mm, respectively, comparable to the values for ice spires on Callisto. Given the scale of the features, it would be safe to ignore the changing dimensions when studying further erosion.

8.3.1 Isostasy

The word **isostasy** comes from the Greek "iso" or equal and "stasis" or standing still, and it is a statement of buoyant equilibrium of lithospheric blocks. The basis of isostasy is the concept that any sufficiently large volume through the outer parts of a planetary body will have the same gravitational force as any other column of the same area and depth. This concept can be expressed mathematically, for a planar geometry, as

$$\sum_{i=1}^{N} \rho_{1,i}h_{1,i} = \sum_{j=0}^{M} \rho_{2,j}h_{2,j} = \text{constant} \tag{8.13}$$

where $\rho_{1,i}$ is the density of each layer of the first column, $h_{1,i}$ is the thickness of each layer in this column, and $\rho_{2,j}$ and $h_{2,j}$ are the densities and thicknesses of the layers in the second column. The summation extends to a sufficient depth – known as the compensation depth (D_c) – that the interior can be considered laterally

Pratt's model

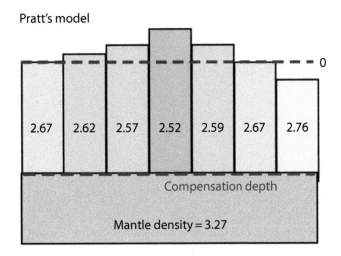

Airy's model

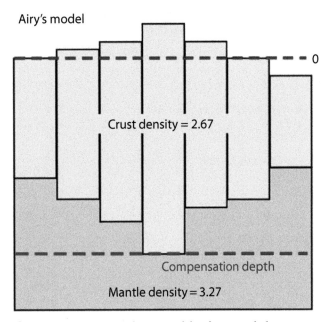

Figure 8.6 The Pratt and Airy models of isostatic balance on a planetary body.

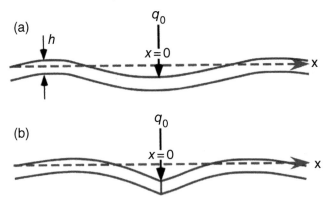

Figure 8.7 Schematic illustration of the flexure of an elastic plate in response to a downward forcing load, q_o. (a) shows the case of an intact plate, and (b) shows the case for a plate that is broken at $x = 0$. Note the forebulges some distance from the load where the plate bows up above the original horizontal. The position of the forebulge can be used to assess flexural properties (including thickness) of the plate. Modified from Turcotte and Schubert (2014).

theory to observed geomorphology is an effective means of determining physical properties of planetary surfaces.

Balancing forces and torques using the principles of Section 8.2 leads to a fourth-order differential equation that guides flexural response:

$$D\frac{d^4w}{dx^4} + P\frac{d^2w}{dx^2} + gw(\rho_m - \rho_{fill}) = q(x) \tag{8.14}$$

This equation describes the vertical displacement, w, of the elastic lithosphere as a function of horizontal position, x. $q(x)$ represents the applied load, which can be a function of x, and P is any applied horizontal force. The third term on the left side represents the restoring force exerted by the fluid asthenosphere against the flexure: g is the acceleration of gravity, ρ_m is the density of mantle rock, and ρ_{fill} is the density of any material that is filling in the basin created by downward flexure of the plate (e.g., water in the ocean or sediments in a sedimentary basin). The mechanical properties of the plate are encapsulated in D, the flexural rigidity:

$$D = \frac{Eh^3}{12(1-v^2)} \tag{8.15}$$

with E and v as defined in Section 8.2 and h representing the elastic thickness of the plate. The solutions to Equation 8.14 depend on the details (boundary conditions) of the situation under consideration and quickly become complex enough to make it impossible to find analytic solutions. Nevertheless, two flexural problems are particularly useful to consider in a planetary context: loading on a plate and support of a periodic topographic load.

Consider a volcanic construct building up on a planetary surface, a plate being thrust on top of an adjacent block, or a ridge building up along a fissure. All of these processes lead to a load weighing down the elastic lithosphere. Figure 8.7 sketches the idealized flexural response

homogeneous. An assumption inherent in isostatic compensation is that the columns can be considered to be independent of each other – i.e., the crust is strengthless at the horizontal scale of the columns considered.

The equality can be achieved by lateral variations in crustal thickness, density, or both. The case of a topographic load on a constant-density crust being compensated by a thickening of the crust (i.e., a "root" extending into the mantle) is the *Airy model*. The case of compensation of a topographic load by lateral density variations is the *Pratt model* (Figure 8.6).

8.3.2 Flexure

Real materials have strength, and stress applied to a plate that is sufficiently strong will cause it to flex rather than to respond buoyantly. As illustrated in Figure 8.3, as long as the stress and strain are not too large, the flexural response can be treated elastically. Application of elastic flexure

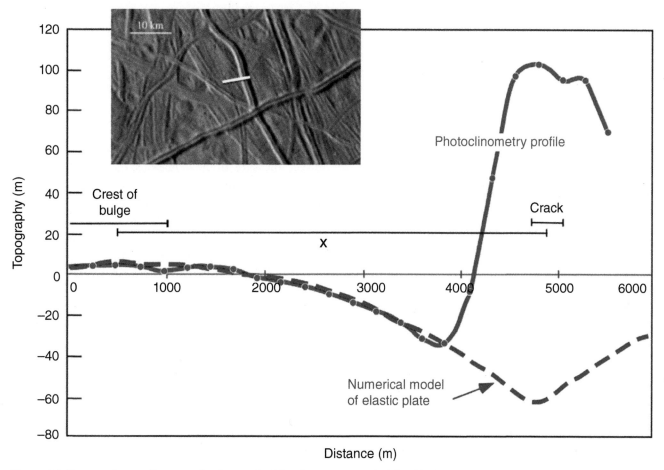

Figure 8.8 Topographic profile perpendicular to a double ridge on Europa (solid red line) along with the modeled flexural plate profile (dashed blue line). The inset image shows the location of the profile. Hurford et al. (2005) used the distance between the center of the load and the crest of the flexural forebulge to estimate the thickness of the elastic plate. Modified from Hurford et al. (2005).

of a plate under a point or line load for the cases of an intact (a) or broken (b) plate. The solution of Equation 8.14 for an intact plate is

$$w = \frac{q_o \alpha^3}{8D} e^{-\frac{x}{\alpha}} \left(\cos \frac{x}{\alpha} + \sin \frac{x}{\alpha} \right) \qquad (8.16)$$

where q_o is the load, and α is the *flexural parameter*: $\alpha = [4D/(\rho_m - \rho_{fill})g]^{1/4}$. The solution for a broken plate is of a similar form:

$$w = \frac{q_o \alpha^3}{4D} e^{-\frac{x}{\alpha}} \cos \frac{x}{\alpha} \qquad (8.17)$$

These solutions reveal that the elastic plate will respond in a series of periodic highs and lows, with an amplitude that decreases exponentially with distance from the load. The distance to the first (largest) flexural high for an intact plate is $x_b = \pi\alpha$ and for a broken plate is $x_b = 3\pi\alpha/4$. We can infer the thickness of the elastic plate by measuring the horizontal distance from a surface load to the adjacent flexural rise.

This straightforward approach of inferring subsurface properties has been applied across the Solar System, from the Hawaiian islands on Earth, to scarps on Mercury and

Venus, to volcanic constructs on Mars, and to various ridges on icy moons and the dwarf planet Pluto in the outer Solar System (e.g., Barnett et al., 2002; Watters, 2003; Hammond et al., 2013; Huppert et al., 2015). Figure 8.8 shows results of such an analysis by Hurford et al. (2005) of double ridges on Europa. They found that Europa's elastic lithosphere at the time the ridges were emplaced was only a few hundred meters, compared to the tens of kilometers common for terrestrial planets.

In our second case, we'll consider periodic loading of a planetary surface, where the emplaced load is given by $q(x) = \rho_c g h_o \sin(2\pi x/\lambda)$. The topographic load is assumed to have the same density as the crustal plate, ρ_c, a topographic amplitude h_o, and wavelength λ. The flexural response of the underlying plate is constrained to have the same periodic response as the topography and can be expressed as

$$w = w_o \sin \frac{2\pi x}{\lambda}, \text{ where } w_o = \frac{\rho_c}{(\rho_m - \rho_c)} \frac{h_o}{\frac{1}{4}\left(\alpha \frac{2\pi}{\lambda}\right)^4 + 1}$$

$$(8.18)$$

The most instructive aspect of this solution is to consider the extremes of very short wavelength

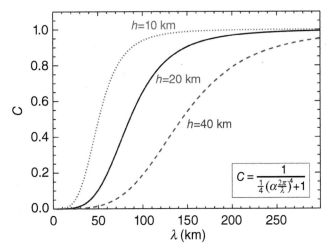

Figure 8.9 Degree of compensation (C) as a function of the length scale (λ) of the topographic load for different values of the elastic thickness of the lithospheric plate (h_e). Small values of C indicate that the weight of the topography is supported by the strength of the lithospheric plate. For $C \sim 1$, the topography is compensated isostatically.

topography and very long wavelength topography. For very short wavelength topography (i.e., for $\lambda \ll a$) the amplitude of plate deflection, w_o, is negligible. In other words, the rigidity of the plate can support loads that have a wavelength much smaller than the flexural parameter. For very long wavelength topography (i.e., $\lambda \gg a$), the solution simplifies to $w_o = \rho_c h_o / (\rho_m - \rho_c)$, which is the isostatic result! Figure 8.9 shows a plot of w_o divided by the isostatic result as a function of wavelength of the periodic load for different values of the elastic thickness (h in Equation 8.15) for a rocky body. Where this degree of compensation approaches 1, the load is supported isostatically, and where it is small, the topography is supported by the rigidity of the plate.

It is common for planetary bodies to exhibit cases of both end members of topographic support. For instance, gravity data (see Section 8.4) of Mars indicate that the southern highlands are in isostatic equilibrium with the northern lowlands. In other words, the lithosphere was not strong enough to support the topographic load when it was emplaced (very early in Mars' history). Using the Airy model, the measured topographic difference between the highlands and lowlands (~3 km), and assuming mantle and crustal densities of 3300 and 2900 kg/m^3, respectively, we find that the southern highland crust is approximately 24 km thicker than that of the northern lowlands. On the other hand, gravity data indicate that the much younger Tharsis volcanic plateau is *not*

isostatically compensated, so the lithosphere must have become much stronger (i.e., thicker) by the time the Tharsis volcanoes were emplaced (e.g., McGovern et al., 2002).

8.4 The Pull of Gravity

The gravity exerted by a planetary body provides a direct measure of how mass is distributed near the surface and in the deep interior. By mapping the gravity field around a body, we can measure the total mass (from which the bulk density can be determined), infer the internal structure (e.g., does the body have a core?), and map near-surface lateral density variations, enabling assessment of the degree of compensation of the topography.

8.4.1 The Geoid

The **geoid** of a planetary body is defined as a surface over which the gravitational acceleration (or gravitational potential) is constant. Since all planetary bodies rotate, the centripetal acceleration from rotation is included in the geoid. On Earth, if the oceans were influenced only by gravity and rotation, they would take the shape of the geoid. The geoid of the Earth is illustrated in Figure 8.10.

The concept of the geoid arises from Newton's law of gravity. The gravitational acceleration felt at any point external to a planetary body can be computed by:

$$g_m(r, \theta, \phi) = \int -G/\vec{b}^2 \, dm \qquad (8.19)$$

where (r, θ, ϕ) are the body-centered coordinates of the external point, G is the universal gravitation constant, $\vec{b}$ is the vector between the infinitesimal mass unit (dm) and the external point, and the integral is throughout the body. For a perfect sphere, the solution of this integral has the relatively simple form

$$g_m(r, \theta, \phi) = -GM/r^2 \qquad (8.20)$$

where $M = 4\pi \int_0^R \rho(r') r'^2 dr'$ is the total mass of the planetary body and r' is measured from the center of the body. The gravitational potential describes the work required to move a unit mass from infinity to a distance r from the center of the body. It is computed by $= \int_\infty^r g_m \, dr$, which, for a sphere, becomes $U = -GM/r$.

Real planetary bodies, however, are not perfect spheres, and the solution for gravitational acceleration and gravitational potential can quickly become very complicated.

Figure 8.10 Geoid of the Earth. The deviations from a reference ellipsoid are quite small compared to the Earth's radius. Modified from delai.gsfc.nasa.gov.

To capture this complexity, gravitational potential is often expressed in terms of spherical harmonics:

$$U = -\frac{GM}{r} + \frac{GM}{r} \sum_{n=1}^{\infty} \left(\frac{R}{r}\right)^n J_n P_n^0(\sin\theta)$$

$$+ \sum_{n=1}^{\infty} \sum_{m=1}^{n} \left(\frac{R}{r}\right)^n P_n^m(\sin\theta)(C_{n,m}\cos m\phi + S_{n,m}\sin m\phi)$$

$$(8.21)$$

In this expression, $P_n^m(\sin\theta)$ represents Legendre polynomials, and J_n, $C_{n,m}$, and $S_{n,m}$ are coefficients that describe the strength of the field at each value of n and m. The goal is to compute these coefficients from measurements of the gravity field for as many values of n and m as possible. Notice that the first term in Equation 8.21 is the spherical solution (corresponding to $n = m = 0$). The second term describes zonal harmonics – the case when $m = 0$ (note that $J_n \equiv C_{n,0}$). The zonal harmonics represent latitudinal gravity inhomogeneities. For example, J_2 describes the amount of rotational flattening (i.e., equatorial bulge) a body experiences from rotation. In the third term, the coefficients with $n = m$ are called *sectorial harmonics*, and these represent longitudinally symmetric inhomogeneities in the gravity field.

Higher degree and order (n and m) describe gravity signatures of ever decreasing size. In order to detect small-scale signatures, many gravity terms are required. The *EGM 2008* global gravity solution for the Earth goes to degree 2190 and order 2159, representing spatial scales of ~2 km). Thanks to NASA's *GRAIL* mission, the gravity field of the Moon is known to degree and order 900 (Lemoine et al., 2014). The higher degree and order terms fall off more quickly with distance from the body, making it very difficult to measure details of the gravity field from large distances.

8.4.2 Gravity Anomalies

Differences between a reference geoid and the measured gravity field are called **gravity anomalies**. The reference geoid for the Earth (World Geodetic System 1984; WGS84) is an ellipsoid of revolution with precisely defined coefficients. Anomalies in the gravity field arise from topography and from an excess or deficit of mass below the surface (i.e., lateral density inhomogeneities) (Figure 8.11). Gravity measurements on Earth have traditionally been made using a gravimeter (a very sensitive accelerometer). In modern times, tracking satellites as they move through Earth's gravity field has provided more consistent, detailed datasets. Similarly, tracking spacecraft as they perform flybys or orbit other planetary bodies provides gravity measurements across the Solar System. The *GRAIL* mission precisely monitored the distance between two spacecraft (*Ebb* and *Flow*) to make detailed measurements of the Moon's gravity field (Zuber et al., 2013).

Gravity data are often reported after one or both of two important corrections are made. The *free air correction* adjusts the data for the elevation or altitude of the measurement above the reference geoid, assuming there is no mass (i.e., just free air) between the instrument and the reference geoid. In other words, any mass from topography is ignored. The *Bouguer correction*, on the other hand, specifically adjusts for the mass of known topography between the measurement and the reference geoid. Since surface topography can be observed, the Bouguer correction highlights subsurface mass variations. Figure 8.12 shows a gravity map of the Moon from *GRAIL*. The large anomalies associated with the lunar maria reveal significant excess mass below those impact basins, providing critical constraints on formation models.

8.4.3 Assessing the Compensation State

If the topography is compensated (e.g., by a root in the Airy model), no anomaly will show up in the free air correction. If the topography is not compensated (e.g., if it is supported by the strength of the lithosphere), a positive gravity anomaly will show up. Isostatically compensated elevated terrain will show up as a gravity low after the Bouguer correction, since the root has a lower density than the surrounding mantle, whereas flexurally supported terrain will not show a Bouguer anomaly (Figure 8.11).

In Section 8.3, we computed the flexural response of an elastic plate to periodic topography of the form $q(x) = \rho_c g h_o \sin(2\pi i/\lambda)$, and we noted that short wavelength (i.e.,

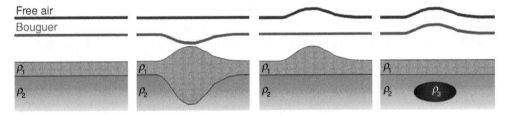

Figure 8.11 Illustration of gravity profiles after free air (blue) and Bouguer (red) corrections. In all frames $\rho_1 < \rho_2 < \rho_3$. (a) If there are no topography or lateral density variations, both gravity profiles are flat. (b) Compensated topography has no free air anomaly, but has a negative Bouguer anomaly. (c) Topography that is not compensated (i.e., is supported by the strength of the lithosphere) shows a positive free air anomaly, but no Bouguer anomaly. (d) A buried mass excess has a positive anomaly with both corrections.

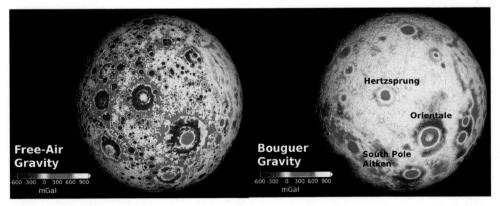

Figure 8.12 Gravity maps of the Moon from the *GRAIL* mission after free air and Bouguer corrections have been applied. Modified from NASA images.

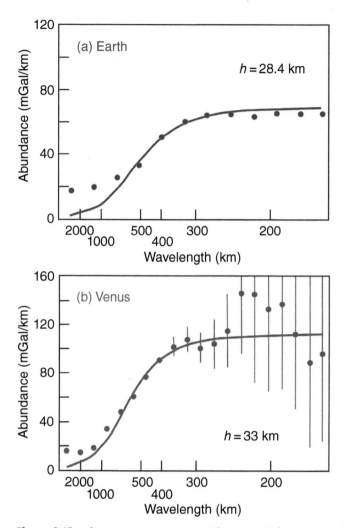

Figure 8.13 Admittance measurements (data points) for regions on the Earth (Hawaii) and Venus (Ulfrun) indicate similar elastic lithosphere thickness on the two bodies. The solid lines are model fits to the data. Modified from Nimmo and McKenzie (1998).

small-scale) topography can be supported by the strength of the plate and long wavelength topography is isostatically compensated. It turns out that the gravity anomaly produced from that topography and the resulting lithospheric deflection is

$$\Delta g_B = \frac{-2\pi\rho_c G}{1 + \frac{1}{4}\left(\alpha\frac{2\pi}{\lambda}\right)^4} e^{-\frac{2\pi z}{\lambda}} h_o \sin\frac{2\pi x}{\lambda} \quad (8.22)$$

Equation 8.22 has two important attributes to note. The first is that, once the topography is divided out, the gravity anomaly has the same form as the degree of compensation plotted in Figure 8.9. The ratio of gravity to topography is called **admittance**. Figure 8.13 shows plots of admittance for Venus and the Earth, illustrating the power of combining measurements of gravity and

topography to assess the elastic thickness on a planetary surface. The second is that the gravity anomaly signature falls off exponentially with altitude and with decreasing wavelength of topography; small-scale gravity anomalies are difficult to detect from large distances.

8.5 Conductive Heat Flow

As we learned in Chapter 6, the transfer of heat is an important driver of many geologic processes throughout the Solar System. Of the three methods of transferring heat (conduction, convection, radiation), conduction is by far the most important in the solid outer layers of planetary bodies with which we interact. The crustal layer of a body is heated and cooled from the top by radiation and may have heat delivered to the bottom through convection, but the thermal energy makes its way through the crust by conduction. The efficiency of conductive versus convective heat transport therefore controls the interior temperature. We can exploit our knowledge of heat conduction to remotely infer surface properties from remote thermal infrared observations. Fracturing of surface rocks from stresses imposed by cyclical heating and cooling has recently been recognized as a potentially important mechanism for breaking down rocks and building regolith on many planetary surfaces.

8.5.1 Fourier's Law and Heat Diffusion

Jean-Baptiste Joseph Fourier, like other famous early scientists, was interested in a wide variety of topics, both natural and philosophical. Fourier made significant advances in understanding heat flow by combining experimentation, mathematical advances, and by breaking his thought from the paradigm of action-at-a-distance, which had been prevalent at the time, reinforced by Newton's brilliant development of the law of gravity. Fourier noted that the flux of heat (energy per time per area flowing through a surface) is directly proportional to the temperature difference immediately on either side of the surface. From this observation, he developed the partial differential equation governing heat flow that now bears the name Fourier's Law:

$$q_z = -k\frac{\partial T}{\partial z} \quad (8.23)$$

or, in three-dimensional vector notation,

$$\vec{q} = -k\,\vec{\nabla}\,T \quad (8.24)$$

In these equations, q is the heat flux (SI units of W/m^2), T is temperature, and k is the thermal conductivity (units of W/m/K). The thermal conductivity describes the amount of energy that can be transported a given distance for a given ΔT. It is typically in the range of 1–4 W/m/K for rocks and ice and 10–80 W/m/K for metals.

A mass of material can store heat as well as conduct it. Conservation of internal energy including both heat storage and conduction adds a time component and leads to the law of heat diffusion:

$$\frac{\partial T}{\partial t} = -\frac{k}{\rho c_p}\frac{\partial^2 T}{\partial z^2} \tag{8.25}$$

or in three-dimensional vector notation

$$\frac{\partial T}{\partial t} = -\frac{k}{\rho c_p}\nabla^2 T \tag{8.26}$$

Here, ρ is density and c_p is the specific heat capacity, which describes the amount of energy required to change the temperature of 1 kg of material by 1 K (units of J/kg/K), and k is assumed to be constant. If there are other energy sources or sinks (e.g., radiogenic heat production), they can be included as additional terms to this equation. The quantity $k/\rho c_p$ is known as the thermal diffusivity (κ; units of m^2/s).

8.5.2 Surface Heat Flux and Temperature Profiles

We can measure the heat flux at the surface of the Earth in detail (Figure 8.14). Earth's average heat flux is ~85 mW/m^2, with q from the continents (~65 mW/m^2) a bit lower than q from the oceanic lithosphere (~100 mW/m^2). It is interesting to note that the heat flux from the Sun at 1 AU is 1367 W/m^2 – a factor of more than 10^4 larger than Earth's internal heat flux. Temperature balance at the surface is therefore dominated by **solar insolation**, but, as we'll see below (and as we know from human experience), the solar contribution does not penetrate deeply into the crust. In a planetary context, the dominance of solar insolation to surface temperature makes it difficult to measure internal heat flux from remote thermal infrared observations. Dramatic exceptions to this are Jupiter's volcanic moon Io, which, due to tidal heating, has $q \sim 4$ W/m^2 (nearly 10 percent of solar insolation) and Saturn's tiny moon Enceladus with a $q \sim 250$ mW/m^2 (about 2 percent of insolation at its distance from the Sun) in its South Polar Terrain from which geysers are erupting (see Figure 10.8a).

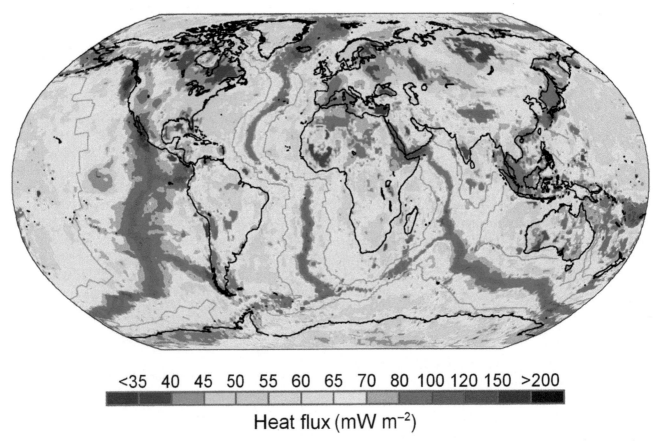

<35 40 45 50 55 60 65 70 80 100 120 150 >200

Heat flux (mW m^{-2})

Figure 8.14 Heat flux at Earth's surface from the interior. Actual heat flow measurements are fairly sparse over much of Earth, and the map has been constructed by incorporating multiple geological and geophysical proxies of heat flow. Modified from Goutorbe et al. (2011).

Integrating Equation 8.23 (or 8.24), assuming a constant heat flux, predicts a linear increase in temperature near the surface of the Earth, with a temperature gradient (geotherm) of ~20–35°/km. Temperature measurements in deep caves and boreholes confirm this approximation close to the surface. In reality, however, decay of radioactive elements in Earth's crust and heat input from the mantle (leftover accretional energy) also affect the heat flow in the lithosphere, and the linear approximation breaks down at depths greater than a few kilometers.

Many small bodies of the Solar System (e.g., asteroids, comets, small moons) never differentiated and can be expected to have already lost all of their accretional energy. The only source of internal heat for these bodies would be radiogenic heating in their rocky components. Assuming steady state ($dT/dt = 0$) and integrating Equation 8.24 in spherical coordinates with a radiogenic term (ρH, where H is the radiogenic heat production in W/kg) predicts an interior temperature profile $T = T_o \frac{\rho H}{6k}(R^2 - r^2)$ and a surface heat flux of $q_o = \rho HR/3$. In these equations, T_o is the surface temperature, R is the radius of the body, and r is the distance from the center at which the temperature is being computed. Assuming radiogenic heat production is the same as measured for chondritic meteorites, the predicted heat flow at the surface of a body the size of the large asteroid Vesta ($R \sim 262$ km) is ~1 mW/m². For the small near-Earth asteroid Bennu ($R \sim 250$ m), the target of NASA's *OSIRIS-REx* sample return mission, the predicted surface heat flow is only ~1 μW/m². These very small heat fluxes would be nearly impossible to distinguish from solar heating in remote observations.

8.5.3 Solar Heating

As mentioned above, the thermal energy balance near the surface of planetary bodies is dominated by solar insolation. Since planetary bodies rotate, the solar energy input could be (loosely) approximated as a time-varying periodic surface temperature: $T_o = \Delta T\cos(\omega t)$, where ΔT is the amplitude of the temperature variation and ω is the rate of variation (i.e., rotation rate). This assumption is oversimplified, but has the virtue of being analytically solvable. Integrating Equation 8.25 with this surface temperature boundary condition gives for the temperature as a function of depth and time:

$$T(z, t) = T_o + \Delta T e^{-\sqrt{\frac{\omega \rho c_p}{2k}}z} \cos\left(\omega t - \sqrt{\frac{\omega \rho c_p}{2k}}z\right)$$

(8.27)

This solution illustrates two important aspects of solar-driven temperature variations: the temperature variation falls off exponentially with depth, and there is a time

(phase) delay between the surface and subsurface temperature cycles. The depth at which ΔT falls off by a factor of $1/e$ is called the *thermal skin depth*, and is given by

$$d_s = \sqrt{\frac{2k}{\omega \rho c_p}}$$

(8.28)

The diurnal thermal skin depth for Earth is between about 5 and 20 cm, depending on the soil properties. The annual thermal skin depth, due to changing seasons, for Earth is about 1–4 m.

In reality, solar heating is not perfectly sinusoidal – there is no solar heat input during night time. In this case, the effective surface boundary condition is on heat flux, not temperature. Flux (energy) balance at the surface can be expressed as

$$\frac{S_o}{r_{AU}^2}(1 - A_B)\cos\theta_i - k\frac{\partial T}{\partial z}\bigg|_{surf} - \epsilon\sigma_B T_o^4 = 0$$

(8.29)

where S_o is the solar flux at 1 AU (1367 W/m²), r_{AU} is the heliocentric distance in AU, A_B is the Bond albedo, θ_i is the solar incidence angle of the surface facet considered, ϵ is the bolometric emissivity, and T_o is the surface temperature. If the specific situation includes other heat sources or sinks (e.g., heating from the atmosphere, volatile sublimation), extra terms can be added. With this boundary condition, Equation 8.25 is no longer solvable analytically – numerical techniques are necessary. Figure 8.15 illustrates temperature versus depths curves for a model surface. Note the decrease in amplitude of temperature variations and phase offset of the temperature wave with depth.

Another parameter that arises from considerations of heat conduction is **thermal inertia**:

$$\Gamma = \sqrt{k\rho c_p}$$

(8.30)

where Γ describes a material's resistance to changes in temperature and has the somewhat cumbersome units of J/m²/K/s$^{1/2}$. Thermal inertia is often used as a proxy for grain size, as described in Section 2.5.4. Small grains (e.g., sand and dust) have low thermal inertias – they heat up and cool down quickly (e.g., the Moon has $\Gamma \sim 50$ in these units). Large grains and bedrock, on the other hand, take a longer time to heat up and cool down. Observations of temperature as a function of time of day can be used to determine thermal inertia. Figure 8.16a shows diurnal temperature curves for surfaces with different thermal inertias. Figure 8.16b plots thermal inertias of asteroids versus their diameters, indicating that large asteroids are covered in fine-grained regolith material, whereas the surfaces of small asteroids appear to be, on average, blockier.

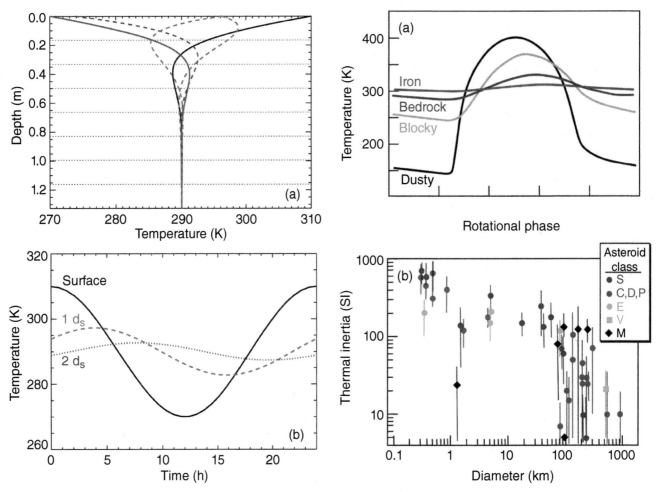

Figure 8.15 Example temperature profiles from Equation 8.27 for a planetary body at 1 AU with a rotation period of 24 h and thermal diffusivity of $10^{-6}\,\mathrm{m^2/s}$. (a) Temperature variation with depth at different times of day. The horizontal dotted lines mark intervals of skin depth down to eight skin depths below the surface. The temperature varies by 40 K at the surface, but that amplitude of temperature variation quickly diminishes, and by five skin depths almost no temperature variation is apparent. The plot also illustrates that the temperature gradient, which is proportional to heat flux (Equation 8.23), changes over the course of the day, so that heat is conducted into the surface in the daytime and out of the surface at night time. (b) The sinusoidal temperature variation assumed in this simple model at the surface and at one and two skin depths below the surface. The diminishing amplitude of the temperature variation is again apparent, as is the changing phase of the temperature variation – the peak temperature moves from 0 h (noon) at the surface to ~4 h at $1d_s$ and ~8 h at $2d_s$.

8.5.4 Thermal Stresses

Heating and cooling of geologic materials leads to expansion and contraction within the material. The strains of expansion and contraction lead, by the constitutive relations of Equations 8.5 and 8.6, to stresses within the material. The amount of strain experienced is described by the thermal expansion coefficient (α): $\varepsilon = \alpha \Delta T$. On Earth,

Figure 8.16 (a) Surface diurnal temperature curves for bodies of different surface types, and therefore thermal inertias. These profiles are computed with a thermophysical model that uses the surface energy balance given in Equation 8.29. Measuring temperatures of a surface at different times of day is a powerful means of determining the thermal inertia of a surface. (b) Thermal inertias of asteroids as a function of diameter. The letters in the key are different asteroid spectral types (which correspond to different inferred compositions). Modified from Delbó et al., (2015).

thermal stresses have long been recognized as a mechanism for spalling of material off of rocks at the surface. Airless planetary bodies (e.g., Mercury, asteroids, comets) can experience even more dramatic temperature variations, and recent thermal cycling experiments on meteorites suggest that thermal fragmentation of surface rocks could, in some cases, surpass impacts as a primary mechanism to form regoliths on these surfaces (Delbó et al., 2014).

8.6 Going with the Flow: Fluid Mechanics

Many processes in planetary science involve the flow of fluids. Fluvial and aeolian processes, those involving flowing liquid and air at the surface of a planetary body,

Table 8.1 Viscosities for some common materials

Material	μ (Pa·s)
Air	1.8×10^{-5}
N_2 (liquid)	1.5×10^{-4}
H_2O (liquid)	9×10^{-4}
Oil (motor, olive)	~0.1
Basaltic, andesitic, rhyolitic lava	~10^4, 10^7, 10^{11}
H_2O (ice, 273 K)	~10^{14}
Earth's mantle	~10^{21}

are covered in detail in Chapters 13 and 14. In this chapter, we focus on the flow of mantle rocks.

A fluid can be defined as a material that deforms continuously under an applied stress. We discussed in Section 8.2 the response of an elastic solid to an applied stress – it will deform (i.e., rearrange its structure at the microscopic level) until it balances the applied stress. For small stresses in most geologic solids, the elastic response is linear ($\sigma = E\varepsilon$). Fluids, however, are never able to balance the stress – their molecules keep slipping past one another. In this case, the applied stress leads to a continuous rate of deformation, or a **strain rate** ($\dot{\varepsilon} = d\varepsilon/dt$; the dot indicates differentiation with respect to time). For many geologic fluids, the relationship between σ and $\dot{\varepsilon}$ is linear; such materials are called Newtonian fluids:

$$\sigma = \mu\dot{\varepsilon} \tag{8.31}$$

The proportionality constant between stress and strain rate is the dynamic viscosity (μ), which has units of Pa·s. Dynamic viscosity describes the stress required to cause a given strain rate in a fluid. Dividing μ by the density of the fluid gives a quantity called the kinematic viscosity (v), which has units of m²/s. Recall that these are the same units as thermal diffusivity, and v similarly characterizes the diffusivity of momentum in a fluid. But here, we will work with the dynamic viscosity, μ. The viscosity spans many orders of magnitude for geologic materials, and we'll see in Section 8.7 that for a given material it depends strongly on temperature, grain size, and in some instances even on the applied stress. Table 8.1 lists typical viscosities for some common materials.

8.6.1 Conservation Laws

Analyses of fluid mechanics problems use the principles of conservation of mass, momentum (i.e., force balance), and energy. Here, we'll focus on mass and momentum conservation. If we think about a small volume in a flow, the conservation of mass is a statement that no mass can be gained or lost from the volume. Any mass leaving the volume must be replaced by new material entering the volume, at the same rate. Similarly, any change in density of the material within the volume must be balanced by material entering or leaving the volume. Considering flow in just one dimension, this conservation law is expressed as

$$\frac{\partial \rho}{\partial t} + \frac{\partial(\rho v_x)}{\partial x} = 0 \tag{8.32}$$

where v_x is the velocity of flow in the x direction.

Conservation of momentum looks at force balance on all sides of the same small volume, including pressure gradients in the flow and buoyancy of the volume relative to the surrounding fluid. The resulting relationship is called the *Navier–Stokes* equation, which with flow in one dimension is

$$\rho\left[\frac{\partial v_x}{\partial t} + v_x\frac{\partial v_x}{\partial x}\right] = \mu\frac{\partial^2 v_x}{\partial z^2} - \frac{\partial P}{\partial x} + \Delta\rho g \tag{8.33}$$

It is useful to look into the physical meaning of each term in the Navier–Stokes equation. The first term on the left side considers how the flow is changing with time (it is zero if flow is constant). The second term characterizes the inertia of the flow; this is the term in which turbulence enters (the term is small if the flow is smooth or laminar). The first term on the right side controls how sluggish the flow is, or how easily momentum is diffused across the flow. The second term describes pressure gradients along the flow, which can help or hinder flow. The last term represents the buoyancy relative to the surrounding flow and should only be included if a component of the flow is along the direction of gravity.

8.6.2 Relaxing Topography

When a load is added to or removed from the lithosphere, the lithosphere bends and the asthenosphere flows to achieve an isostatic balance. The best-known example of this process on Earth is post-glacial rebound. During the last ice age, great sheets of ice weighed down the lithosphere, flexing it into the asthenosphere. Ever since that ice melted, the fluid-restoring force of the asthenosphere has been pushing the lithosphere back into place.

Because the rebound is controlled by the viscous flow of the mantle, Equation 8.33 can be solved to describe the subsequent isostatic rebound of the surface. The full solution combines the flexural response of the lithosphere and the fluid response of the asthenosphere and is generally done numerically. Nevertheless, it has been shown that the vertical displacement (w) of the topography recovers with an exponential timescale:

$$w = w_o e^{-\frac{t}{\tau}}, \qquad \text{where } \tau \approx \frac{4\pi\mu}{\rho g L} \tag{8.34}$$

Here, w_o is the original vertical displacement, and L is the horizontal length of the original load. This solution holds for the common case of $w \ll L$. By dating paleo-shorelines (i.e., elevated beaches) since the end of the ice age, geologists have been able to determine the timescale of rebound and, from that, estimate the viscosity of the asthenosphere (Figure 8.17).

On other planetary bodies, impacts act to remove mass very quickly from the lithosphere, leaving craters. These craters undergo isostatic rebound as well. In some cases, particularly on icy bodies, the craters can relax completely, so that no negative topography is left. Because the crater rims have a much shorter horizontal length scale, the crater rims still often rise above the surface after the crater itself has completely rebounded (e.g., Ganymede in Figure 8.17). Images of the surface of the dwarf planet Ceres (diameter ~960 km) from the *Dawn* mission revealed large impact craters that had rebounded. This observation is a clear indication that Ceres has, or at least had for some period of time after the large craters formed, a ductile asthenosphere.

As we'll see in Section 8.7, viscosity is a strong function of temperature. As temperature increases, viscosity decreases, and topography relaxes more quickly. Crater relaxation can therefore be used to uncover changes in heat flow within a planetary body. Enceladus is an excellent example. As will be described in Section 9.5.3, the south pole of Enceladus contains tectonic fractures with high heat flow, but other parts of the surface are geologically old, as evidenced by high densities of impact craters. Many of these craters, it turns out, have experienced significant amounts (up to 90 percent) of relaxation. Numerical modeling of the process by Bland et al. (2012) indicates that heat flows comparable to that occurring at the south pole must have occurred in these other regions at some time in Enceladus' past.

8.6.3 Convection

A fluid layer heated from below and cooled from the top (a common occurrence in planetary bodies radiating their heat to space from their surfaces) is gravitationally unstable. The hot fluid at the base is less dense due to thermal expansion, and it therefore wants to rise buoyantly. Viscous forces in the fluid layer fight against this buoyancy; convection can only occur if the buoyancy force is larger than the viscous forces.

Mathematically, the thermal buoyancy is included in the buoyancy term in the Navier–Stokes equation (Equation 8.33). The change in density driving the buoyancy is controlled by the thermal expansion coefficient and the temperature. In other words, inclusion of thermal buoyancy adds a temperature term to the Navier–Stokes equation. The temperature is solved using the heat diffusion equation (Equation 8.25), but a term must be added to that equation as well. As the hot parcel rises, it carries heat with it – heat is being transported directly by the movement of the fluid. This *advection* adds a term that includes velocity to the heat diffusion equation. The appearance of temperature in the flow equation and velocity in the heat equation requires both to be solved simultaneously to investigate convection. This coupling makes quantitative studies of convection difficult, requiring numerical techniques.

To examine the conditions under which convection gets started, it is useful to define a parameter called the *Rayleigh number*. The Rayleigh number compares the thermal diffusion timescale across the layer to the timescale for the warm fluid to move through the layer and is given by

$$\mathrm{Ra} = \frac{g\rho_o \alpha_v (T_1 - T_0) b^3}{\kappa \mu} \tag{8.35}$$

Here, ρ_o is the standard density of the fluid (before heating), α_v is the volumetric thermal expansion coefficient, T_1 and T_0 are the temperatures at the base and top of the layer, respectively, κ is the thermal diffusivity, and b is the thickness of the layer. Convection can get started if Ra is greater than some critical value that depends strongly on whether the top layer is free to move horizontally (mobile lid) or is fixed in place (stagnant lid) (see Chapter 9 for more detail).

Convection cells can be envisioned somewhat like a conveyer belt, with warm fluid rising, moving horizontally while it cools to the surface, and the cool fluid falling back down. The horizontal motion imposes a stress on the (typically rheologically brittle) surface. If the stress on the lithosphere from convection is larger than its strength, the lithosphere will break, and the resulting plates can be moved by the convection conveyer. On the other hand, if the lithosphere is strong enough to withstand convective stresses, it will remain intact and will not move. As we know, the Earth's lithosphere is broken into tectonic plates that move and subduct along with the convection cells. No other planetary bodies are known to support plate tectonics; their lithospheres are apparently strong enough to withstand the stresses of the convection roiling beneath the surfaces.

8.7 Rheology

In this chapter, we have introduced several of the more common processes that geodynamicists investigate. For each process, we have made some assumptions about how geologic materials behave under stress. We have treated rocks as solid materials that respond in a linear, elastic way to stress – deformation is instantaneous, recoverable when the stress is removed, and scales

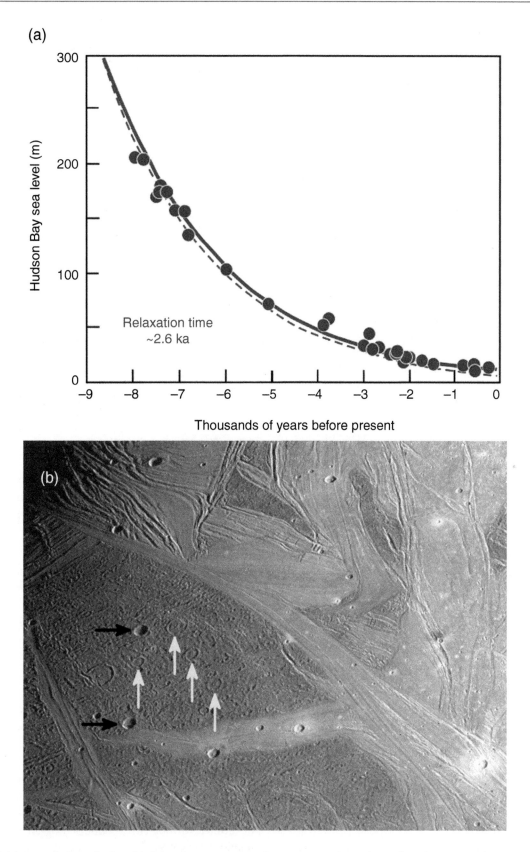

Figure 8.17 (a) Rate of rebound of Earth's lithosphere around Hudson Bay since the melting of ice sheets. Paleo-beach marks are used to measure rebound as a function of time. Using these data in Equation 8.34 yields a viscosity of ~10^{21} Pa·s for Earth's mantle. (b) Image of Ganymede (one of Jupiter's large moons). The yellow arrows indicate craters whose topography has relaxed. The rims, because they are smaller than the craters themselves, have not yet relaxed. The black arrows point to younger craters that have not relaxed, perhaps indicating a change in heat flow with time. NASA image.

linearly with stress as described by Young's modulus (E). We have treated ductile layers as fluids that also respond linearly to stress – deformation builds up with time, remains when the stress is removed, and the rate of deformation scales linearly with applied stress (i.e., Newtonian fluid) as described by dynamic viscosity (μ). Real geologic materials are more complex and often don't fit cleanly in one of these two categories.

8.7.1 Visco-Elastic Rheology

Rocks and ices near the surface of Earth or other planetary bodies are generally brittle – they respond elastically to stress until their strength is overcome, at which point they fracture. When rocks and ices are deep enough that the confining pressure is comparable to or greater than their strength, they can be ductile – they flow to achieve an equilibrium stress. In reality, rocks and ices in almost any situation exhibit a combination of these behaviors – a linear elastic response at first, followed by ductile flow, then fracturing if the stress overcomes the strength and confining pressure (see Figure 8.3). The dominant behavior we see in a specific situation often depends on the timescale involved.

Looking at this **visco-elastic** model as a function of time, we see that when a stress is first applied, elastic strain is incurred immediately. As that stress is maintained on the (nominally solid) object, strain continues to increase linearly with time as the object deforms ductilely (Figure 8.18). In other words, all solid geologic materials behave elastically over short timescales but, given sufficient time, will flow. A question that follows is: Over what timescales should we treat a given material as a solid versus as a fluid? A common way to answer this question

is to find the time that it takes for the viscous (ductile) strain to equal the initial elastic strain. We can calculate this time by dividing the elastic strain by the viscous strain rate. Assuming linear elastic behavior as a solid and Newtonian (linear) behavior as a fluid, we find that this timescale is

$$\tau_M = \frac{\mu}{G_s} \tag{8.36}$$

where G_s is the shear modulus (or modulus of rigidity) as described in Section 8.2. This quantity is called the Maxwell time, after the same Maxwell who established the famous equations for electromagnetism. It seems that he was conducting long-term pendulum experiments, and those experiments were compromised by slow extension of the pendulum wire. The Maxwell time for the Earth's mantle is around 100 years, which explains why it transmits short period seismic waves like a solid, but is rebounding from the melting of ice sheets as a fluid. The Maxwell time of H_2O ice near its melting temperature is on the order of an hour; warm ice flows fairly readily under sustained stress.

8.7.2 Non-Newtonian Rheology

Some materials do not follow the linear Newtonian flow law defined in Equation 8.31, and the viscosity of almost all materials is temperature-dependent. The generalized relationship between stress and strain rate has the form

$$\dot{\varepsilon} = A\sigma^n d^{-m} e^{-\frac{E_a + PV_a}{RT}} \tag{8.37}$$

Here, d is grain size, E_a is an activation energy, V_a is an activation volume, A is a constant, R is the universal gas constant, and T is temperature. E_a, V_a, A, n, and m are determined from experiments. Compared to Equation 8.31, we can write an effective viscosity as

$$\mu_{\text{eff}} = \frac{\sigma}{\dot{\varepsilon}} = \frac{\sigma^{1-n} d^m}{A} e^{-\frac{E_a + PV_a}{RT}} \tag{8.38}$$

In this form, the viscosity of a material depends on the grain size of the material, the stress applied, the temperature, and several parameters intrinsic to the material.

Solid materials flow as reorganization occurs within the crystal lattice. This reorganization can happen in different ways, depending on the material, the stress applied, the temperature, etc. At relatively low stress, diffusion of defects or vacancies through the volume of the grains or along grain boundaries enables flows to occur. Both of these diffusion pathways tend to lead to Newtonian behavior ($n = 0$), but with different grain size dependences ($m = 2$ and 3, respectively). At higher pressures, dislocations within the crystal lattice migrate, leading to non-Newtonian behavior and different grain size dependences.

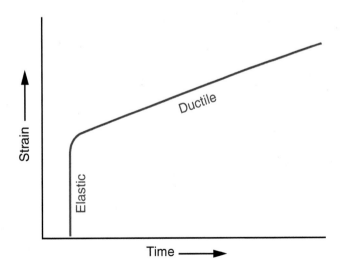

Figure 8.18 Illustration of visco-elastic behavior. When a constant stress is applied and maintained, elastic strain occurs immediately, and viscous or ductile strain accrues linearly with time.

The temperature, stress, and grain size dependencies of flow behavior can make it difficult to apply insights gained from terrestrial geology to other planetary bodies. Jupiter's large moon Europa is a prime example. Planetary geologists would like to know whether convection in Europa's ice shell contributes to the complex geology on the surface. In fact, it has been suggested that plate tectonics may even occur on Europa, if it can be driven by convection in the ice shell (Kattenhorn and Prockter, 2014). Unfortunately, the effective viscosity of ice under the conditions at Europa are not well constrained. Even if the flow law itself were known, there is no way with current measurements to know the grain size in the ice shell. The effective viscosity is uncertain by orders of magnitude just from the uncertainty in grain size alone. Nevertheless, models do favor convection in Europa's ice shell, and these models will be tested in the next decade by NASA's *Europa Clipper* mission.

Summary

Geologic materials under stress strive for balance – they want to relieve that stress as effectively as they can. The way to relieve stress is through strain, and the strain that a material will undergo when stress is applied depends on its rheology. We have seen how planetary surfaces support topography and respond to surface loads. Detailed measurements of a body's gravity field provide a window into the interior, at least in terms of how mass is arranged below the surface. The observation from gravity measurements that the giant Tharsis volcanoes on Mars are not in isostatic balance indicates that the lithosphere must have been extremely thick and strong when they formed, and remains so now. Yet, relaxation of even relatively small craters on some regions of Saturn's small moon Enceladus reveals high internal heat flow in those regions in the past.

Planetary surfaces are solid and brittle, and heat transfer through them is by conduction. Surface temperatures are controlled primarily through solar heating and therefore change systematically over the course of a day and an orbit. Measurements of temperature as a function of time, which can be done remotely by measuring thermal flux emitted from the surface, provide a powerful means to determine properties of the surface, such as grain size and induration. Planetary interiors, however, often behave, at least over geologic timescales, as a fluid. Understanding fluid mechanics is therefore critical for assessing the dynamics of planetary interiors. One consequence of more fluid-like behavior is that heat may be transferred by convection. When convection occurs, it induces stress in the overlying lithosphere, the effects of which may be measurable by planetary geologists.

Review Questions

1. What does it mean to say that something is a linear elastic material? How are stress and strain related for a linear elastic material? How does strain in one dimension affect strain in another dimension?

2. Imagine a sedimentary basin on some planetary body that is confined on all sides and filled to a depth of 1 km. The horizontal strains, ε_{xx} and ε_{yy} are both zero, and the vertical stress at any depth z in the basin is given by $\sigma_{zz} = \rho z g$. Use the constitutive relations in Equation 8.4 to find expressions for the strain in the vertical direction (ε_{zz}) and the stresses in the horizontal directions (σ_{xx} and σ_{yy}). Compute these quantities at depths of 0.1, 0.5, and 1.0 km for a sedimentary basin on Earth, assuming $\rho = 2100 \, \text{kg/m}^3$, $E = 50 \, \text{GPa}$, and $v = 0.2$. Now compute the same quantities for an icy sedimentary basin on Titan, where $g = 1.3 \, \text{m/s}^2$, $\rho = 917 \, \text{kg/m}^3$, $E = 10 \, \text{GPa}$, and $v = 0.3$. Compare the results for the two bodies.

3. The South Pole-Aitken (SPA) impact basin, located on the far side of the Moon, is about 8 km deep and appears to be in isostatic balance. Describe how you would use the principle of isostasy to compute the thickness of the crust surrounding the basin, under the assumption that the basin-forming impact excavated all the way to the mantle. For crustal and mantle densities of $2800 \, \text{kg/m}^3$ and $3400 \, \text{kg/m}^3$, respectively, compute the crustal thickness and discuss the reliability of this estimate.

4. Describe how to estimate the thickness of the lithosphere from topography near a big surface load. Use this technique to compute the elastic thickness (h) of the lithosphere on Europa from the data in Figure 8.8, assuming the plate is broken beneath the load, $g = 1.32 \, \text{m/s}^2$, $E = 10 \, \text{GPa}$, $v = 0.3$, $\rho_\text{m} = 1000 \, \text{kg/m}^3$, and there is no fill in the flexural basin ($\rho_\text{fill} = 0$). Compare your results to values of elastic thickness given in Figure 8.13 for the Pacific plate beneath the Hawaiian islands and the Ulfrun region of Venus.

5. How can topography and gravity measurements be combined to determine the compensation state of surface features?

6. Describe diurnal temperature variations at the surface and near-surface (top few meters) of planetary bodies.

7. Why are some craters on Ganymede, Enceladus, and other icy bodies not topographic lows? Why do they still have raised rims?

8. What does it mean to say that something is a Newtonian fluid? What are the various factors that affect the effective viscosity of geologic materials? Which of these factors imposes the largest uncertainty in estimating the viscosities of the mantles of other planets (i.e., besides Earth)? Compare and contrast the relative importance of each in the mantles of planetary bodies.

SUGGESTIONS FOR FURTHER READING

Gerya, T. (2010) *Introduction to Numerical Geodynamic Modelling.* Cambridge: Cambridge University Press. An excellent resource for delving more deeply into the computational techniques necessary for modern geophysics/geodynamics.

Stacey, F. D., and Davis, P. M. (2008) *Physics of the Earth.* Cambridge: Cambridge University Press. This book presents a complete history of the Earth from a geophysical perspective.

Turcotte, D., and Schubert, G. (2014) *Geodynamics*, 3rd edition. Cambridge: Cambridge University Press. The classic textbook for learning geodynamics for terrestrial and planetary applications. The third edition includes chapters and examples for computational modeling of geodynamical processes.

REFERENCES

Barnett, D. N., Nimmo, F., and McKenzie, D. (2002). Flexure of Venusian lithosphere measured from residual topography and gravity. *Journal of Geophysical Research* 107(E2). DOI: 10.1029/2000JE001398.

Bland, M. T., Singer, K. N., McKinnon, W. B., et al. (2012) Enceladus' extreme heat flux as revealed by its relaxed craters. *Geophysical Research Letters,* 39, L17204.

Delbó, M., Libourel, G., Wilderson, J., et al. (2014) Thermal fatigue as the origin of regolith on small asteroids. *Nature,* 508, 233–236.

Delbó, M., Mueller, M., Emery, J. P., et al. (2015) Asteroid thermophysical modeling. In *Asteroids IV*, eds. Michel, P., DeMeo, F., and Bottke, W. F. Tucson, AZ: University of Arizona Press, pp. 107–128.

Goutorbe, B., Poort, J., Lucazeau, F., et al. (2011) Global heat flow trends resolved from multiple geological and geophysical proxies. *Geophysical Journal International,* 187, 1405–1419.

Hammond, N. P., Phillips, C. B., Nimmo, F., and Kattenhorn, S. A. (2013) Flexure on Dione: investigating subsurface structure and thermal history. *Icarus* 223, 418–422.

Huppert, K. L., Royden, L. H., and Perron, J. T. (2015) Dominant influence of volcanic loading on vertical motions in the Hawaiian islands. *Earth and Planetary Science Letters,* 418, 149–171.

Hurford, T. A., Beyer, R. A., Schmidt, B., et al. (2005) Flexure of Europa's lithosphere due to ridge loading. *Icarus,* 177, 380–396.

Kattenhorn, S. A., and Prockter, L. M. (2014) Evidence for subduction in the ice shell of Europa. *Nature Geoscience,* 7, 762–767.

Lemoine, F. G., Goossens, S., Sabaka, T., et al. (2014) GRGM900C: a degree 900 lunar gravity model from GRAIL primary and extended mission data. *Geophysical Research Letters,* 41, 3382–3389.

McGovern, P. J., Solomon, S. C., Smith, D. E., et al. (2002) Localized gravity/topography admittance and correlation spectra on Mars: implications for regional and global evolution. *Journal of Geophysical Research,* 107, E12.

Nimmo, F., and McKenzie, D. (1998) Volcanism and tectonics on Venus. *Annual Reviews of Earth and Planetary Science*, **26**, 23–51.

Turcotte, D., and Schubert, G. (2014) *Geodynamics*, 3rd edition. Cambridge: Cambridge University Press.

Watters, T. R. (2003) Lithospheric flexure and the origin of the dichotomy boundary on Mars. *Geology*, **31**, 271–275.

Zuber, M. T., Smith, D., Watkins, M., et al. (2013) Gravity field of the moon from the Gravity Recovery and Interior Laboratory (GRAIL) mission. *Science*, **339**, 668–671.

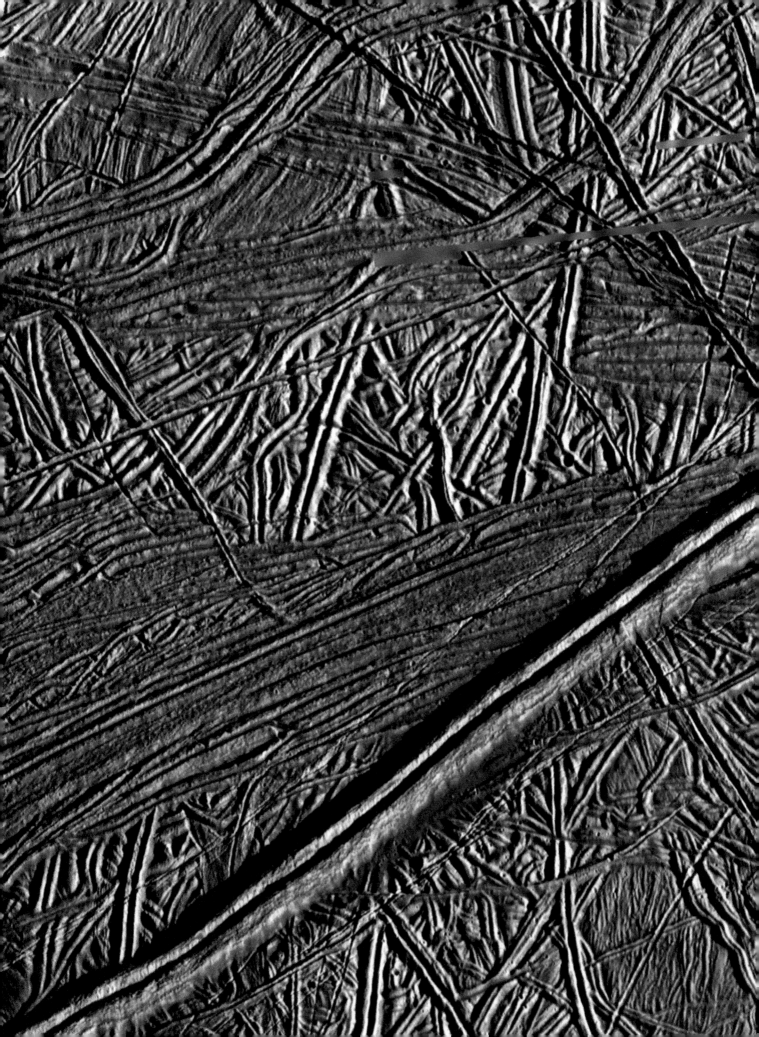

9

Planetary Structures and Tectonics

The surfaces of terrestrial planets and icy satellites have enjoyed deformation marked by faults and folds. We use these geologic structures not only to characterize the morphology of the surfaces, but also to describe the motions, stresses, and deformation processes that created the structures. Ultimately, we can sum these structural data and interpretations to infer the tectonic deformation for large portions of, or even entire, planets and satellites. As we will see, understanding deformation at this large tectonic scale enables us to investigate what is driving overall planet or satellite development. We will also learn that while the expected will happen, conundrums exist too. For example, the rocks of terrestrial planets deform quite differently from the icy shells of the satellites of the gas giants, yet the magnitude of these differences and their causes can surprise us. On the other hand, Venus and Earth are quite similar in many planetary characteristics but have strikingly different tectonic histories, which challenges us to understand why.

9.1 Active-Lid versus Stagnant-Lid Planets and Satellites

If the lithospheres of planets and satellites were continuous and attached to underlying mantles and cores, this chapter would likely be a short description of structures related to long-term cooling shrinkage of the bodies and impact cratering. Such bodies with immobile and intact **stagnant lids** are abundant in the Solar System (Figure 9.1a). Yet, we also have bodies with **active lids**, such as the Earth and Europa, with lithospheres that are able to develop faults/shear zones and deform significantly in response to driving stresses related to agents such as mantle convection or eccentric orbits. In between, we have bodies such as Mars, which through much of its history has been stagnant lid, but with greater tectonic deformation than expected due to Tharsis, or Venus,

which is presently stagnant lid, but has evidence to indicate that possibly it was episodically an active lid in the past.

Considering planets and satellites from the perspective of structural geology and tectonics, the contrast between stagnant and active lids is fundamental. For example, a stagnant-lid body such as Mercury dominantly shrank through its history due to cooling, so that contractional structures on its surface record a very modest shortening strain. In contrast, the active-lid Earth has individual thrust faults in single mountain belts that each accommodated displacements that are 2–5 times greater than the total shortening recorded for Mercury. Consequently, active-lid and possibly episodically active-lid bodies display a much richer abundance and range of tectonic structures representing more complex histories than found for stagnant-lid bodies (Figure 9.1b).

Given the range of behaviors from simple stagnant-lid to fully active-lid bodies in the Solar System, understanding what controls the occurrence of particular behaviors is essential to grasping their tectonic development. A necessary requirement for active-lid behavior is that the driving stress from the agent powering the deformation exceeds lithospheric strength (see Chapter 8 for quantitative consideration of the driving stresses and onset of deformation). For example, if the agent is mantle convection, then the stresses generated by mantle circulation are sufficient to cause breakage of the overlying lithosphere so that segments of the lid can move relative to each other, creating active-lid behavior (O'Neill et al., 2007).

Additionally for the case of mantle convection, the thermal and deformation history can play a role in whether active- or stagnant-lid behavior is present. For example, transitioning from stagnant lid to active lid is more difficult than from active to stagnant for two reasons. First, stagnant lids tend to be thicker than active

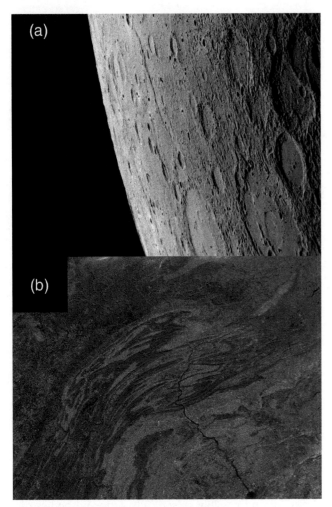

Figure 9.1 (a) Surface image of Mercury, which experienced stagnant-lid tectonics. Abundant craters of various sizes accumulated over time and very few topographic features are related to faults or folds. This image does contain a prominent north–south trending tectonic lobate ridge in its middle. Image from *MESSENGER*. (b) Image of a planetary surface that experienced active-lid tectonics. Portion of the Appalachian mountain belt in North America that formed due to collision between two lid segments, creating abundant faults and folds. NASA *Aqua Satellite* image.

lids, with greater vertical thermally inefficient, conductive cooling paths, so heat is trapped in the underlying mantle, reducing its viscosity (strength) (Weller and Lenardic, 2012). Second, the thicker stagnant lid will have greater yield strength, as we will see in the next section about lithospheric material properties, which reduces the ability of mantle convection to trigger lithospheric failure.

9.2 Lithospheric Materials, Deformation Behaviors, and Strengths

While a number of factors influence the occurrence of active- versus stagnant-lid behavior, the crux of the matter is the interplay of the magnitudes of the driving forces from causative agents versus the lithospheric material strength.

9.2.1 Materials

The compositions of planetary and satellite lithospheres are critical to determining their material strengths. As we are focusing on terrestrial planets and the Moon, and on icy satellites of the gas giants, we are dealing with lithospheres composed of rock or ice, respectively. While these materials can be quite complex as a function of being polymineralic, having different grain sizes and containing a variety of impurities or fluids, we will keep our considerations simple, so as to focus on the major aspects of their deformation behaviors.

For most rocky bodies, we are dealing with exterior mafic igneous rocks (basalt, gabbro) and internal ultramafic rocks (peridotite) in the lithosphere to a first order, so the suite of minerals commonly includes a primary mix of olivine, pyroxene, amphibole, plagioclase, and garnet. Ideally, we would construct our material understanding from these rocks using their mineral suites allowing for compositional variation, grain size variation, etc. At this time, our experimental and simulation capabilities for rocks limit our ability to characterize and usefully predict the deformation behavior of these natural materials. So, we will use a single mineral that is common to these rocks to serve as a proxy for their behavior: olivine, the ferromagnesian silicate containing a structure of isolated silica tetrahedra with orthorhombic symmetry. The robustness of using this proxy is based on over 50 years of actual deformation experiments and 25 years of computer-simulated deformation including polycrystalline aggregates interpreted in terms of comparisons to naturally deformed rocks from Earth.

For icy satellites, we are restricted to information about their exteriors for nominating a proxy of their deformation behavior. Almost all exteriors have ice compositions dominated by water although methane or ammonia may be in the mix. Thus, water ice serves as our proxy, and in particular Phase I water ice (henceforth, just water ice). Other phases of ice typically require confining pressures that would place their occurrence at several hundred kilometers depth within satellites, and the consensus is that the ice/water layer for these bodies does not normally reach that thickness for the satellites with visible tectonic deformation. An important limitation is that useful ice deformation experiments and simulations are much fewer than for rocks and only date back about 30 years. Further, we lack naturally deformed ice samples from the satellites, and ice from Earth's glaciers and ice sheets deforms at much higher temperatures than found on the satellite exteriors. So, our understanding of material behavior for ice-bearing lithospheres is more speculative than for rock-bearing lithospheres.

9.2.2 Deformation Behaviors

Materials deform in three basic ways: elastically, brittlely, or ductilely. During an elastic deformation when the causative stress state is relaxed, the material returns to its original condition and the deformation is recoverable (further explained in Section 8.2). By contrast, brittle and ductile deformations are permanent and not recovered when the causative stress state is relaxed. For brittle deformation, the material breaks and grain attrition processes such as wear and fracture occur to accommodate displacement, whereas for ductile deformation the material flows without breakage. Two end member types of flow are viscous deformation where the flow rate is a function of the differential stress magnitude, and plastic deformation where flow does not occur until reaching a threshold or yield stress.

Olivine and water ice display combinations of these behaviors during deformation. For example, in the upper lithosphere where temperatures and pressures are lower, small deformations are typically elastic and larger and/or faster deformations are brittle. Likewise, very small deformations are elastic and larger/faster ones are permanent in the lower lithosphere. In this case, the permanent deformation will be the interplay of viscous flow often related to deformation processes involving atomic diffusion (creep processes) and plastic flow related to overcoming yield strengths for glide systems within atomic lattices (dislocation glide) or on grain boundaries (grain-boundary sliding).

An important point to consider when comparing the behaviors of ice-bearing and rock-bearing lithospheres is whether olivine and water ice are similar or different as deforming materials. Elastically, they differ as the stiffness (Young's modulus) of olivine is about 50 GPa, whereas in water ice it is about 9 GPa, so olivine is a much stronger elastic material than ice. Frictionally, they are quite similar initially in that both have coefficients of friction, which is the resistance to sliding on a surface, of about 0.45 to 0.55. So, once sliding surfaces form, both materials display similar resistance to slip. Ductilely, olivine and water ice are again quite different. Ice has a much lower melting temperature and therefore at much lesser depths in a lithosphere nears its melting temperature, unlike olivine. In fact, for olivine and considering the Earth, a current debate is whether the base of oceanic lithosphere at the asthenosphere is due to a small amount of melting or melting is unneeded because thermally driven deformation has become very efficient (e.g.,

Boettcher et al., 2007; Hansen et al., 2016). So, the olivine-rich oceanic lithosphere of the Earth may not reach its melting temperature at any depth, making it stronger when compared to ice! Further, as water ice is 10^4–10^8 times less viscous than olivine, ice flows much more easily for the same driving stress and is thus much weaker as a ductile material. In summary, water ice with respect to olivine is weaker elastically and easier to break, similarly strong initially as a brittle sliding material, and much weaker as a ductile material due to a lower melting temperature and lesser viscosity. Simply, water ice is more susceptible to deformation than olivine (rock), and we will see the implications of this difference when we consider Europa.

9.2.3 Lithospheric Strength as a Function of Depth

Given knowledge of our materials and their deformation behaviors, we can identify the depth at which the lithosphere is strongest, which is where the driving stress must be greatest to trigger active-lid tectonics. Assuming a compositionally homogeneous lithosphere with a constant grain size for simplicity, we first consider the Earth as a representative for rocky bodies. The strength of the lithosphere is a function of depth, where depth serves as a proxy for temperature and confining pressure (Figure 9.2a). In a stressed material, different deformation behaviors compete as a function of conditions such as pressure, differential or shear stress, and confining temperature, and the most efficient behaviors tend to dominate, setting the local value for lithospheric strength. Overall, in the lithosphere, this competition is between brittle frictional versus ductile viscous and/or plastic processes. The frictional processes require shear stresses to exceed the coefficient of friction and the normal stress, which is a function of overburden and the tectonic stresses. The viscous and plastic processes that are inversely and nonlinearly related to temperature and hence, the reservoir of heat energy, require sufficient differential stresses to drive deformation processes such as atomic diffusion, lattice sliding, grain-boundary sliding, grain-boundary nucleation, and grain-boundary migration. Consequently, the cooler upper lithosphere is dominantly elastofrictional, whereas much of the lower hotter lithosphere deforms viscously and/or plastically. The depth at which the lithosphere transitions from elastofrictional to viscous/plastic is the **Brittle–Ductile Transition** (BDT in Figure 9.2). This depth is also the position of the greatest strength in the lithosphere because it has the greatest normal stress on a potential translithospheric fault for the brittle regime, and the least available amount of heat energy for deformation in the ductile regime.

(a)

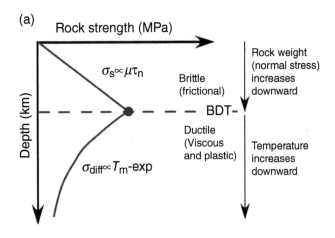

(b)

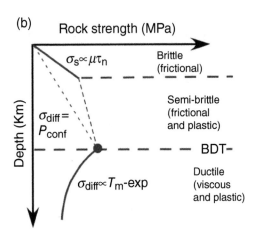

Figure 9.2 (a) Graph of depth versus rock strength for (a) idealized olivine lithosphere, and (b) idealized olivine lithosphere where brittle and plastic deformation mechanisms are both active above the Brittle–Ductile Transition (BDT). Large blue dot marks the depth of greatest rock strength, where normal stress is greatest in the brittle regime and temperature is lowest in the ductile regime. σ_s is shear stress, μ is coefficient of friction, σ_n is normal stress, σ_{diff} is differential stress, T_m is melting temperature, τ_n and P_{conf} is confining pressure.

Ice lithospheres may differ from olivine-dominated lithospheres by having an elastic zone between the brittle and ductile lithospheric regimes, if driving stresses are insufficient for brittle failure in the lower elasto-plastic lithosphere. Another variation of lithospheric strength with depth occurs with olivine due to the competition of deformation mechanisms above the ductile regime. There, both frictional and dislocation-related plastic processes may be occurring with near equal efficiency in the rock so that the lithosphere has a composite semi-brittle behavior (Figure 9.2b). Yet, in any of these cases, the top of the ductile regime is the strongest position in the lithosphere and that is the strength value that must be exceeded by driving stresses for active-lid tectonics to initiate.

9.3 Energy Sources and Driving Stresses

Tectonic deformation requires energy, which can be derived from a variety of sources. These include response to temperature changes, unstable density configurations, tidal distortions and, on icy bodies, true polar wander.

9.3.1 Thermal Sources

Temperature and, more importantly, cooling and heating, have quite opposite roles as an energy source for driving tectonic deformation. The long-term cooling of a planet or satellite results from the body releasing heat to the cold of space. If the body is not generating sufficient replacement heat through processes such as radioactive decay, buoyancy effects with potential energy releases or major impacts, it cools. Within the body, heat release is a combination of outward heat conduction and convection that leads to thermal contraction of solids, solidification of liquids, and phase changes to denser minerals (see Sections 8.5 and 8.6 for further explanation). These processes typically are operating over most of the history of a body, excluding the early heavy bombardment period, up to the present day. The primary physical effect of this cooling is a reduction in the radius of the planet or satellite. As the surface area of such a body is dependent on its radius, a radial reduction requires an areal reduction triggering horizontally directed, contractional tectonic deformation on the surface of a body. As we will see, Mercury is an excellent example of this phenomenon.

The other role relates more to heat flow than heat release and for terrestrial bodies concerns the vigor of mantle convection and its ability to affect lithospheric behavior and deformation. We know from our consideration of material strength for the lithosphere that mantle convection initiates active-lid tectonics by creating sufficient stress to overcome the lithospheric strength at the BDT. Consequently, the velocity of mantle convection and the viscosity of the convecting mantle are key to determining whether convection generates sufficient stress to trigger failure. Another important attribute is the degree of traction between the base of the lithosphere and the underlying mantle, which is assumed to be strong even if an asthenosphere with a very small amount of melt is present. However, it is worth noting that such an assumption is more problematic for a convecting, denser water ocean underneath an icy lithosphere.

It is likely that all four terrestrial planets have some intensity of mantle convection at present as part of their suite of processes for achieving heat release. The vigor of these convections is dependent on the temperature gradient across the mantles, which is a function of both their radial thickness and their prior cooling histories. Consequently, thinner mantles such as for Mercury and Mars will have lesser convective vigor because they have released a greater proportion of their heat through time due to relatively greater surface areas versus volumes, as compared to Venus and Earth. Therefore, we would predict that active-lid tectonics driven by mantle convection is only a possibility for Venus and the Earth. It is also worth noting that if we can trigger lithospheric failure and feed it directly into the mantle circulation, any descending lithosphere will enhance convection by materially strengthening and increasing the magnitude of thermal downwelling (Figure 7.4).

9.3.2 Density Inversion Sources

Inside a planet or satellite, a system is gravitationally more stable if more dense materials are closer to the body center than less dense materials. The classic example from the Earth is the inward vertical stacking of the least dense quartzofeldspathic crust on an ultramafic mantle on the most dense, iron-rich core (pictured in Figure 7.2). Another important example of this type of gravitationally stable stacking for icy satellites is the presence of less dense icy lithosphere on either an underlying water ocean or rocky mantle that are both more dense. Yet, not all planets and satellites have this form of gravitational stability throughout their histories. When absent, the body is gravitationally unstable, which may trigger deformation driven by the potential energy release of lowering the more dense material closer to the body center. A good example from the early Solar System of such a release of potential energy was the differentiation of mantles and cores for terrestrial bodies that was achieved by the iron metal and sulfides sinking to body centers (Section 6.2.3). This process primarily focuses on the negative buoyancy of the iron-rich materials as it is mainly about core formation.

More typically, nearer the surface of planets and satellites, we focus on cases of positive buoyancy where the upward rise of less dense materials with related "floundering" or burial of more dense material downward releases potential energy and achieves gravitational stability (Section 8.7). A smaller-scale example that is peculiar to the Earth is the formation of salt diapirs where mobile, less dense, weak salts exploit geometric irregularities in stratigraphic sequences to create vertical paths upward, displacing denser overlying sediments downward. Perhaps a more significant case with respect to lithospheric-scale deformation is the rise of hotter, less dense mantle rocks as plumes or less dense magmas within a planet or satellite.

9.3.3 Tidal Sources

If they do not spin or orbit around other bodies, planets and larger satellites tend to adopt spherical shapes achieving gravitational equilibrium (Section 8.4). When they spin, they develop an oblate spheroid shape where an equatorial diameter is greater than the polar diameter as a function of rotation rate, object size, object mass, and the material behavior(s) of the body. When a spinning body orbits around a larger body, the two bodies are gravitationally attracted to each other. This attraction further distorts body shapes, creating an ellipsoid shape due to the formation of two tidal bulges (Figure 9.3a) where the longest axis (diameter) points toward the other body and the shortest axis (diameter) is still the polar one. Ideally, a less massive body will orbit around a more massive body and have a circular orbit so that the shape of the bodies remain constant while changing orientation as a function of orbital position, which produces tides (Figure 9.3b).

If we are dealing with a less massive orbiting satellite, such as the Moon for the Earth or a satellite of Jupiter, another important effect occurs. While it is true that the tidal bulge migrates around the satellite as it orbits, this change in shape is not instantaneous because the satellite rock and/or ice actually takes time to deform. Consequently, the bulge is not exactly where it should be and as a result, the gravitational attraction of the host planet exerts a torque for the purpose of moving the actual bulge back to its ideal position (Figure 9.3c). Ultimately, this torque and the related work are eliminated in the system by locking the orbiting satellite so the tidal bulge occupies the same material points at all times, which means that the satellite takes exactly the same amount of time to complete one spin on its axis as to complete one orbit around its host. This tidally locked behavior applies to the Moon and major satellites of the gas giants in the Solar System. In the case of two bodies with similar masses, both bodies become tidally locked to each other, as with Pluto and Charon.

To this point, we are discussing gravitational and motion effects that do not generate the stresses needed to initiate active-lid tectonics. We can change that state of affairs by considering the major moons of Jupiter and Saturn, which are massive hosts compared to their satellites, and where we are changing from a "two-body system" to a "multi-body system." Having several satellites around a planet leads to the satellites having elliptical orbits because each satellite not only experiences gravitational attraction from the planet but also the other satellites (Greenberg et al., 1998). Put differently, the gravitational attraction of the other satellites changes the shape of any satellite's orbit from circular to elliptical. For example, consider Europa orbiting Jupiter with

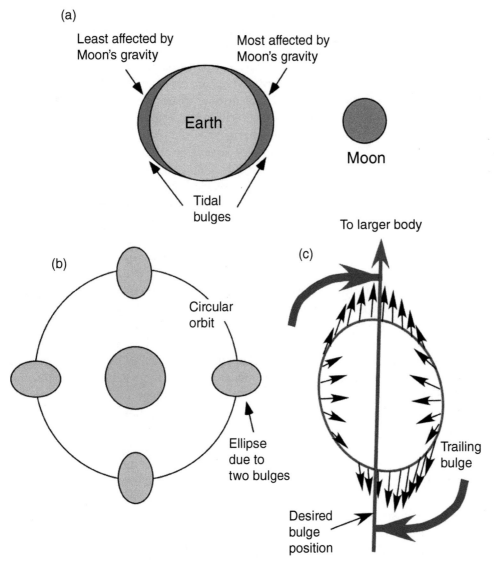

Figure 9.3 (a) Two tidal bulges created by gravitational attraction on one body by another, illustrated using gravitational attraction of the Moon on Earth's oceans because they are more responsive than the solid Earth. (b) Orientation of tidal bulges for less massive body in a circular orbit about a more massive body. (c) Torque on bulge flexing the orbiting, less massive body to re-align the actual bulge to the "instantaneous" bulge orientation.

gravitational interactions with Io and Ganymede such that its orbital eccentricity is 0.01, which while small is sufficient to generate gravitationally driven stresses that trigger stresses sufficient to cause lithospheric deformation, and by flexing the satellite generates heat (Section 6.2.4). The stresses are created by two effects: (1) tidal stress is greater when Europa is closest to Jupiter at pericenter and less when it is furthest at apocenter; and (2) while Europa's rotational speed is constant during an orbit, it should vary such that it is faster near pericenter and slower at apocenter. The first effect changes the amplitude of the tidal bulge and the second effect causes the tidal bulge to "wobble" such that the bulge is ahead of its expected position leaving pericenter and behind leaving apocenter.

These variations create a "diurnal tide" that is a daily (one orbit of about 3.5 Earth days around Jupiter), and hence rapid phenomenon for Europa. As a result, stresses are created with the potential to deform the lithosphere and, just as importantly, work done flexing the satellite generates heat, creating the water-dominated ocean beneath the satellite's icy lithosphere and above its rocky interior.

The existence of an off-center bulge and a water ocean beneath the icy lithosphere introduces an important effect that adds not only to the stress sources for lithospheric deformation, but also the complexity of the possible structural geometries: **nonsynchronous rotation**. For Europa and some other icy satellites such as Enceladus around Saturn, a water ocean decouples the icy lithosphere from the rocky interior so that it can operate with a degree of mechanical independence. For Europa, this independence means that when the icy lithosphere experiences a gravitational torque attempting to pull the tidal bulge back into an ideal position, the icy lithosphere spins just a little faster than the rocky interior, creating nonsynchronous rotation (Figure 9.4a). This rotation means that a point on the surface of the lithosphere moves perpendicular to longitude and does not have a fixed position with respect to Jupiter. Further, through time, it will move onto a tidal bulge experiencing tensile stresses as it expands and off of a bulge experiencing contraction and compression (Figure 9.4b). Several lines of physical evidence support the existence of nonsynchronous rotation: (1) the crater abundance on the leading hemisphere for Europa is not greater than for the trailing hemisphere as would be expected for a fixed lithosphere; (2) Europa contains many tectonic structures that are not predicted to occur where they do if only diurnal tidal stresses exist; and (3) comparison of the change in position of structures to comparable Europan terminators from images for the *Voyager* and *Galileo* missions allow for this type of motion. As we will see, the combination of diurnal tides and nonsynchronous rotation creates rich structural suites on icy satellites such as Europa and Enceladus.

9.3.4 True Polar Wander as a Source

The existence of decoupled icy lithospheres for some satellites of the gas giants allows for another type of absolute motion of the icy lithosphere relative to the satellite center that can create lithospheric deformation: **true polar wander**. This can occur when polar ice is much thicker than equatorial ice, causing the lithosphere to move perpendicular or obliquely to latitude across the tidal bulge (axis) to compensate by relocating the thicker lithosphere at the tidal bulge. Also, major density variations in the icy lithosphere or even large impacts can trigger some true polar wander because of the relative ease of motion due to lithospheric decoupling above the ocean. Consequently, unlike nonsynchronous rotation, true polar wander is not restricted to motion perpendicular to longitude. It could in fact be perpendicular to latitude (Figure 9.4c). For example, a recent large-scale topographic analysis indicates that true polar wander may have occurred for Enceladus, repositioning which parts of the lithosphere were located at both the poles and equator (Tajeddine et al., 2017).

9.4 Structures and Tectonics for Stagnant Lids

Stagnant lids are foreign to our terrestrial experience, but they are far more common than active lids. Let's see how deformations are expressed under these conditions.

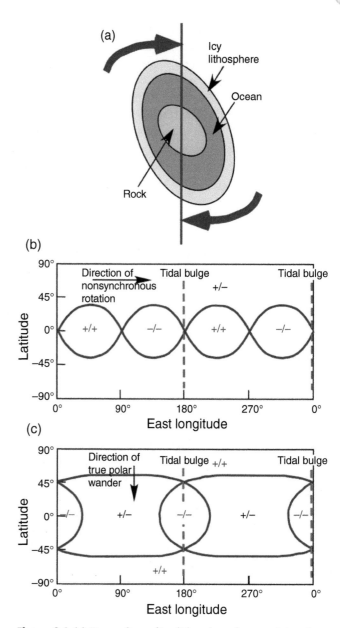

Figure 9.4 (a) Decoupling of icy lithosphere from rock interior by the intervening ocean that allows the icy lithosphere to move separately relative to the interior, creating nonsynchronous rotation. (b) Map view of deformation due to an eastward-moving nonsynchronous rotation of icy lithosphere over tidal bulges. Around the equator, it extends and contracts, leaving bulges. (c) Map view of deformation due to southward-moving true polar wander of icy lithosphere over tidal bulges, where portions extend approaching the bulges and contract moving away. Plus and minus identify surface extension and contraction; in pairs, first symbol is east–west trending and second is north–south. Modified from Collins et al. (2010).

9.4.1 Simple Stagnant Lids (Mercury, Callisto)

The signature surface appearance of a satellite or terrestrial planet that has experienced stagnant-lid tectonics is a great abundance of impact craters rather than a great abundance of tectonic structures (Figure 9.1a). As is described and illustrated in Chapter 11, these impact craters tend to have structures that either concentrically ring the craters, particularly larger ones, or are radial about the crater centers. Thus, these features are quite recognizable and will not be part of our consideration of the tectonic structures and histories of bodies in the Solar System.

While the overall appearance of a stagnant lid can appear to be quite similar between terrestrial bodies (e.g., Mercury, Figure 9.1a) and icy satellites (e.g., Callisto, Figure 9.5), key differences do exist because the icy satellites have weaker icy lithospheres, an underlying water ocean in many cases, and the gravitational interactions between many orbiting satellites around giant planets. Thus, Callisto has no systematic tectonic structures, whereas Oberon, a heavily cratered moon of Uranus, has some extensional normal fault systems that may relate to early thermal expansion of the interior of the satellite, but also otherwise lacks evidence of tectonic structures. Overall, no consistent pattern of tectonic structures exists among the many heavily cratered icy satellites with stagnant lids, although some satellites have unique structural patterns related to their particular accretion, cooling, and/or orbital histories.

By contrast, the terrestrial bodies with stagnant lids – Mercury, the Moon, and Mars – share a common pattern of tectonic deformation that timewise is entwined with the history of cratering, particularly after Late Heavy Bombardment at about 3.8 Ga. In all cases, they experienced shrinkage due to body cooling and the development of contractional structures accommodating the related reduction in circumference and surface area. These structures, in terms of increasing size, are **wrinkle ridges, lobate scarps,** and **long-wavelength undulations.** Wrinkle ridges and lobate scarps are interpreted to be fault-related folds in the hanging walls of thrusts (Figure 9.6). Wrinkle ridges on Mercury have topographic reliefs of less than 500 m and lengths of less than 100 km, whereas lobate scarps have topographic reliefs of 500–1500 m, and even up to 2000 m, with lengths that are 100 to more than 500 km. Both types of structure have asymmetric topography with relatively longer, shallower slopes that are thought to be above the underlying thrust faults. Thus, the asymmetric shape and positive topographic relief represent contractional displacement accumulation in the hanging walls above and along these thrusts at the surface of a terrestrial body. By making assumptions about the dip and shape (e.g., planar, listric, etc.) of these unexposed subsurface faults, the amount of crustal shortening achieved by individual or groups of wrinkle ridges and lobate scarps can be determined from the shapes and sizes of their topographic slopes (Figure 9.6c). Similarly, the larger undulations are treated

Figure 9.5 Image of Callisto showing abundant craters. NASA image.

as folds with amplitudes of 1–3 km and wavelengths of 800–1300 km, so that their shortening can be determined by comparing their wavelength to their curvilinear length, including amplitude.

Considering this cooling contractional deformation with *MESSENGER* data for Mercury (Byrne et al., 2014; Crane and Klimczak, 2017), almost 6000 contractional structures were identified and characterized in terms of size, orientation, shortening magnitude, and relative age where a tectonic structure and impact crater rim intersected. To determine the planet-scale cooling contraction, eight great-circle traces were positioned on the planet surface and the total shortening for the intersected contractional structures was calculated. This analysis determined a 7 km reduction in radius for Mercury with a related ~44 km reduction in planetary circumference. Turning to relative ages for the contractional structures and using the crater-related stratigraphy for Mercury, the oldest Pre-Tolstojan/Tolstojan craters are almost all cut by younger faults, whereas over half of the younger Mansurian craters overprint older faults (Figure 9.7). Thus, almost all preserved contractional structures postdate 3.8 Ga and many are older than 1.0 Ga, indicating that after Late Heavy Bombardment, crustal contraction was faster earlier in the history of Mercury than later. Such a result is consistent with models for thermal cooling of Mercury and other stagnant-lid tectonic bodies. They lost heat faster in the past when they were hotter because radioactive decay rates were greater, so they would

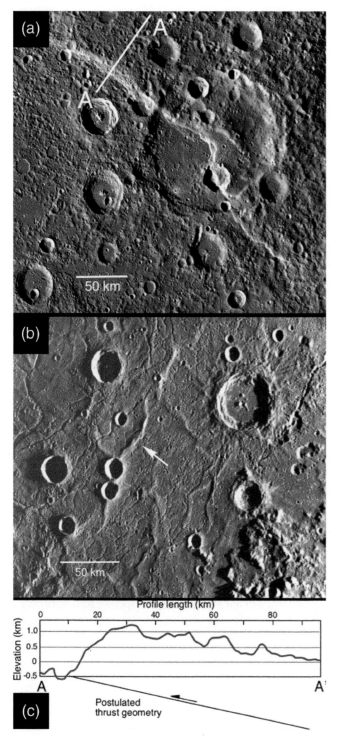

Figure 9.6 Contractional structures on Mercury. (a) Lobate scarp deforming a large crater. (b) Smaller wrinkle ridge. (c) Example of geometric assumptions for determining magnitude of crustal shortening (Byrne et al., 2014, supplemental materials). NASA *MESSENGER* images.

have contracted more then. Still, one should remember that the total shortening for Mercury due to cooling-driven contraction is about 0.3 percent, which is trivial compared to the shortening for any mountain belt on Earth.

9.4.2 A Loaded Stagnant Lid (Mars)

Mars shares a history of contraction due to cooling with Mercury and the Moon. Yet, Mars has a more complex tectonic history due to the presence of the Tharsis bulge, an igneous-related feature with a volume of about 3×10^8 km that has a footprint of about 20 percent of the planetary circumference. Additionally, Mars has a third important feature – its topographic dichotomy between the northern lowland plains over thin crust and the southern uplands over thick crust. However, as the predicted structural patterns for the global-scale stress field due to this dichotomy are absent, this important feature is generally thought to pre-date the formation of most tectonic structures on Mars, so it is not a factor in our considerations.

In terms of types of structures, Mars has the features that we have seen on Mercury and Moon, but with greater abundances (Figure 9.8):

1. Grabens: These pairs of facing extensional faults are separated by distances of kilometers, with topographic reliefs of hundreds of meters, lengths of tens to hundreds of kilometers, individual displacements of less than 100 m, and often occur as clusters (Figure 9.8a).
2. Wrinkle ridges: These asymmetrically sloping topographic ridges have reliefs of hundreds of meters, widths of tens of kilometers, and lengths of tens to hundreds of kilometers (Figure 9.8b). They are interpreted as resulting from subsurface contractional faults with displacements of less than 100 m. They typically occur on relatively uncratered lava plains.
3. Lobate scarps. These larger asymmetric topographic ridges have reliefs of hundreds of meters to several kilometers, lengths of hundreds of kilometers, and interpreted underlying thrust faults with displacements of hundreds of meters. They are more abundant in more intensely cratered regions, particularly in the southern uplands, and locally wrinkle ridges in plains continue into a lobate scarp in more intensely cratered regions.

The big difference with Mars compared to Mercury or the Moon is the arrangement of these structures on the surface of the planet (Golombek and Phillips, 2010). Many extensional grabens are arranged radially around the Tharsis uplift and many wrinkle ridges and lobate scarps are arranged concentrically. Such a geometric relationship between these structures and this major martian province favors the potential for a genetic relationship between them. As it happens, such a relationship can exist, but it derives from a stress driver that we have not previously considered: vertical lithospheric loads.

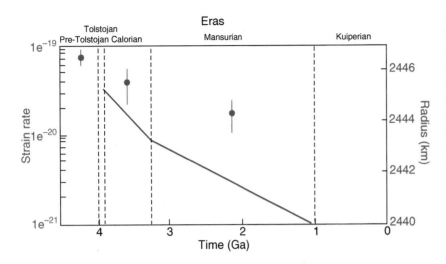

Figure 9.7 Graph showing decrease in magnitude of contraction by cooling through time for Mercury. Blue line shows change in radius with time, and red dots show average strain rates for particular times. From the right axis, one can see that the planetary radius decreased by up to 7 km. Modified from Crane and Klimczak (2017).

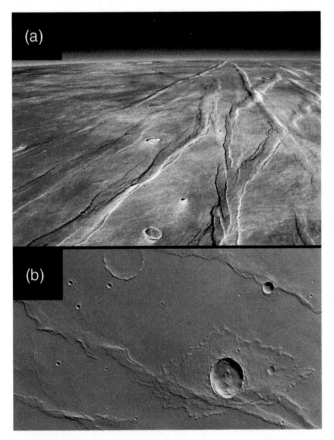

Figure 9.8 Common types of tectonic structures on Mars. (a) Complex rift system of parallel normal faults. (b) Parallel wrinkle ridges. NASA images.

The outer lithosphere can be treated as an elastic shell with flexural rigidity that means that it has a bending strength and will support loads over geologically long periods of time (Section 8.3). While this bending behavior can create suites of structures with concentric and/or radial geometries around the causative load (e.g., advancing thrust system, intruded igneous province, emplaced wedge of sedimentary rocks), these tectonic outcomes are still relatively local in character and hence not typically part of our consideration of behaviors for entire planets and satellites. Yet, given the size of Tharsis, the tectonic response of Mars to this massive load is actually planet-scale (Figure 9.9). Such a lithospheric loading creates radial horizontally directly maximum compression outward, which provides the driving stress to form the wrinkle ridges and lobate scarps that are tangential to the load geometry (Figure 9.9a). Likewise, the minimum horizontal stress is tangential to the load geometry, so that grabens can form radially. These grabens may be singular pairs of faults, many clustered pairs, which is typical, and even one of the largest grabens in the Solar System, Valles Marineris. Using the age relationships of these faults to craters and sedimentary units, structural development due to Tharsis-related loading began in the Noachian, peaked in the Early Hesperian and continued into the Amazonian (post-2.5 Ga). So, the primary emplacement of the igneous-related load in the Tharsis region is interpreted to be completed prior to 3.0 Ga, but to have continued with lesser contributions through much of martian history.

Still, we need to remember that Mars experienced contractional cooling as well. Evidence for this typical stagnant-lid behavior is the occurrence of populations of wrinkle ridges in the lava plains of the eastern martian hemisphere away from Tharsis and on the northern lowland plains, where ridge trends do not match the predicted geometries for genesis by Tharsis. Second, in regions around Tharsis, lateral-slip faults, which are near vertical faults with horizontal displacements and are rare for stagnant-lid tectonics, occur with lengths of up to a few hundred kilometers and displacements of 5–10 km (Figure 9.10) (Andrews-Hanna et al., 2008). They are unlike grabens and thrust-related folds because they have

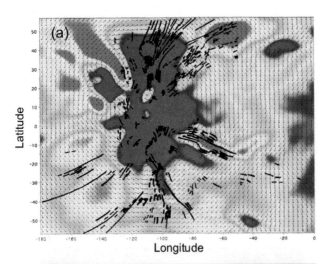

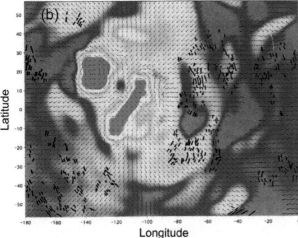

Figure 9.9 Maps showing the results of modeling the strains for flexural loading of Mars by Tharsis. (a) Major volcanoes are located in the purple region. Arrow pairs show the maximum extension directions from the load and are mostly perpendicular to the black bold traces of the major normal faults. (b) Major volcanoes are located in the red regions. Arrows show the directions of maximum contraction from the load and are mostly perpendicular to the short bold traces of wrinkle ridges. After Golombek and Phillips (2010).

both the minimum and maximum stresses trending horizontally, which does not happen with simple cooling contraction where minimum stress is typically vertical or with flexural loading. So, their occurrence implies that the stress fields created by lithospheric loading from Tharsis and cooling contraction operated at the same time for some period, such that their addition created the situation in which both the minimum and maximum stresses were horizontal. This outcome is a very important lesson: Structures may result from more than one driving stress operating simultaneously.

So, in summary, while in comparison to other bodies with stagnant lids Mars can be seen to have had more tectonic deformation for a longer period of its history, the

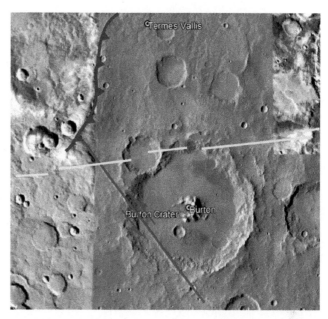

Figure 9.10 Example of evidence for a lateral-slip fault (red line) on Mars. A wrinkle ridge (blue line ornamented with triangles) and a graben system (yellow lines) are present. After Andrews-Hanna et al. (2008).

overall deformation is still very modest, with total strains of only a few percent. Mars in this context can be seen as "active," but it was certainly not a body with an active lid as we will see.

9.5 Structures and Tectonics for Active Lids

Active lids may seem synonymous with plate tectonics. As we will see, however, lithospheres (or more properly, cryospheres) can actively deform in other ways.

9.5.1 Active Lid with Plate Tectonics (Earth)

The signature active-lid tectonic system belongs to Earth and is plate tectonics. Given that a great deal has been written about plate tectonics for the Earth, we simply focus on a few observations that relate to our overall consideration of stagnant-lid versus active-lid tectonics. The cooling Earth provides sufficient energy through mantle convection to create translithospheric faults/shear zones, so that the lithosphere is a series of pieces that move from spreading centers where new lithosphere forms to subduction zones where lithosphere is returned to the mantle. Because the mantle is "self-heating" due to radioactive decay, more heat flows out of the top of the mantle and to the Earth's surface than into the bottom of the mantle from cooling of the core due to its solidification. As a result, downwelling is more important in mantle convection and is enhanced by the insertion of the colder, stronger subducting lithosphere.

BOX 9.1 THARSIS-DRIVEN TECTONIC DEFORMATION: WHEN, WHERE, AND HOW DO WE KNOW?

The formation of Tharsis as a very large igneous province flexurally loaded the martian lithosphere at a planetary scale. While this loading is elastic for small deformations (Section 8.3) and modeled elastically to predict expected strains and structures (Figure 9.9), the magnitude and duration of the Tharsis load created permanent deformation with an impressive suite of related structures. Thus, the formation of Tharsis and its related structures was not simply an elastic event. Loading yielded abundant populations of normal faults and wrinkle ridges, and we would like to document how these structures accumulated through billions of years to better understand the history of loading and, ideally, the formation of the Tharsis bulge.

An approach is to consider the geometries of all related tectonic structures around Tharsis, particularly in the western hemisphere of Mars. Using *Viking*-derived data, Anderson et al. (2001) were the first to use this approach and mapped a total population of about 24,500 structures, of which ~20,000 are normal fault systems and ~4500 are wrinkle ridges. Their analysis was superseded by the study by Bouley et al. (2018), who used data from many recent obiter missions and derivative products including a new Mars geologic map (Tanaka et al., 2014). Their mapped structures are illustrated on a Google Mars basemap in Figure 9.11. Rather than just counting numbers of faults within particular stratigraphic intervals, Bouley et al. determined the lengths of the fault traces, as a more complete proxy for the magnitude of deformation, and considered deformation intensity in terms of fault trace-length as a function of area with respect to time (Figure 9.12). Following these results, we would first determine the age of each structure by correlating this age to the youngest age of the rocks deformed by it. With these age determinations, we then define six stages of structures (Figure 9.12): Early/Mid-Noachian, Late Noachian, Early Hesperian, Late Hesperian, Early Amazonian, and Middle Amazonian to the present (refer to Figure 3.7 for the martian timescale). In terms of structural abundance (Figure 9.12), fault trace-length intensity peaked in the Early Hesperian for both radial extensional faults and concentric contractional faults (wrinkle ridges). If structure abundance equates to the magnitude of deformation, then Tharsis loading has caused extensional faulting from Early Noachian time onward, whereas contractional faulting was concentrated in Late Noachian through Hesperian time. We should also note that the rate of fault formation would have decreased markedly in the Amazonian, given the long duration of that time period (Bouley et al., 2018).

As determined by both Anderson et al. (2001) and Bouley et al. (2018), the center of igneous activity and the related development of extensional and contractional faults in Tharsis moved within the region through time (Figure 9.11). Consequently, questions remain for future researchers as to how the change in load center positions correlates with igneous eruptions and mantle circulation, and why crustal contraction related to Tharsis loading is focused in time, unlike crustal extension.

Figure 9.11 Mars paleotectonic map, centered on Tharsis, showing populations of normal (radial extensional) faults and wrinkle ridges (concentric contractional structures). Base image is from Google Mars, and fault traces are from a .KMZ file in the Supplemental Materials of Bouley et al. (2018). Colors represent the different time stages for the sequence of structure formation (periods given in Figure 9.12).

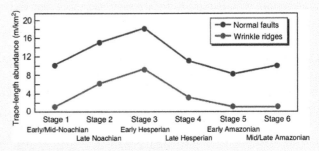

Figure 9.12 Graph of the changing abundance of fault traces for different time stages and the ages of the host units for the faults. Data are modified from Bouley et al. (2018).

A signature of plate tectonics is the importance of horizontal displacements in the lithosphere along major lateral-slip transform faults, but also at the top of subduction zones, along the major thrust faults of mountain belts and the major normal faults of extensional terranes. Individually, these faults may have up to hundreds of kilometers and collectively thousands of kilometers of horizontal displacement. While it is possible to have active-lid tectonics without these major horizontal displacements, they are a requirement of active lids with plate tectonics.

Another important aspect of active-lid tectonics for the Earth is that lithosphere is recycled and replaced. While continental lithosphere due to its buoyancy may survive on the outside of the planet for billions of years, oceanic lithosphere is typically recycled in fewer than 250 million years. Thus, even though the lithosphere has some mechanical independence from the Earth's interior due to the presence of underlying asthenosphere, it is part of the integrated heat-flow system for the planet. It assists with heat outflow at spreading centers and heat return at subduction zones. So, again, while it is possible to have active-lid tectonics without lithospheric recycling, the recycling is a requirement of plate tectonics. These distinctions between the existence of active lids with and without plate tectonics are important, as we will see when considering active icy satellites.

9.5.2 Active Lid without Plate Tectonics (Europa)

Probably the body with the greatest activity but no plate tectonics in the Solar System is Io. However, as its development is mainly the result of essentially vertical tectonics, including a daily topographic tidal range of 100 m, related to igneous-dominated rather than tectonically dominated processes, we will not consider its development.

Instead, we will consider two satellites with icy lithospheres: Europa in the Jovian system and Enceladus in the Saturnian system. Both are active bodies, but they differ in structural and tectonic development due to factors such as distance from their host planet, the number of other satellites with which they interact gravitationally, and the thickness variation of their lithospheres.

Europa is clearly a recently active if not a currently active tectonic body, given a very limited number of craters indicating an average surface age of 40–90 Ma. It is also covered by structures (Figures 3.10, 9.13). In fact, these structures have sufficient complexity to have spawned a special vocabulary exceeding 40 terms just for naming their types. If the driving stresses for deformation mainly yielded simple effects due to the diurnal

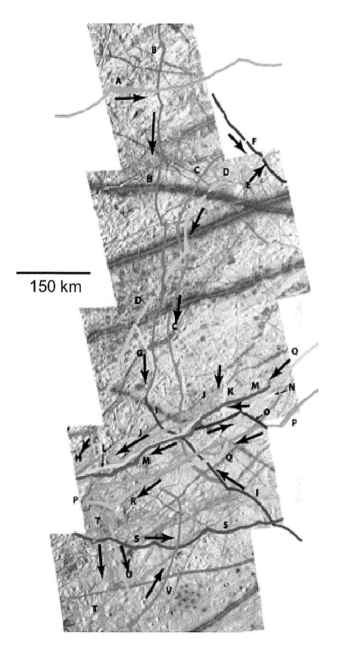

150 km

Figure 9.13 Overlapping cycloid traces in the Bright Plains area of Europa (see also Figure 3.10). Arrows are interpreted propagation directions for each trace. Modified from Groenleer and Kattenhorn (2008).

tides, the dominant structures would likely be **cycloidal traces** (Figure 9.13). They consist of multiple convex segments, where each segment has a length of about 50–200 km. A segment initiates when the diurnal tidal stress generates a sufficient tension to overcome ice strength of about 25 kPa. Segments have a direction of curvature that is a function of its propagation direction and the rotation of the stress field from the ongoing diurnal tides, a propagation velocity of about 1–3 km/h, and a depth of about 50 m up to a few hundred meters.

Propagation only occurs for the portion of each orbit when the critical ice strength is exceeded, and propagation typically terminates with the formation of a tailcrack. In the next orbit, when the critical strength is again exceeded, propagation exploits that tailcrack, creating an apex and the next curved segment. Consequently, traces in Figure 9.13 show sequences of segments recording from four to more than ten orbits when fracturing occurred (Groenleer and Kattenhorn, 2008).

The array of cycloidal traces in Figure 9.13 has a great range of trends that should not be possible for a simple stress history involving only waxing and waning diurnal tidal stresses. This assemblage of different trace trends requires that the icy lithosphere of Europa is changing position with respect to its tidal bulges. This repositioning is achieved by nonsynchronous rotation, and this population of traces has been interpreted to represent almost two entire rotations of the icy lithosphere about the satellite center that each took about 10–25 million years. This interpretation could be further complicated by the occurrence of true polar wander and local stress variations due to thermal/buoyant upwellings that are both thought to occur in Europa. Consequently, at any moment for a point on Europa, the stress state is the summation of several causes, and as that summation is more complex than, for example, we considered for Mars, we observe a more complex structural suite than found on Mars.

The complexity results from the facts that structures of different types can be occurring simultaneously at different locations because stress conditions vary from point to point, and further, structures at particular points reactivate as different structures because stress conditions at those locations also change through time. For example, ridges and **bands**, which are two very important types of Europan structures, can simply form or result from reactivation of cycloid traces. Then, they can be reactivated as lateral-slip faults or even reverse faults. Ridges are the most common structures on Europa (e.g., Figure 3.11). They can occur singly, or commonly as double ridges or ridge complexes, where this increasing geometric complexity is also thought to be a developmental sequence. Consequently, they record the complexity of a stress field that represents several drivers and is changing every orbit (3.55 Earth days). For example, the Bright Plains region records at least two complete lithospheric rotations with respect to the satellite center due to nonsynchronous rotation, while experiencing at least one billion diurnal tidal stress cycles (Figure 3.11) (Kattenhorn, 2002). These ridges represent material accumulation from the lithospheric interior onto the Europan surface, but their cause is not well understood. The material may derive from reactivation of the ridges during diurnal tidal events as faults with lateral slip,

Figure 9.14 Example of band structure on Europa. The band is a location where lithospheric dilation occurred and new icy lithosphere was added at the surface (Collins, 2010). NASA image.

if that slip exceeds 10 cm, generating sufficient heat to weaken and even melt ice, and forming material that rises to the surface. Alternately, some or all of the material may come out of the lithosphere from where a ridge fracture connects to upwelling diapiric ice or even an intruding water "magma" from below.

As common as ridges are on Europa, one can argue that bands are more important in terms of their tectonic implications (Figure 9.14). Bands are up to 30 km wide and hundreds of kilometers long, with a central trough, matching hummocky textural zones across the trough, boundaries that may include narrow parallel ridges and troughs representing normal faults, and most importantly older structures that fit back together when the band is removed. Consequently, bands are dominantly dilational structures that are places where younger lithosphere was inserted frequently as weak ice or water that froze to preserve the structural opening. Given that their widths can be up to tens of kilometers and that locally they represent 40 percent of the satellite's surface area, they indicate that Europa has experienced significant additions of icy lithosphere to its surface.

Such major additions immediately create a problem because as Europa is not an expanding body, some process or processes must shorten and/or remove lithosphere from the satellite surface. Large folds with wavelengths of about 25 km but modest amplitudes occur, but they do not represent nearly enough shortening to match the dilation recorded by bands. Similarly, thrust faults and reverse faults that reactivated other structures occur, but their effects are quite modest and insufficient. Recently, another process has been proposed to remove lithosphere from the surface, which is **subsumption** with subsumption bands

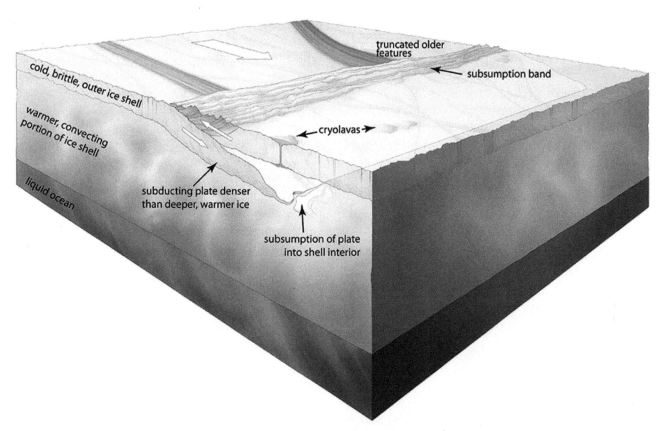

Figure 9.15 Conceptual block diagram of the process of subsumption that recycles brittle, icy lithosphere into the underlying ductile lithosphere, with creation of cryolavas that create volcanogenic landforms in the overriding brittle lithosphere. Reprinted by permission from Springer Nature: Nature Geoscience, Evidence for subduction in the ice shell of Europa, Simon A. Kattenhorn and Louise M. Prockter, Copyright (2014).

and cyrolava bodies (Figure 9.15) (Kattenhorn and Prockter, 2014). Subsumption bands contain elongate hummocks oriented subparallel to the margins, smooth regions with pits, and margin-parallel linear features; they lack central troughs and bilateral symmetry of structures within the bands, and have somewhat elevated relief as compared to adjacent regions. Most importantly, they juxtapose regions with different features that do not match across them. Given that the bands lack the relief and structure complexity to account for "missing lithosphere," this lithosphere is proposed to return by subsumption into the lithosphere. Essentially, the outer, cold, brittle lithosphere is returned, warmed and converted to either ductile lithosphere that remains in the lithosphere or provides melts that rise to form cyrolava bodies in the overriding lithosphere. Given this proposed "pairing" of bands with lithospheric addition and subsumption bands with lithospheric removal, Europan tectonics would bear kinematic similarities to plate tectonics. The key difference of this proposed system is that the addition and subtraction of lithosphere would work entirely within the lithosphere of Europa, whereas plate tectonics on Earth involve the lithosphere and the entire underlying mantle.

9.5.3 Partially Active Lid without Plate Tectonics (Enceladus)

Unlike Europa, Enceladus only has major tidal engagement with one other satellite around its host planet, and has about 1/216th of the volume and 53 percent of the density. So, Enceladus has not experienced all of the structural complexity of Europa, but the effects of diurnal tides, nonsynchronous rotation, and true polar wandering have structurally marked the surface of this satellite. Further, Enceladus is currently active – plumes of vapor and ice crystals were observed escaping from linear features ("**tiger stripes**") in its South Polar Terrain by *Cassini* (Figure 9.16).

Perhaps the easiest way to recognize the difference in tectonic histories between Europa and Enceladus is that whereas Europa has a complex array of structures across its entire surface, Enceladus has four quite different terrains. They range from a cratered terrain with crater abundances indicating a mean maximum surface age of a few billion years, to the South Polar Terrain that is devoid of craters, so that it is very young (Figure 9.17). The cratered terrain does contain troughs, pit chains, and merged pit chains that are parallel and indicate at least modest recent extension as they deform surface regolith, possibly with a tensional driving stress (Nahm and Kattenhorn, 2015). However, their occurrence does not disturb the determination of an ancient maximum age for the heavily cratered terrain.

The leading (western) and trailing (eastern) tectonized terrains contain more and a greater variety of structures than the cratered terrain (Figure 9.17). These structures include bands, **chasma**, ridges, scarps, and troughs with and without pits. The majority of these structures have been recently interpreted to result from extension without involving lithospheric surface additions, unlike Europa.

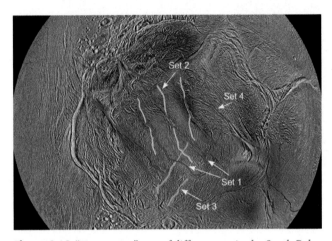

Figure 9.16 "Tiger stripe" sets of different age in the South Polar Terrain of Enceladus. By implication, the South Polar Terrain has not experienced any significant true polar wander during the development of this structural sequence. Relative age information from Patthoff and Kattenhorn (2011).

The South Polar Terrain covers that polar region up to about 55 °S latitude, where a variety of topographic features and structures create a boundary between it and the other terrains. This terrain contains abundant double ridges and subdued double ridges, and the subdued ridges are thought to be older (Patthoff and Kattenhorn, 2011). All double ridges are found to have orientations modes where the youngest mode is the present tiger stripes (Figure 9.16). A total of four modes exist and are interpreted to represent a sequence of events in which a fracture set forms parallel to the tidal bulges. As time passes, that fracture population, with its ancient tiger stripes, rotates due to nonsynchronous rotation from being parallel to the bulges and ceases to fracture. Then, a new fracture set with tiger stripes forms to accommodate deformation and plume occurrence. This interpretation has identified about 153° of nonsynchronous rotation with decreasing angular increments between modes that may represent greater tidal heating, weaker and/or thinner ice lithosphere, or change in orbital eccentricity, for example. The duration of this rotation is unknown, but in any case the combination of diurnal tides and nonsynchronous rotation indicates the presence of a global underlying ocean. Yet, this ocean, if present under all of Enceladus, has not enabled Enceladus to be tectonically active everywhere and eliminate the cratered terrains. In summary, while Enceladus has one terrain, the Southern Polar Terrain, that is possibly even more active than any location on Europa, it does not have evidence for body-scale lithospheric recycling or even satellite-wide tectonic resurfacing. Consequently, the

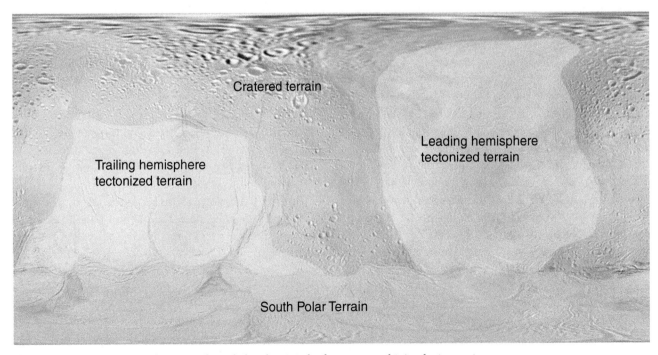

Figure 9.17 Surface topographic map of Enceladus showing the four topographic/geologic terrains.

strength of the icy lithosphere for Enceladus has been sufficient to prevent tectonic deformation driven by diurnal tides, nonsynchronous rotation, true polar wandering, density inversion, convection, and other causes over much of its surface, as compared to Europa.

9.6 Stagnant Lid Possibly Active in the Past? (Venus)

Despite having seen through the clouds of Venus using the radar data of the *Magellan* mission, the planet is still quite enigmatic. Venus has clear similarities to the other terrestrial bodies with stagnant lids, such as Mercury and Mars. It is covered in basalt, has abundant wrinkle ridges, has abundant small grabens and linear extensional fractures, lacks structures like major strike-slip faults associated with active tectonics, and has an abundance of volcanogenic surface features that range from only 1 km in width to volcanic complexes that span 1500 km (Figure 3.9). These features can be interpreted to mean that Venus is another body with thermally driven vertical tectonics shelled by an igneously built stagnant lid (McGill et al., 2010) . Further, the present atmosphere and very hot surface of Venus lack water, which implies that the current lithosphere of Venus is quite dry compared to Earth, and hence strong, which is appropriate for a stagnant lid.

Yet, Venus in terms of size and mass is very similar to Earth, the home of plate tectonics. Further, Venus has a number of unique features that are absent on Mercury, the Moon, or Mars:

1. Volcanic rises: About ten of these features occur on Venus with volumes of 10^4–10^6 km^3 and diameters of 1400–2500 km that overlie positive gravitational anomalies, which are consistent with active underlying mantle plumes.
2. Coronae (Figure 9.18a): These circular features with diameters of 50–500 km may have raised or depressed centers. They are too small and in many cases too close together to be plume-related volcanogenic features. They are commonly associated with significant volcanic flows that were typically deposited prior to concentric topographic rims that are commonly found around coronae.
3. Chasmata (Figure 9.18a): Venus has five large, complex normal fault systems with lengths of thousands of kilometers and reliefs of up to 7 km, along with a number of other, somewhat smaller systems that are preferentially located in equatorial highland regions.
4. Tesserae (Figure 9.18b): These terrains are mostly located in the crustal plateaus that topographically compose about 10 percent of the planet surface, although a few **tesserae** occur in the abundant plains of Venus. They are

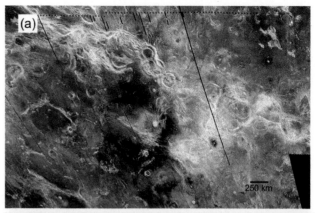

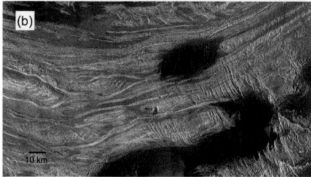

Figure 9.18 Structural features on Venus. (a) Coronae (circular structures) and chasmata (complex white linear traces from the upper left to middle right of the image). Note the scale and hence large sizes of these structures. (b) Tesserae, showing broad ridges with open folds cut by north-northwest-trending grabens. NASA *Magellan* radar images.

commonly overlapped by the adjacent terrains so that the tesserae are the oldest features. Further, they are structurally complex with sinuous topographic antiforms dissected by one or more sets of grabens and fractures.

5. A limited population of large craters: Venus has fewer than 1000 impact craters, which are all larger than 2 km in diameter because the venusian atmosphere destroys smaller possible impactors. This population is randomly distributed across the planet surface and analysis shows that their presence means that, on average, the present surface of Venus is less than one billion years old, so the surface is young compared to other terrestrial bodies with stagnant lids. Further, the planet may have been completely repaved with igneous rocks about one billion years ago.

Still, the occurrence of volcanic rises, coronae, and chasmata would be consistent with stagnant-lid behavior, if Venus is considered to still be thermally active internally, which is appropriate for a planet that is similar in size, mass, and composition to the Earth. Where the applicability of the stagnant-lid model becomes more uncertain is when considering the tesserae, the relatively young overall age of the venusian surface as compared to other terrestrial bodies with stagnant lids, and, surprisingly, both the existence of the stagnant lid and the lack of present-day water.

The Earth loses about 70 percent of its heat at spreading centers and subduction zones while constantly replenishing surface liquids and gases from mantle volatile escape. By comparison, Venus lacks such an efficient means for releasing mantle heat or mantle volatiles. Thus, heat could be accruing beneath the present venusian lithosphere that could trigger another surface repaving event by igneous flows that would again leave isolated tesserae and result from thermally driven vertical tectonics without the need for active tectonics. Alternately, it is possible that this trapping of heat triggers a phase transition in which the lithosphere is not a distinct layer but rather compositionally continuous with the underlying mantle, allowing whole-mantle circulation to entrain the lithosphere as a part of an actively convecting system in which active tectonics temporarily operates and resurfaces the planet. Further, this possibility could have been more likely in the past because the present deuterium/hydrogen ratios for Venus indicate that it likely had about 100 times more water than it presently has. The past presence of water would have favored a weaker lithosphere and a more mobile mantle that would have supported active tectonics driven by mantle convection. Still, Venus is an enigmatic planet: It lacks erosion so a long-term complex history is preserved on its surface; it provides little data about rock compositions and ages as a function of terrains, structures, and volcanogenic features; it lacks geophysical data about its interior; and it remains shrouded in clouds.

Summary

Terrestrial planets and icy satellites are distinguishable by whether they function or functioned as stagnant- or active-lid bodies. This choice of behavior is driven by whether stress-generating agents such as mantle convection, diurnal tides, or nonsynchronous rotation are able to overcome the strength of the exterior shell (lithosphere) to trigger active-lid tectonics. If strength prevails, stagnant-lid behavior occurs. In that case, the planet or satellite is typically intensely ornamented by craters with radial and concentric structures, plus wrinkle ridges and lobate scarps that accommodate the very modest shrinkage of particularly the terrestrial bodies such as Mercury. This deformation pattern is enhanced when planetary-scale lithospheric loading, such as for Tharsis on Mars, or continuing active mantle heat circulation, as for Venus, occurs. In these cases, structures such as wrinkle ridges and graben fault systems are more abundant, with distributions that relate to the lithosphere load or the history of lithospheric responses to continuing mantle convection and plumes. Yet, when the driving stress prevails, active-lid tectonics enhances the structural complexity on the body surface. For example, Europa has a complex array of structures reflecting the interaction of diurnal tides, nonsynchronous rotation, and true polar wander in both space and time around the satellite, and a very limited population of young impact craters. Alternately, the Earth has some ancient preserved continental regions that lack craters due to complex geologic histories and an active atmosphere, but more importantly returns more than about 70 percent of its lithosphere into the interior every 250 million years as a function of convection-driven plate tectonics. Even where only partially active, the structure complexity increases greatly, as with the tiger stripes and related structures in the South Polar Terrain of Enceladus. Finally, the temporary occurrence of active-lid tectonics in the past might even explain the potentially episodic replacement of the complex geologic surface of Venus.

Review Questions

1. Considering terrestrial planets and satellites with radii greater than 200 km, are surfaces with stagnant-lid or active-lid tectonics more abundant? Why do you believe that one behavior has been more abundant than the other?

2. Why is it reasonable to say: "From a material perspective, an icy lithosphere is more likely to show evidence for active-lid tectonics than a rocky lithosphere."

3. Contrast the roles of thermal and tidal sources as driving stresses for triggering active-lid tectonics.
4. What are the similarities and differences between stagnant-lid surfaces for icy satellites and rocky planets?
5. Why is Europa interpreted to be a very tectonically active icy satellite, whereas Enceladus is a selectively active satellite?
6. Develop two lists: (a) evidence favoring Venus having always been a stagnant-lid body; and (b) evidence favoring Venus having one or more episodes of active-lid behavior in its past and/or future. Which explanation do you favor and why?

SUGGESTION FOR FURTHER READING

Watters, T. R., and Schultz, R. A., eds. (2010) *Planetary Tectonics*. Cambridge: Cambridge University Press. Chapter 3 presents information about the structural geology and tectonics of Venus, whereas Chapter 5 does the same for Mars, and Chapter 7 considers the icy satellites. Chapter 9 considers the strength and deformation of lithospheres.

REFERENCES

Anderson, R. C., Dohm, J. M., Golembek, M. P., et al. (2001) Primary centers and secondary concentrations of tectonic activity through time in the western hemisphere of Mars. *Journal of Geophysical Research*, **106**(E9), 20563–20585.

Andrews-Hanna, J. C., Zuber, M. T., and Huack, S. A. (2008) Strike-slip faults on Mars: observations and implications for global tectonics and dynamics. *Journal of Geophysical Research*, **113**. DOI: 10.1029/2007JE002980.

Boettcher, M. S., Hirth, G., and Evans, B. (2007) Olivine friction at the base of oceanic seismogenic zones. *Journal of Geophysical Research: Atmospheres*, **112**, 1205–1218.

Bouley, S., Baratoux, D., Paulien, N., et al. (2018) The revised tectonic history of Tharsis. *Earth and Planetary Science Letters*, **488**, 126–133.

Byrne, P. K., Klimczak, C., Sengor, A. M. C., et al. (2014) Mercury's global contraction much greater than earlier estimates. *Nature Geoscience*, **7**, 301–307.

Collins, G. C., McKinnon, W. B., Moore, J. M., et al. (2010) Tectonics of the outer planet satellites. In *Planetary Tectonics*, eds. Watters, T. M., and Schultz, R. A. Cambridge: Cambridge University Press, pp. 264–350.

Crane, K. T., and Klimczak, C. (2017) Timing and rate of global contraction on Mercury. *Geophysical Research Letters*, **44**, 3082–3089.

Golombek, M. P., and Phillips, R. J. (2010) Mars tectonics. In *Planetary Tectonics*, eds. Watters, T. M., and Schultz, R. A. Cambridge: Cambridge University Press, pp. 183–232.

Greenberg, R., Geissler, P., Hoppa, G., et al. (1998) Tectonic processes on Europa: tidal stresses, mechanical response, and visible features. *Icarus*, **135**, 64–78.

Groenleer, J. M., and Kattenhorn, S. A. (2008) Cycloid crack sequences on Europa: relationship to stress history and constraints on growth mechanics based on cusp angles. *Icarus*, **193**, 158–181.

Hansen, L. N., Conrad, C. P., Boneh, Y., et al. (2016) Viscous anisotropy of textured olivine aggregates: 2. Micromechanical model. *Journal of Geophysical Research: Solid Earth*, **121**. DOI: 10.1002/2016JB013240.

Kattenhorn, S. A. (2002) Nonsynchronous rotation evidence and fracture history in the Bright Plains region, Europa. *Icarus*, **157**, 490–506.

Kattenhorn, S. A., and Prockter, L. M. (2014) Evidence for subduction in the ice shell of Europa. *Nature Geoscience*, **7**, 762–767.

McGill, G. E., Stofan, E. R., and Smrekar, S. E. (2010) Venus tectonics. In *Planetary Tectonics*, eds. Watters, T. M., and Schultz, R. A. Cambridge: Cambridge University Press, pp. 81–120.

Nahm, A. L., and Kattenhorn, S. A. (2015) A unified nomenclature for tectonic structures on the surface of Enceladus: implications for Enceladus's tectonics and ice shell. *Icarus*, **258**, 67–81.

O'Neill, C., Jellinek, A. M., and Lenardic, A. (2007) Conditions for the onset of plate tectonics on terrestrial planets and moons. *Earth and Planetary Science Letters*, **261**, 20–32.

Patthoff, D. A., and Kattenhorn, S. A. (2011) A fracture history on Enceladus provides evidence for a global ocean. *Geophysical Research Letters*, **38**, L18201. DOI: 10.1029/2011GL048387.

Tajaddine, R., Soderlund, K. M., Thomas, P. C., et al. (2017) True polar wander of Enceladus from topographic data. *Icarus*, **295**, 46–60.

Tananka, K. L., Skinner, J. A., Dohm, J. M., et al. (2014) Geologic map of Mars. US Geological Survey Scientific Investigations Map 3292.

Weller, M. B., and Lenardic, A. (2012) Hysteresis in mantle convection: plate tectonic systems. *Geophysical Research Letters*, **39**. DOI: 10.1029/2012GL051232.

10

Planetary Igneous Activity

All planets, and many moons and asteroids, have experienced igneous activity. Magma compositions on Earth vary widely, reflecting different melting mechanisms in the various tectonic settings, different source compositions, and the effects of magmatic processes like fractional crystallization and assimilation. Most magmas are emplaced in plutons rather than erupt on the surface. On other planets, we study volcanic constructs and rocks, because plutonic rocks are not commonly exposed. Eruptive styles vary with each planet, depending on neutral buoyancy zones in the subsurface, the amount of volatiles in magmas, gravity, atmospheric pressure, and other factors. Basalts are ubiquitous on all rocky bodies, and fractional crystallization of basaltic magmas has produced cumulates and fractionated residual melts. The formation of abundant, highly evolved felsic magmas, as far as we can presently discern, has been restricted to Earth. On some icy bodies cryovolcanoes erupt cold brines and gases. Volcanism mostly ceased on the Moon when melting retreated to the deep interior, on Mercury when global contraction closed pathways for magma ascent, and on asteroid Vesta when the radiogenic heat sources were exhausted. Magmatic activity continues on Earth and Io, and recent (possibly ongoing) activity occurs on Venus and Mars. Where sufficient information is available to judge, magma compositions appear to have evolved with time, in a manner unique to each body.

10.1 Magmas, Everywhere You Look

Magmatism is a common thread running through the geologic evolution of planets large and small. With the exception of chondritic asteroids and perhaps a few icy moons, all Solar System bodies have experienced some degree of melting. The processes that affect molten magmas once they are generated within these bodies – segregation from the unmelted residue, ascent,

emplacement or eruption, interaction with enclosing rocks, crystallization – are alike in many ways. But that is where the similarities end. Planetary igneous rocks represent the results of nature's experiments conducted under a panoply of physical conditions and involving a wide array of compositions.

We will begin by briefly reviewing magmatic activity on Earth, a logical but perhaps unfortunate choice because terrestrial igneous activity appears to be so much more complex than on other planets. This partly reflects more complete sampling on Earth, not only of volcanic rocks but also of plutonic rocks exposed by tectonic forces; we are mostly restricted to the study of surface volcanism on other planets. But, as we will see, there are also major differences in magma compositions on our world with its jostling tectonic plates, relative to other bodies with stagnant crustal lids.

10.2 Magmatic Activity on the Planet We Know Best

Let us begin by considering why the Earth's interior melts. The increase in both temperature and pressure with depth (the **geothermal gradient**, often abbreviated as "geotherm") is illustrated schematically by a dashed line in Figure 10.1. Note that nowhere does the geotherm cross the dry solidus for peridotite – the line where mantle rock begins to melt. So why does melting occur? Three mechanisms for melting are illustrated by colored arrows. An increase in temperature (labeled "+T"), which we will call hot plate melting, moves a point on the geotherm horizontally across the dry solidus; this occurs when hot material in an ascending plume underplates some portion of the mantle or crust. Crystallization of the stalled magma is exothermic, and the heat generated can cause melting of the overlying rock. Local increases in radioactive material could also form hot spots in the

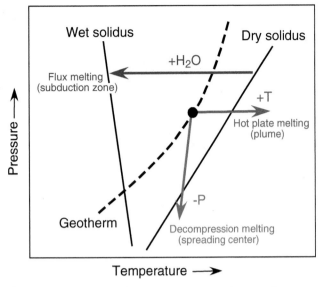

Figure 10.1 Schematic pressure–temperature diagram showing a typical terrestrial geotherm (dashed line) and the mantle solidus (solid line), with and without water present. Three mechanisms for melting are illustrated by arrows.

mantle. Another mechanism is decompression melting (labeled "–P") that occurs when pressure is reduced with little or no change in temperature, also causing that point to cross the dry solidus; this happens at a spreading center, where rifting thins the crust and thus diminishes the pressure on the rocks below. Rapidly ascending mantle plumes can also melt by decompression. The third mechanism displaces the solidus itself, rather than some point on the geotherm. Adding water (labeled "+H_2O") lowers the melting temperature, shifting the melting curve as shown by the wet solidus in Figure 10.1, and results in flux melting; this occurs in a subduction zone, where water is driven out of the descending plate and invades the hot, overlying mantle wedge.

The relative volumes of magmas generated by these mechanisms in different tectonic environments vary considerably, as illustrated by the boxes in Figure 10.2. Spreading centers (mostly mid-ocean ridges, although nascent divergent margins also occur on some continents) account for nearly two-thirds of the Earth's present-day magma production. Subduction zones account for most of the rest, with hot spots (plumes) a distant third. During certain periods, however, plume volcanism has been much more important, producing vast volcanic outpourings over just a few million years. This occurs when plume heads approach the surface. Regardless of the tectonic setting, most ascending magmas never actually erupt onto Earth's surface, instead crystallizing as plutons. On average, the ratio of the volumes of plutonic to volcanic rocks is estimated at ~5:1 (White et al., 2006); other planets may have different ratios.

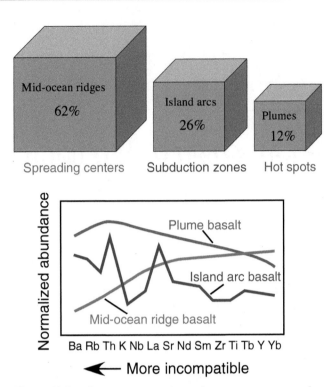

Figure 10.2 Volumetric proportions of magmas on present-day Earth vary with tectonic setting, as illustrated by the sizes of the boxes. The lower diagram shows how different tectonic settings impart distinctive trace element patterns to magmas, reflecting differences in the compositions of mantle source regions and melting mechanisms.

Mantle-derived magmas formed by all three melting mechanisms are basaltic. That is not surprising, since partial melting of ultramafic rocks always yields basalt. However, the composition of the basalt varies in alkalis ($Na_2O + K_2O$) and SiO_2, depending on the degree of partial melting, pressure (depth), and composition of the mantle source. Tholeiitic basalts are lower in alkalis and higher in silica than alkali basalts, and form at higher degrees of melting and/or at shallower depths. The mantle source regions in different tectonic settings also impose distinctive trace element patterns in basalts, as illustrated in the lower part of Figure 10.2. This diagram plots the abundances of trace elements, normalized to chondrites, and arranged according to their incompatibility in major minerals. Mid-ocean ridge basalts, formed at spreading centers, are depleted in the most incompatible elements. This pattern mimics the composition of their upper mantle source regions, which were previously partially melted and depleted in incompatible elements that are now sequestered in the continental crust. Conversely, plume basalts are enriched in **incompatible elements**, reflecting low degrees of partial melting in the lower mantle, which did not participate in the formation of continental crust. Island arc basalts, formed at

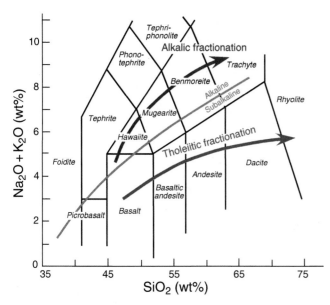

Figure 10.3 Alkalis versus silica diagram, commonly used to classify volcanic rocks, showing the boundary between alkaline and subalkaline compositions and typical fractionation paths (arrows) for tholeiitic and alkali basaltic magmas.

subduction zones, show a spiked pattern reflecting the depletion or enrichment of elements according to their solubilities in water, as appropriate for flux melting.

Once primary magmas form and separate from their source regions, they commonly undergo chemical changes. The segregation of early-formed crystals from the melt (fractional crystallization) modifies the composition of the residual liquid. In this way, a variety of igneous rocks can arise from basaltic magmas, as illustrated by the fractionation arrows in Figure 10.3. This alkali–silica diagram is used to classify volcanic rocks on Earth, and it works for planetary lavas as well. The green line in this figure separates tholeiitic and alkali basalts and their fractionation products – the subalkaline and alkaline trends. Fractional crystallization also increases the iron contents of most residual melts, because early-formed olivine and pyroxene are magnesium-rich. However, fractionation of basalt in a subduction zone occurs without progressive iron enrichment, because a higher oxidation state causes early crystallization of iron oxides. This produces the so-called calc-alkaline trend on diagrams containing iron.

Magmas can also assimilate or melt the surrounding rocks, producing hybrid compositions. Assimilation or melting of continental crust by invading basaltic magma produces andesite, dacite, and rhyolite, and their plutonic equivalents (diorite, granodiorite, and granite, respectively). Derivation of magmas from different parts of the mantle, as well as their interactions with the continental crust, imprint melts with distinctive ratios of radiogenic

to nonradiogenic isotopes, such as $^{87}Sr/^{86}Sr$, $^{143}Nd/^{144}Nd$, $^{207}Pb/^{204}Pb$, and $^{208}Pb/^{204}Pb$.

This brief summary does not exhaust the full range of terrestrial igneous rocks, although it encompasses most of them. Ultramafic magmas (komatiites and kimberlites) represent higher degrees of melting than basalts and come from extreme mantle depths. Carbonate melts, formed by liquid immiscibility from CO_2-rich silicate magmas, form carbonatites. Lavas choked with vesicles (pumice) and pyroclastic rocks composed of mineral grains, rock fragments, and glass shards are produced by explosive eruptions of magmas rich in exsolved volatiles (H_2O, SO_2, CO, and CO_2).

10.3 Planetary Volcanism and Eruptive Styles

The morphologies of volcanoes on Earth reflect their styles of eruption. Basaltic magmas with relatively low viscosity produce shield volcanoes or erupt as flood basalts from fissures that generate broad plains. By analogy with Hawaiian lavas, the flow morphologies are characterized as pahoehoe or aa. Silica-rich lavas with high viscosity commonly also contain large amounts of dissolved gases. The production of gas bubbles during ascent results in explosive eruptions of pyroclastic material that form large composite volcanoes and smaller cinder cones. Volcanic explosions also occur if dry magma encounters the water table during ascent, producing maars. The pyroclasts ejected during explosive eruptions are deposited from ground-hugging surges or as airfalls from materials ejected high into the atmosphere. High density in the interiors of fire fountains produces liquid droplets that coalesce into melt ponds, whereas the exteriors of fountains cool to form glasses that are dispersed into mantling deposits. The sheer diversity of terrestrial igneous rock compositions distinguishes them from their counterparts on other planets. Do these features have counterparts on other planets?

10.3.1 Moon

Lunar volcanism following the early magma ocean stage (described in Section 6.3) consisted primarily of lavas filling large impact basins on the nearside, producing the **maria**. Mare lavas cover 17 percent of the lunar surface but constitute only 1 percent of the volume of the crust. Maria formed from multiple thin flows of low-viscosity basaltic magmas. Although it has been argued that the Moon's lower gravity should have produced thicker lava flows, other factors such as magma viscosity may mitigate gravity's effect. Sinuous **rilles** are channels that begin in crater-like sources and fade downslope into smooth mare surfaces. The time delays between the formation of impact basins and their filling by mare basalts indicate that the impacts did not directly cause the melting. Instead, magmas formed through heating of the mantle by radioactive isotope decay and ascended using the extensive fracture systems beneath impact basins.

Large, regional pyroclastic units, commonly called "dark mantle deposits," and small cinder cones constitute lunar eruptions powered by high contents of CO or H_2. Pyroclastic eruptions that would have generated localized cones on Earth were dispersed over a much larger area on the Moon, owing to its lower gravity and lack of an atmosphere (Lopes et al., 2013). The largest lunar pyroclastic deposit, on the Aristarchus Plateau, covers ~50,000 km^2. On a smaller scale, fire fountaining has produced glass beads, like the *Apollo 17* orange glass described in Section 10.4.1.

Magmatism on the Moon largely ceased several billion years ago. However, new observations of uncommon, irregularly shaped mare patches with few impact craters have been interpreted to represent basaltic eruptions within the past 100 Ma (Braden et al., 2014).

10.3.2 Mercury

Smooth volcanic plains (Figure 10.4a) cover about 40 percent of the surface of Mercury, mostly in the northern polar region and within the huge Caloris basin (Denevi et al., 2013). These lavas form rolling plains with wrinkle ridges and host fewer impact craters than other units. These magmas are inferred to have erupted from sunken vents, rather than from volcanic edifices. Crater counting suggests that all these lavas were emplaced at 3.7–3.8 Ga,

indicating vast outpourings at that time. Contraction of the planet's radius by as much as 7 km as it cooled, evidenced by a global network of thrust faults, shut off volcanic eruptions and forced magmas to intrude as extensive sills.

Older units on Mercury are described as intercrater plains and heavily cratered terrain, and these too appear to be volcanic in origin (Murchie et al., 2015). They uniformly have ages of ~4.1 Ga, so global volcanism at that time buried the pre-existing surface.

Although Mercury was long thought to lack the volatiles that would have powered explosive eruptions, deposits of pyroclastic ash – the telltale signature of explosive volcanism – have been imaged by the *MESSENGER* spacecraft (Kerber et al., 2011). The pyroclastic deposits occur as bright, spectrally reddish materials around volcanic vents (Figure 10.4b). Crater counts suggest these eruptions occurred between 3.5 and 1.0 Ga, so they continued long after effusive eruptions ceased. The volatiles driving the explosive eruptions on Mercury could have been oxidized sulfur and carbon.

10.3.3 Venus

Venus holds the record for the number (nearly 1200 volcanic centers) and diversity of volcanic landforms. These eruptive features, clearly recognizable in radar images, include large shield volcanoes (Figure 10.5a) with calderas and extensive flows, fields containing hundreds of smaller volcanoes (Figure 10.5b), and vast, low-relief wrinkled plains probably emplaced by flood volcanism (Head et al., 1992; Crumpler et al., 1997). Lava channels and rilles meander for great distances, some for thousands of

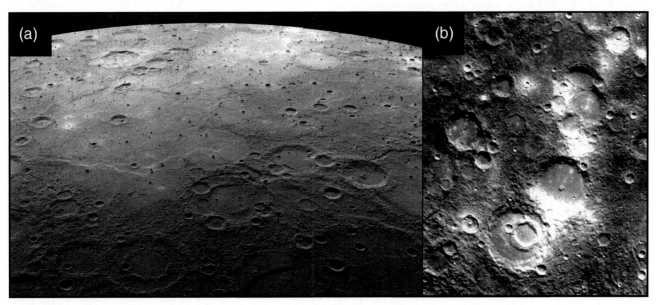

Figure 10.4 Volcanic terrains on Mercury. (a) Smooth volcanic plains occupy much of Mercury's northern hemisphere. (b) Bright pyroclastic deposits. NASA *MESSENGER* images.

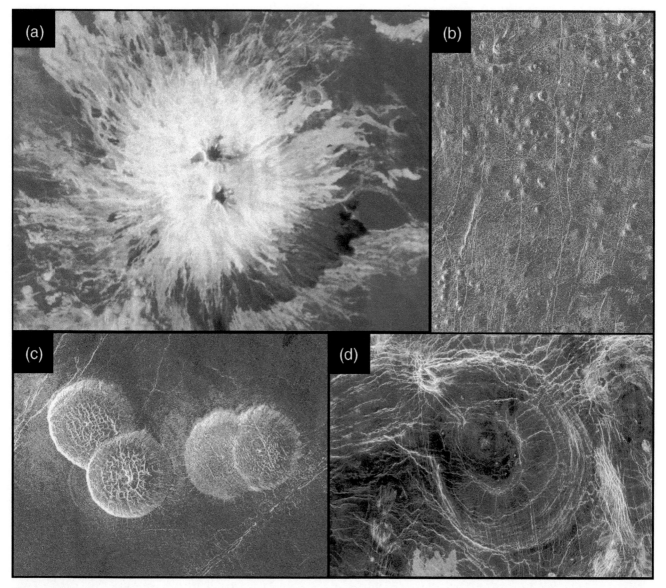

Figure 10.5 Volcanic features on Venus. (a) Aerial view (vertically exaggerated) of 8 km high shield volcano, with lava flow in the foreground. (b) Field of numerous small volcanoes. (c) Pancake domes with fractured tops. (d) Corona. NASA *Magellan* radar images.

kilometers, possibly related to Venus' high surface temperature. Hundreds of pancake-shaped domes (Figure 10.5c) are apparently built of more viscous material. The most unique features are large circular patterns of ridges and troughs called "**coronae**" (Figure 10.5d) that likely formed by upwelling of hot mantle. The coronae are not igneous landforms *per se*, but they are the loci of many volcanic flows and constructs. They represent the major mechanism for heat loss on Venus.

What is not present on Venus is any evidence of pyroclastic activity (Head et al., 1992). The high pressure of Venus' dense atmosphere, likely coupled with the absence of water, has prevented explosive eruptions.

The ages of volcanic features on Venus are difficult to determine from crater counts, because small incoming meteors are screened out by the dense atmosphere. One model posits that volcanism subsided after a planet-wide resurfacing event between half and one billion years ago. In that view, Venus is now volcanically quiescent, and will remain so until it catastrophically erupts at some point in the future. However, new infrared measurements have revealed hot spots (520–830 °C) on the planet's surface, suggesting ongoing magmatism. These spots occur in a region with recognizable extensional tectonics and mantle upwelling, processes associated with volcanism on Earth. They may represent present-day eruptions (Shalygin et al., 2015), spatially contained by Venus' high surface pressure. An observation of a sharp rise in atmospheric SO_2, followed by gradual decrease over the following several years, also hints at modern volcanic activity.

10.3.4 Mars

The most notable volcanic features on Mars are its massive shield volcanoes, including Olympus Mons (Figure 10.6a) and other volcanoes in Tharsis, Elysium, and Syrtis Major (Carr, 2006). The summit of Olympus Mons shows nested calderas and wrinkle ridges (Figure 10.6b). Arcuate lobes extend outward from Olympus Mons, formed by gravity-induced failure of the volcano flanks, as seen in Hawaiian volcanoes. Smaller volcanoes, called "**tholi**," on Tharsis and Elysium may be shields whose lower slopes were buried by later lava flows. Individual flows from Tharsis volcanoes have been mapped using orbital data, and high-resolution images of Arsia volcano show pahoehoe and aa textures (Figure 10.6c). The massive size of these volcanoes (Olympus Mons is the largest volcano in the Solar System at 550 km across and 21 km high) points to a plume origin, without plate movements that would otherwise have produced a chain of smaller volcanoes. Tharsis has been the site of active volcanism for perhaps 4 Ga, and the locations of its eruptive centers have migrated around over that time (Anderson et al., 2001).

Paterae are older, dish-shaped features, typically located in the southern highlands of Mars (Carr, 2006). The paterae are often deeply channelized, suggesting that they are composed of more easily erodible, pyroclastic material (Figure 10.6d). The low gravity and atmospheric pressure on Mars would allow ascending magmas to vesiculate at greater depths, so violent eruptions releasing large amounts of ash are expected. Crater counts on calderas in 20 volcanoes suggest that martian eruptions transitioned from explosive to effusive during Hesperian time (Robbins et al., 2011).

Vast volcanic plains are widespread on Mars and likely erupted from fissures. The plains may exhibit evidence for a volcanic origin in the form of overlapping flow lobes, and supporting evidence in the form of wrinkle ridges consistent with subsidence in a layered volcanic substrate. Collapsed lava tubes on the plains form chains of pit craters.

10.3.5 Io

More than 150 active eruptions have been observed on Io by passing spacecraft and telescopic observations (Lopes et al., 2013), making Io the most volcanically active body in the Solar System. Three styles of eruption have been noted on Io (Lopes et al., 2013): effusive, explosive, and those confined to paterae (Figure 10.7a). The paterae are large vents resembling calderas, but they are not atop volcanoes, and it is not known whether they form by

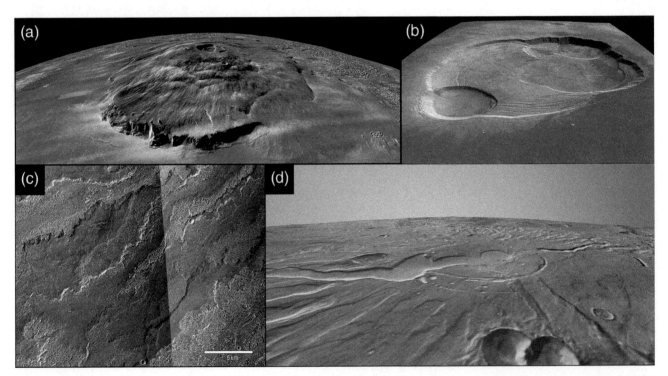

Figure 10.6 Volcanic features on Mars. (a) *Mars Odyssey* THEMIS perspective view of Olympus Mons (840 × 640 km wide). (b) ESA image of nested craters on the summit of Olympus Mons. (c) *Mars Reconnaissance Orbiter* CTX image of smooth pahoehoe and rough aa flows emanating from Arsia Mons. (d) THEMIS perspective view of highly eroded Tyrrhenus Mons (40 km diameter), likely composed of pyroclastic materials.

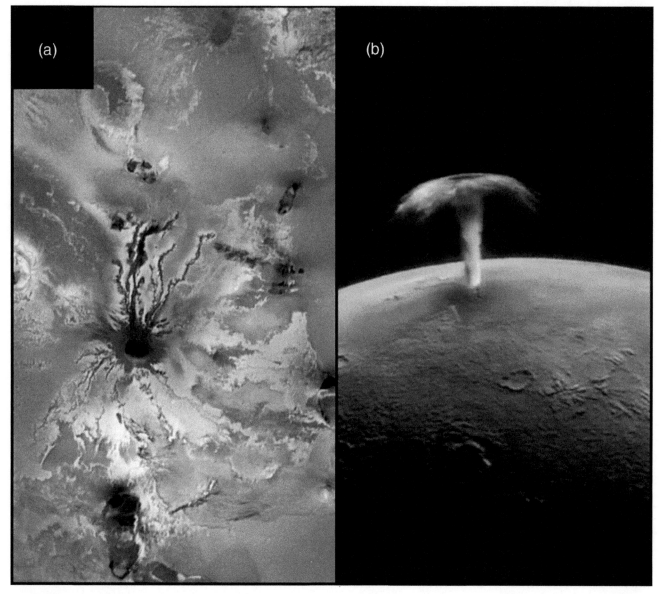

Figure 10.7 Eruptions on Io. (a) Ra Patera, showing a vent with dark lava flows. (b) Prometheus plume, rising 75 km above the surface. NASA *Galileo* images.

collapse or by exhumation of sills to form lava lakes. Dark flows emanate from the paterae. Flow fields exhibit compound flows that extend for tens to hundreds of kilometers, and eruptions can persist for years or decades.

On the other hand, explosive eruptions on Io have short durations, lasting only days or weeks. These eruptions are especially hot and bright. The most spectacular explosions produce umbrella-shaped plumes of gas and dust (Figure 10.7b). Prometheus-type plumes are mostly dust, less than 100 km tall, and collapse to produce bright, circular deposits. Larger Pele-type plumes contain more gas and can reach heights of 500 km; they form red and black pyroclastic deposits, suggesting high sulfur contents.

10.3.6 Comparisons of Eruptive Style

The most prominent volcanic features on Earth, Venus, and Mars are huge shield volcanoes, constructed from many small flows. These eruptions emanate from shallow magma storage chambers, a conclusion reinforced by the occurrence of calderas that collapse when underlying magma bodies are emptied. Such magma chambers form when ascending melts reach a zone of neutral buoyancy. Because the lower gravity on Mars means that rocks at a given depth will be less compacted than on Earth or Venus, ascending martian magmas are likely to encounter neutral buoyancy zones at greater depths. Deeper magma storage regions on Mars also lead to higher rates of eruption, which in turn affect the size of constructed edifices and the lengths of lava flows (Wilson and Head, 1994).

Conversely, lavas on the Moon, Mercury, and Io have erupted from fissures or vents, without producing large volcanic edifices. This implies that shallow neutral buoyancy zones do not occur in the crusts of these smaller bodies. Instead, ascending lavas produce steadily erupting outpourings. Venus exhibits both large volcanoes and clusters of small edifices, depending on whether voluminous magmas formed shallow chambers or slower magma replenishment did not allow shallow reservoirs. The high ratio of plutonic to volcanic rocks on the Earth and inferred for other large planets may not apply to these smaller worlds.

Another difference is the occurrence of pyroclastic deposits. Composite volcanoes that erupt violently are prominent features on Earth, owing to abundant water and to subduction that recycles it into mantle source regions. Ancient Mars, Mercury, and the Moon also hosted pyroclastic eruptions, in each case likely powered by different volatiles. The depths of magma vesiculation and fragmentation differ, depending on gravity's effect on internal pressures. On Venus, however, explosive eruptions are precluded by the high atmospheric pressure. The spectacular volcanic plumes on Io are in a class by themselves and illustrate the effect of very high volatile (sulfur) abundances.

10.4 Planetary Igneous Petrology and Geochemistry

Besides the Earth, we have igneous samples of three other explored bodies – a satellite (Earth's Moon), a planet (Mars), and an asteroid (Vesta). Analyses of these samples provide petrologic and geochemical details that are not otherwise obtainable from remote sensing data. We will focus on the Moon, Mars, and Vesta before turning to Mercury, Venus, and Io, for which we can infer some things about their petrology and geochemistry from remote sensing data. Achondritic meteorites provide additional igneous samples from other, unknown asteroids. Although interesting because they expand the range of magma compositions and conditions, they are not described here; the interested reader can find descriptions of achondrites in Mittlefehldt (2004).

10.4.1 Moon

The ancient (~4.5 Ga) lunar crust (Papike et al., 1998) is composed mostly of ferroan (iron-rich) anorthosite, formed by flotation of plagioclase in an early magma ocean (as described in Section 6.3). Although the anorthosites are cumulates, they rarely preserve igneous textures because impacts have pulverized them, producing breccias (Figure 10.8a). Clasts of **KREEP** basalts and monzodiorites, representing the last dregs of the magma ocean, have also been incorporated into crustal breccias.

Crust formation continued after solidification of the magma ocean, with intrusion of the so-called magnesian and alkali suites, consisting of norites, troctolites, and dunites. These plutons apparently intruded on the anorthositic crust between 4.5 and 4.1 Ga.

Mare basalts (Figure 10.8b), sampled by six of the nine *Apollo* and *Luna* missions (refer to Figure 1.4), consist of pyroxenes and plagioclase, sometimes with olivine or silica and a variety of opaque minerals. The mare basalts are classified based on their bulk TiO_2 contents, readily apparent in the abundance of ilmenite. Radiogenic isotopes in these rocks indicate mare volcanism peaked at 3.8–3.6 Ga and then experienced a smooth decline to 3.2 Ga, finally changing to sporadic activity that persisted until ~1 Ga and perhaps even until the last 100 Ma. High-Ti basalts tend to be older than low-Ti basalts, and experiments indicate that the low-Ti basalts melted at higher pressure and thus come from greater depths. This suggests that mare magmas retreated to greater depths over time, finally ceasing to erupt when the lithosphere became too thick for magmas to transit.

The mineralogy of lunar rocks (Papike et al., 1998) is distinct from terrestrial rocks because of the low oxidation state and very low water contents. Iron occurs as metal (Fe^0) and Fe^{2+} but not Fe^{3+}, forming ulvöspinel (Fe_2TiO_4) instead of magnetite. Titanium (and chromium) also occur partly in low oxidation states (Ti^{3+} and Cr^{2+}), producing armalcolite ($FeTi_2O_5$), named for the *Apollo 11* astronauts: Armstrong, Aldrin, and Collins (Figure 10.8c). No hydrous minerals like amphibole or mica occur, and only apatite contains any water. The plagioclase in lunar rocks is calcic (anorthite), reflecting the depletion of sodium and other volatiles in the Moon.

Volcanic beads (Figure 10.8d) are formed by fire fountains of hot lava erupting into the lunar vacuum. The *Apollo 17* orange glasses, as well as green glasses found in soils at other sites, were derived from deeper in the lunar mantle than the crystalline mare basalts. The lunar glasses have more primitive compositions than the basalts, and may have been parental magmas that fractionated to produce basaltic magmas. They clearly contained volatiles, including water.

10.4.2 Mars

Orbital mineralogical and geochemical studies of igneous terranes on Mars indicate that rocks of basaltic composition are widespread (McSween and McLennan, 2014).

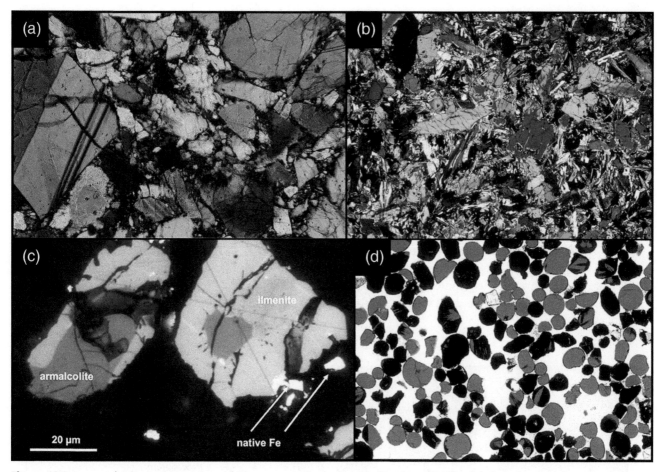

Figure 10.8 Lunar samples in thin section. (a) Ferroan anorthosite breccia, X-polars. (b) Mare basalt, consisting of pyroxene, olivine, and plagioclase, X-polars. (c) Reflected light photomicrograph showing ilmenite grains with cores of armalcolite. Bright white grains are metallic iron. (d) Pyroclastic orange glass beads from the *Apollo 17* site. The black beads are partly crystallized, plane-polarized. NASA images.

A much more informed picture of martian magmatism comes from volcanic rocks analyzed on the surface by rovers. The volcanic rocks studied by the Spirit rover in Gusev crater consist of high- and low-calcium pyroxenes, olivine, and plagioclase, with minor oxides and apatite. These rocks (McSween, 2015) are alkaline basalts and related lavas (trachybasalts and tephrites), commonly porphyritic and sometimes containing vesicles (Figure 10.11a), and pyroclastic rocks. Basaltic volcanic rocks analyzed by the Curiosity rover in Gale crater are strongly alkalic, consisting of pyroxenes, plagioclase, nepheline, olivine, orthoclase, oxides, and apatite. Some feldspar-rich monzonites and diorites were identified in Gale (Figure 10.11b), and feldspathic rocks have been inferred elsewhere from orbiting spectroscopy. However, granites have not been found anywhere on Mars, indicating that fractionation of magmas was limited to feldspar enrichment and did not lead to silica-rich residual melts. The ages of the volcanic surfaces in Gusev and Gale, based on crater counting, are estimated to be ~3.65 Ga

and ~3.7 Ga, respectively, so rocks at both sites represent ancient crust.

The majority of **martian meteorites** (McSween, 2015) are shergottites (Figure 10.12a), composed of pyroxenes (pigeonite and augite), plagioclase, oxides (magnetite and ilmentite), and sometimes olivine or silica. The shergottites are subdivided into basaltic (olivine-free basalts), olivine-phyric (basalts containing olivine phenocrysts), and lherzolitic (cumulates of olivine and orthopyroxene) varieties. The radiometric ages of shergottites range from 170 to 475 Ma. Nakhlites (Figure 10.12b) are 1.3 Ga old cumulates of augite with minor olivine, and chassignites are dunites of the same age. The rare, 2.4 Ga augite basalts are composed mostly of augite and plagioclase, without pigeonite. The only ancient martian meteorites are ALH 84001, a 4.1 Ga old orthopyroxene cumulate, and the 4.4 Ga NWA 7034 breccia and a handful of meteorites paired with it. This breccia contains clasts of alkaline rocks including alkali basalt and mugearite.

BOX 10.1 COMBINING REMOTE SENSING AND SAMPLE GEOCHEMISTRY ON THE MOON

The *Lunar Prospector* orbiter carried a gamma-ray spectrometer that analyzed a handful of elements. This instrument was especially sensitive to thorium and iron. Global maps of the distributions of these two elements (total iron is expressed as FeO) were used to identify and map compositionally distinct terranes, as shown in Figure 10.9 (Jolliff et al., 2000). The rock types comprising these terranes were then determined by comparing their compositions with laboratory analyses of Th and FeO in returned *Apollo* samples (Figure 10.10). The compositional fields for the various terranes in Figure 10.10 extend beyond the *Apollo* rock compositions because they reflect mixing of other rocks within the large gamma ray analysis footprints.

The Feldspathic Highlands Terrane consists of ferroan anorthosite and forms the ancient lunar crust. The mare basalts show up as high-FeO spots in Figure 10.10; high-Ti and low-Ti basalts cannot be distinguished in this map, but their global distributions have been mapped in other studies using reflectance spectra from the *Clementine* orbiter (Giguere et al., 2000). The Procellarum KREEP Terrane is characterized by relatively high Th (Figure 10.10), an incompatible element concentrated in KREEP, and the KREEP end member is mixed with both highlands and mare materials. The South Pole-Aitken (SPA) Terrane fills a gigantic basin (Figure 10.9) with higher FeO content than typical anorthosite, and likely contains exposed mantle rocks. This example indicates that there is great value in combining orbital geochemical data with samples analyzed in the laboratory.

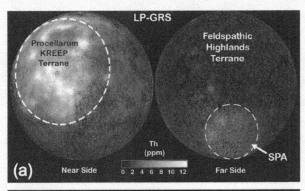

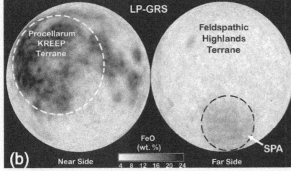

Figure 10.9 Global lunar maps of thorium and iron (expressed as FeO), based on gamma ray measurements from *Lunar Prospector*. These maps allow recognition of compositionally distinct terranes (Jolliff et al., 2000). SPA is South Pole-Aitken Terrane. Figure courtesy of B. Jolliff.

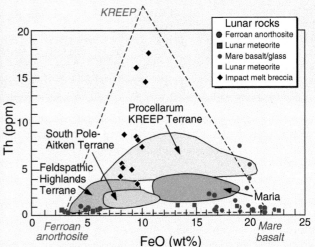

Figure 10.10 FeO and Th contents of lunar terranes in Figure 10.9 and of lunar rocks (*Apollo* samples and lunar meteorites) measured in the laboratory. This comparison allows the assignment of specific rock types to each terrane, except for the South Pole-Aitken (SPA) Terrane, for which we apparently have no samples. Modified from Jolliff et al. (2000).

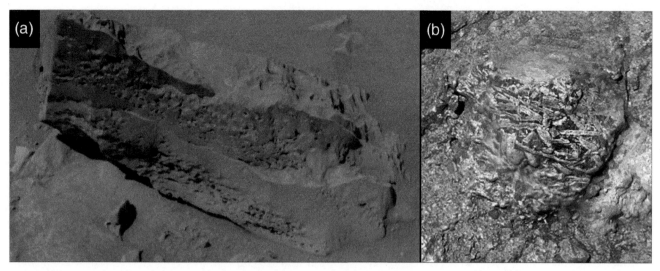

Figure 10.11 Martian igneous rocks. (a) Spirit rover image of vesiculated basalt in Gusev crater. (b) Curiosity rover image of igneous rock with K-feldspar phenocrysts in Gale crater. NASA images.

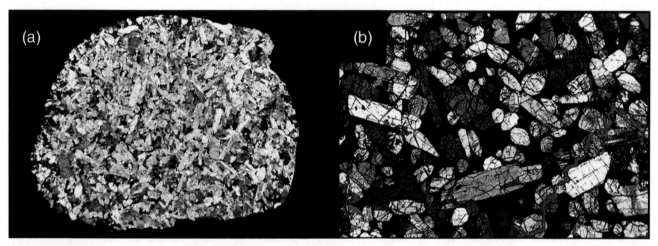

Figure 10.12 Photomicrographs of martian meteorites, under X-polars. (a) Basaltic shergottite (Zagami), and (b) nakhlite (Nakhla). Images courtesy of Dante Lauretta, from A Color Atlas of Meteorites in Thin Section (2005). Reprinted with permission.

The compositions of martian meteorites and of igneous rocks analyzed by rover APXS in Gusev and Gale craters are compared in Figure 10.13. Basalts dominate, but the ancient rocks analyzed by rovers are mildly to strongly alkaline, whereas the younger martian meteorites are tholeiitic. However, one Gusev sample (Bounce rock) is compositionally like the shergottites, and one meteorite (NWA 7034 breccia) is compositionally like Gusev rocks (Figure 10.13). Rocks at the *Mars Pathfinder* site were reported to have andesitic compositions, but this analysis probably represents a weathered rind rather than an igneous composition. Orbital gamma ray spectrometer (GRS) measurements covering most of the martian surface have low SiO_2 contents, confirming that the basalts found at rover sites are representative. This instrument analyzed K_2O but not Na_2O, so the alkalis sum was estimated by assuming the average Na/K ratio in either Gusev rocks or martian meteorites (colored boxes in Figure 10.13).

The various igneous rock compositions in Gusev crater can be related through fractional crystallization at different depths of a common hydrous basaltic magma found at the site (McSween et al., 2006), or through different degrees of partial melting (Schmidt and McCoy, 2010). The igneous rocks in Gale crater likewise formed by fractionation of alkali basalt, but at more extreme depths (Stolper et al., 2013). These two sites represent the best-characterized igneous provinces on the planet. The martian meteorites sampled over at least a half-dozen locales each contain a specific lithology. Although the specific

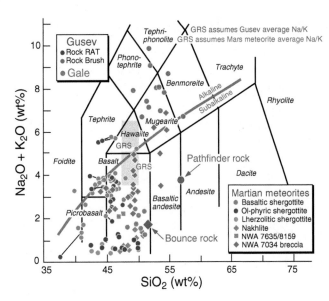

Figure 10.13 Alkalis–silica diagram comparing the rover-analyzed compositions of Mars surface rocks from Gusev and Gale craters with laboratory analyses of martian meteorites. Compositions of broad areas of the surface measured by the gamma ray spectrometer (GRS) on *Mars Odyssey* are shown as colored boxes, with different assumptions about Na/K. After McSween (2015), with permission.

launch sites for the meteorites are unknown, some very young impact craters with rays are plausible sources (Tornabene et al., 2006).

10.4.3 Asteroid Vesta

More than 1000 meteorites are linked to asteroid Vesta (Figure 10.14) based on spectra, oxygen isotopes, and geochemistry (McSween et al., 2013). Eucrites are basalts or cumulate gabbros composed of pigeonite and plagioclase; diogenites are cumulates of orthopyroxene with or without olivine; and howardites are breccias composed of eucrite and diogenite fragments. Taking the first letters of these meteorite names, they are collectively called **HED meteorites**. The crystallization ages of HEDs are very old, all near 4.5 Ga.

The global distributions of eucrites, diogenites, and howardites on Vesta (Figure 10.15) have been mapped by neutron absorptions and visible/near-infrared spectra measured by instruments on the *Dawn* orbiter (Raymond et al., 2017). Most of Vesta's surface is covered by pulverized howardite, but eucrites are concentrated near the equator and diogenites are exposed in ejecta from the huge Rheasilvia basin near the south pole. Most HED

Figure 10.14 Photomicrographs under X-polars of igneous meteorites from asteroid Vesta: basaltic eucrite, cumulate eucrite, and diogenite (left to right).

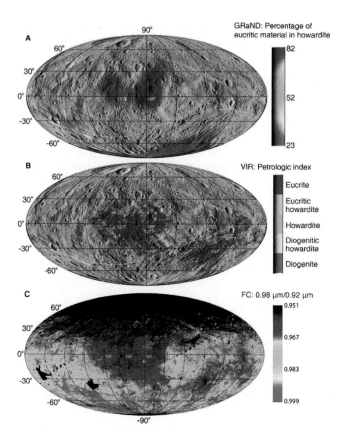

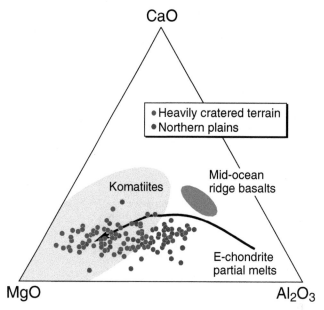

Figure 10.16 Compositions of basaltic terrains on Mercury, compared to terrestrial komatiites and mid-ocean ridge basalts. Arrow shows melts produced by increasing degrees of partial melting of enstatite chondrite. Modified from Weider et al. (2012).

Figure 10.15 Global distributions of eucrite (blue), diogenite (red/yellow), and howardite (green) on asteroid Vesta, mapped by neutron absorption (GRaND) and visible and near-infrared spectra (VIR and FC) on the *Dawn* orbiter. From Raymond et al. (2017), with permission.

meteorites were probably excavated and launched from this basin. Smaller Vesta-like asteroids have orbits that extend to a nearby resonance with Jupiter, and this escape hatch provides an explanation for the high abundance of meteorites from this body.

The origin of the igneous rocks on Vesta is surprisingly complex for such a small body. Most workers have favored an early magma ocean, which allowed the formation of a large metal core and crystallized to form a diogenite mantle and a eucrite crust. However, the details of eucrite and diogenite geochemistry are hard to reconcile with crystallization from a single, evolving global magma. An alternative model suggests that partial melting produced magmas that rapidly ascended, precluding a global magma ocean. In this model, diogenites formed by fractional crystallization in separate, subsurface magma chambers.

10.4.4 Bodies without Samples

Mercury: The *MESSENGER* X-ray spectrometer (XRS) and gamma ray spectrometer (GRS) measured elemental abundances on the planet's surface. The absence of iron

absorption bands in silicates indicates that this element is severely depleted in crustal materials. The younger smooth volcanic plains have lower magnesium, calcium, and sulfur and higher aluminum than the older terrains (Weider et al., 2012). The Mg, Ca, and Al contents are compared to some terrestrial magma compositions in Figure 10.17. The northern smooth plains lavas are basalts and the older units resemble komatiites. The different compositions may indicate distinct source regions or degrees of melting or fractionation.

The mineralogy of surface rocks is dominated by enstatite and plagioclase, with a few percent of unusual sulfides like oldhamite (CaS). Calcium does not usually form sulfide and its occurrence requires a very low oxidation state. GRS measurements of sodium suggest that plagioclase on Mercury may contain a substantial albite component. Dark areas of the surface contain a few percent of graphite, which is likely to have floated to the top in Mercury's magmas.

The mineralogy and reduced oxidation state of Mercury igneous crust are similar to the enstatite chondrites. Partial melting of E-chondrite material at varying temperatures can produce magmas that are geochemically similar to Mercury's lavas (arrow in Figure 10.17). The iron contents of these magmas are too low, but iron in the source regions was likely sequestered in the planet's massive core.

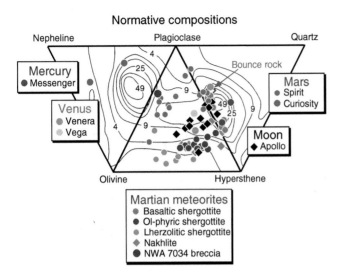

Figure 10.17 Comparison of normative mineral abundances in basalts from different planets. The relative abundances of terrestrial basalts are illustrated by contours. Modified from McSween et al. (2009).

Venus: Most of the surface consists of extensive volcanic plains, with landforms characteristic of basaltic lava (Bruno and Taylor, 1995). The very long channels on Venus' surface imply magmas with low viscosity, and carbonate–sulfate melts have been suggested. The pancake domes imply high viscosity and could represent either silicic magma or basalt with high contents of crystals or gas bubbles. Highlands terrae exhibit complex tectonics which might have induced crustal melting and yielded silicic magmas, although that could be difficult without water.

Major element abundances of three rocks in the Venus lowlands were measured by X-ray fluorescence spectrometers on the Soviet *Venera 13* and *14* and *VEGA 2* landers (Kargel et al., 1993). The scalding surface of Venus offers a very challenging environment for geochemical measurements. Basaltic compositions were reported for these rocks, but the analytical uncertainties were so large as to allow silica contents ranging from foidite to andesite, so the identification of basalts should be viewed with caution (Treiman, 2007). The rocks at various sites show either enrichments or depletions of incompatible elements (K, U, Th), which could reflect different degrees of partial melting or fractionation. The potassium, uranium, and thorium measurements by GRS were imprecise, except for one high-K sample with ~4 wt% K_2O. The mineralogy of these volcanic rocks is unknown and may have been altered by reaction with the hot, venusian atmosphere.

Io: The garish colors of lavas on Io were originally thought to indicate molten flows of elemental sulfur. Later measurements of magma temperatures (900–1300 °C, far higher than the boiling point of sulfur) have shifted thinking to silicate (basaltic or perhaps ultramafic) lavas, and this idea has been bolstered by the spectral identification of orthopyroxene. However, sulfur and SO_2 are the primary constituents of Io's plumes.

10.5 Petrologic Comparisons and Magmatic Evolution

Although the igneous rocks on planets, moons, and meteorite parent bodies are diverse, nowhere is as petrologically diverse as Earth. This is not just a sampling problem – our world is different. However, the varied magmatic rocks found throughout the Solar System can be understood as having experienced many of the same igneous processes as on Earth.

10.5.1 Planetary Igneous Rocks

The recurring magmatic theme for rocky Solar System bodies is basaltic volcanism. Differences in basalt geochemistry can be understood using the basalt tetrahedron, which uses calculated mineralogy (norms) as its apices. A simplified version of the tetrahedron, collapsed onto a plane, is shown in Figure 10.17. Basalts containing normative hypersthene are silica-saturated (tholeiites), and rocks containing normative nepheline are silica-undersaturated (alkali basalts and nephelinites). Rocks containing olivine do not coexist with quartz, so the plagioclase–hypersthene boundary, separating olivine-bearing and quartz-bearing tholeiites, is also significant. To calculate normative mineralogy, we must know all the major element abundances, which are not available from most orbital remote sensing measurements. The normative mineralogies of the available basalt compositions from different bodies are compared with terrestrial basaltic rocks (contours) in Figure 10.17. Basalts on Earth define two concentrations, representing tholeiitic and alkaline compositions. As far as we know now, only Earth and Mars contain both kinds of basalts.

Fractional crystallization of basaltic magma produces a variety of other melt compositions and can eventually yield silica-rich (rhyolite) residual melts (see Figure 10.3). However, this process is very inefficient and produces only small quantities of highly fractionated magma; tiny clasts of lunar "granite" (actually monzodiorite), the ultimate fractionation product of the lunar magma ocean, are examples. The efficiency of the process is increased if water is present, accounting for the prevalence of calc-alkaline magmas in terrestrial subduction zones, but no calc-alkaline rocks are known from other planets. Fractionation paths depend on many other variables besides the starting magma composition, including pressure and oxidation state, so each planet may have compositionally

BOX 10.2 CRYOVOLCANISM ON ICY BODIES

The discovery of **cryovolcanism** in the outer Solar System prompts the need to expand the definition of volcanoes as mountains from which molten rock and exsolved gas emanate. Cryovolcanoes erupt liquid water, ammonia, methane, or slurries of other vapor-phase volatiles at cold temperatures. *Cassini* imaged numerous jets emanating from lineaments (called "tiger stripes") at the south pole of Enceladus (Figure 10.18a) and extending to heights of 500 km. It is possible that these eruptions are curtains rather than discrete geysers. Beneath its icy crust, Enceladus apparently hosts a global ocean that is tapped to form the jets. Several of these eruptions were sampled as *Cassini* flew through the clouds, detecting mostly water vapor but also nitrogen and hydrocarbons, as well as entrained solid particles. Cryovolcanic eruptions on Enceladus are thought to be the source of particles for Saturn's E ring.

Flows emanate from large conical features interpreted to be ice volcanoes on Titan, the largest moon of Saturn. These volcanoes are thought to erupt liquid methane, although water, ammonia, or other hydrocarbons cannot be ruled out. Eruptions of methane could account for the CH_4 in Titan's atmosphere, which breaks down fast enough that it must be replenished.

The *New Horizons* spacecraft imaged peaks towering nearly 6 km above the surface of Pluto (Figure 10.18b). Summit depressions support the idea that these are cryovolcanoes. The inferred viscosities suggest flows of water–ammonia mixtures. A large dome imaged by *Dawn* on dwarf planet Ceres is thought to be a cryovolcano (Figure 10.18c).

All of these bodies have icy crusts and are located in the frigid outer regions of the Solar System. The ices can melt or vaporize at relatively low temperatures, likely aided by dissolved salts that further lower the melting points of ices. Melting of ices may result from tidal forces or decay of long-lived radionuclides deep in the bodies' interiors. The ascent and eruption of cryomagmas are not well understood, because water-dominated fluids are negatively buoyant relative to water ice crust; possible mechanisms to overcome this problem include exsolution of volatiles, pressurization of liquid chambers during freezing, and incorporation of denser silicates into the ice shell (Lopes et al., 2013).

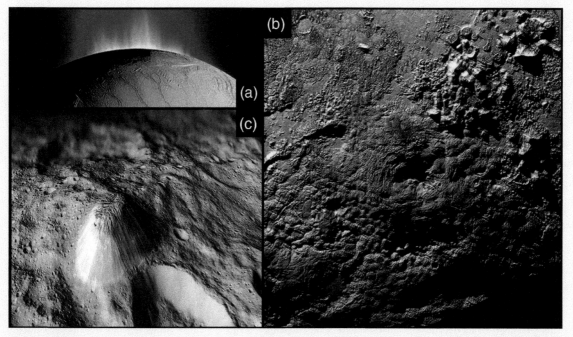

Figure 10.18 Cryovolcanism on icy bodies. (a) Erupting jets on Enceladus. (b) Volcanic mountain ~150 km across with a central depression on Pluto. (c) Ahuna Mons, a probable volcanic dome on Ceres. NASA *Cassini*, *New Horizons*, and *Dawn* images.

distinctive lavas. Another way to make silica-rich magmas is to partially remelt basalts or their fractionation products (crust rocks) rather than ultramafic (mantle) rocks. This certainly occurs on Earth and might have occurred on Venus, although there is as yet no clear evidence for venusian granite.

10.5.2 Planetary Magmatic Evolution through Time

Planetary magmatic activity clearly evolves over time. The most glaring change is when volcanism ceases altogether, largely controlled by how a planetary body loses its internal heat. Figure 6.1 illustrates the effect of planet size on the duration of magmatic activity driven by radioactive decay, reflecting the insulating effect of lots of rock. Small bodies like Vesta melted early, but igneous activity lasted for only a few million years. The Moon's magmatic activity was mostly confined to its first several billion years, with the zone of melting retreating into the deep interior over time. Volcanism on Mercury appears to have been truncated as the planet cooled and shrank, closing the pathways for magma ascent. However, pyroclastic eruptions powered by volatiles continued well into the latter half of Mercury's history. Eruptions on Mars appear to show the opposite trend, transitioning from explosive to effusive, and volcanism has extended to relatively recent time. Magmatism on Venus may be globally episodic and perhaps ongoing. Earth is obviously volcanically active. Exceptions to this pattern occur if the source of heating is tidal interactions, which do not dissipate with time unless the orbits change.

Magma compositions also appear to evolve with time. The younger volcanic plains on Mercury appear to be more fractionated than the older, heavily cratered terrains. We know very little of magma compositions on Venus. Sudden resurfacing events probably involve such large degrees of melting that little variation would exist at any one time. Magmatism on Mars appears to have evolved from early alkalic magmas (analyzed by rovers) to later tholeiitic magmas (martian meteorites). Speculations attribute this evolution to differences in oxidation states or water contents. Ancient igneous rocks on the Earth – the Archean TTG suite and komatiites described in Chapter 6 – are distinct from the mid-ocean ridge basalts and calc-alkaline rocks erupted at modern plate boundaries. The evolutionary progression of magma compositions is distinctive for each planet.

Summary

Igneous rocks are the primary constituents of planetary crusts. Their eruptive styles and the resulting volcanic landforms reflect differences in magma composition and volatile content, barriers to ascent, gravity, and other factors. Most planetary magmas are basaltic, a consequence of partial melting of ultramafic mantle sources. They undergo fractional crystallization and assimilate enclosing rocks, as do terrestrial magmas, producing a range of lava compositions and cumulate rocks. Icy bodies erupt frigid volatile liquids and gases, instead of silicate magmas. Each planet, moon, or asteroid has produced distinct igneous rocks, but the methods of volcanology, petrology, and geochemistry developed by studying rocks on Earth serve to characterize and interpret magmatic activity everywhere.

Review Questions

1. How and why do melting mechanisms, relative volumes of magmas, and magma compositions relate to plate tectonic settings on Earth?
2. Describe the kinds of volcanic landforms that occur on the terrestrial planets.
3. What kinds of eruptions are missing on Venus, and why?
4. Why are basalts so ubiquitous, and how does basalt chemistry vary among the planets?
5. How has planetary volcanism changed over time on the Earth and terrestrial planets?

SUGGESTIONS FOR FURTHER READING

Jolliff, B. L., Wieczorek, M. A., Shearer, C. K., et al. (2006) *New Views of the Moon*. Chantilly, VA: Mineralogical Society of America. A masterful set of review papers describing what has been learned about the Earth's Moon; chapter 2 describes lunar volcanic features, and chapter 3 focuses on the petrology and geochemistry of igneous rocks.

Lopes, R. M. C., Fagents, S. A., Mitchell, K. L., et al. (2013) Planetary volcanism. In *Modeling Volcanic Processes: The Physics and Mathematics of Volcanism*, eds. Fagents, S. A., Gregg, T. K. P., and Lopes, R. M. C. Cambridge: Cambridge University Press, pp. 384–413. This chapter details discoveries about volcanism on Mars, Io, and cryovolcanic satellites; it does not cover Venus and Mercury. The book itself is mostly focused on modeling volcanic processes on Earth.

McSween, H. Y. (2015) Petrology on Mars. *American Mineralogist*, **100**, 2380–2395. A comprehensive review of the mineralogy, textures, geochemistry, and origins of martian igneous rocks.

Papike, J. J., ed. (1998) *Planetary Materials*. Chantilly, VA: Mineralogical Society of America. This book provides comprehensive descriptions of the petrology and geochemistry of lunar rocks, achondrites, and martian meteorites.

REFERENCES

Anderson, R. C., Dohm, J. M., Golombek, M., et al. (2001) Primary centers and secondary concentrations of tectonic activity through time in the western hemisphere of Mars. *Journal of Geophysical Research*, **106**(E9), 20563–20585.

Braden, S. E., Stopar, J. D., Robinson, M. S., et al. (2014) Evidence for basaltic volcanism on the Moon within the past 100 million years. *Nature Geoscience*, **7**, 787–791.

Bruno, B. C., and Taylor, G. J. (1995) Morphologic investigation of venusian lavas. *Geophysical Research Letters*, **22**, 1897–1900.

Carr, M. H. (2006) Volcanism. In *The Surface of Mars*. Cambridge: Cambridge University Press, pp. 43–76.

Crumpler, L. S., Aubele, J. C., Senske, D. A., et al. (1997) Volcanoes and centers of volcanism on Venus. In *Venus II*, eds. Bougher, S. W., Hunten, D. M., and Phillips, R. J. Tucson, AZ: University of Arizona Press, pp. 697–756.

Denevi, B. W., Ernst, C. M., Meyer, H. M., et al. (2013) The distribution and origin of smooth plains on Mercury. *Journal of Geophysical Research*, **118**, 891–907.

Giguere, T. A., Taylor, G. J., Hawke, R. B., et al. (2000) The titanium contents of lunar mare basalts. *Meteoritics & Planetary Science*, **35**, 193–200.

Head, J. W., Crumpler, L. S., Aubele, J. C., et al. (1992) Venus volcanism: classification of volcanic features and structures, associations, and global distribution from Magellan data. *Journal of Geophysical Research*, **97**, 13153–13197.

Jolliff, B. L., Gillis, J. J., Haskin, L. A., et al. (2000) Major lunar crustal terranes: surface expressions and crust-mantle origins. *Journal of Geophysical Research*, **105**(E2), 4197–4216.

Kargel, J. S., Komatsu, G., Baker, V. R., et al. (1993) The volcanology of Venera and VEGA landing sites and the geochemistry of Venus. *Icarus*, **103**, 253–275.

Kerber, L., Head, J. W., Blewett, D. T., et al. (2011) The global distribution of pyroclastic deposits on Mercury: the view from MESSENGER flybys 1–3. *Planetary and Space Science*, **59**, 1895–1909.

Lopes, R. M. C., Fagents, S. A., Mitchell, K. L., et al. (2013) Planetary volcanism. In *Modeling Volcanic Processes: The Physics and Mathematics of Volcanism*, eds. Fagents, S. A., Gregg, T. K. P., and Lopes, R. M. C. Cambridge: Cambridge University Press, pp. 384–413.

McSween, H. Y. (2015) Petrology on Mars. *American Mineralogist*, **100**, 2380–2395.

McSween, H. Y., and McLennan, S. M. (2014) Mars. In *Treatise on Geochemistry* Vol. 2, 2nd edition, eds. Holland, H. D., and Turekian, K. K. Amsterdam: Elsevier, pp. 251–300.

McSween, H. Y., Ruff, S., Morris, R., et al. (2006) Alkaline volcanic rocks from the Columbia Hills, Gusev crater, Mars. *Journal of Geophysical Research*, **111**, E09S91.

McSween, H. Y., Taylor, G. J., and Wyatt, M. B. (2009) Elemental composition of the martian crust. *Science*, **324**, 736–739.

McSween, H. Y., Binzel, R. P., De Sanctis, M. C., et al. (2013) Dawn; the Vesta–HED connection; and the geologic context for eucrites, diogenites, and howardites. *Meteoritics & Planetary Science*, **48**, 2090–2104.

Mittlefehldt, D. W. (2004) Achondrites. In *Treatise on Geochemistry*, Vol. 1, ed. Davis, A. M. Amsterdam: Elsevier, pp. 291–380.

Murchie, S. L., Klima, R. L., Denevi, B. W., et al. (2015) Orbital multispectral mapping of Mercury with the MESSENGER Mercury Dual Imaging System: evidence for the origins of plains units and low-reflectance material. *Icarus*, **254**, 287–305.

Papike, J. J., ed. (1998) *Planetary Materials*. Chantilly, VA: Mineralogical Society of America.

Raymond, C. A., Russell, C. T., and McSween, H. Y. (2017) Dawn at Vesta: paradigms and paradoxes. In *Planetesimals*, eds. Elkins-Tanton, L., and Weiss, B. Cambridge: Cambridge University Press, pp. 321–340.

Robbins, S. J., Di Achille, G., and Hynek, B. M. (2011) The volcanic history of Mars: high-resolution crater-based studies of the calderas of 20 volcanoes. *Icarus*, **211**, 1179–1203.

Schmidt, M. E., and McCoy, T. J. (2010) Evolution of a heterogeneous martian mantle: clues from K, P, Ti, Cr, and Ni variations in Gusev basalts and shergottite meteorites. *Earth and Planetary Science Letters*, **296**, 67–77.

Shalygin, E. V., Markiewicz, W. J., Basilevsky, A. T., et al. (2015) Active volcanism on Venus in the Ganiki Chasma rift zone. *Geophysical Research Letters.* DOI: 10.1002/2015GL064088.

Stolper, E. M., Baker, M. B., Newcombe, M. E., et al. (2013) The petrochemistry of Jack_M: a martian mugearite. *Science*, **341**. DOI: 10.1126/science.1239463.

Tornabene, L. L., Moersch, J. E., McSween, H. Y., et al. (2006) Identification of large (2–10 km) rayed craters on Mars in THEMIS thermal infrared images: implications for possible martian meteorite source regions. *Journal of Geophysical Research*, **111**, E10006.

Treiman, A. H. (2007) Geochemistry of Venus surface: current limitations as future opportunities. *American Geophysical Union, Geophysical Monograph Series*, **176**, 7–22.

Weider, S. Z., Nittler, L. R., Starr, R. D., et al. (2012) Chemical heterogeneity on Mercury's surface revealed by the MESSENGER X-ray spectrometer. *Journal of Geophysical Research*, **117** (E12). DOI: 10.1029/2012JE004153.

White, S. M., Crisp, J. A., and Spera, F. J. (2006) Long-term volumetric eruption rates and magma budgets. G^3, **7**. DOI:10.1029/2005GC001002.

Wilson, L., and Head, J. W. (1994) Mars: review and analysis of volcanic eruption theory and relationships to observed landforms. *Reviews of Geophysics*, **32**, 221–264.

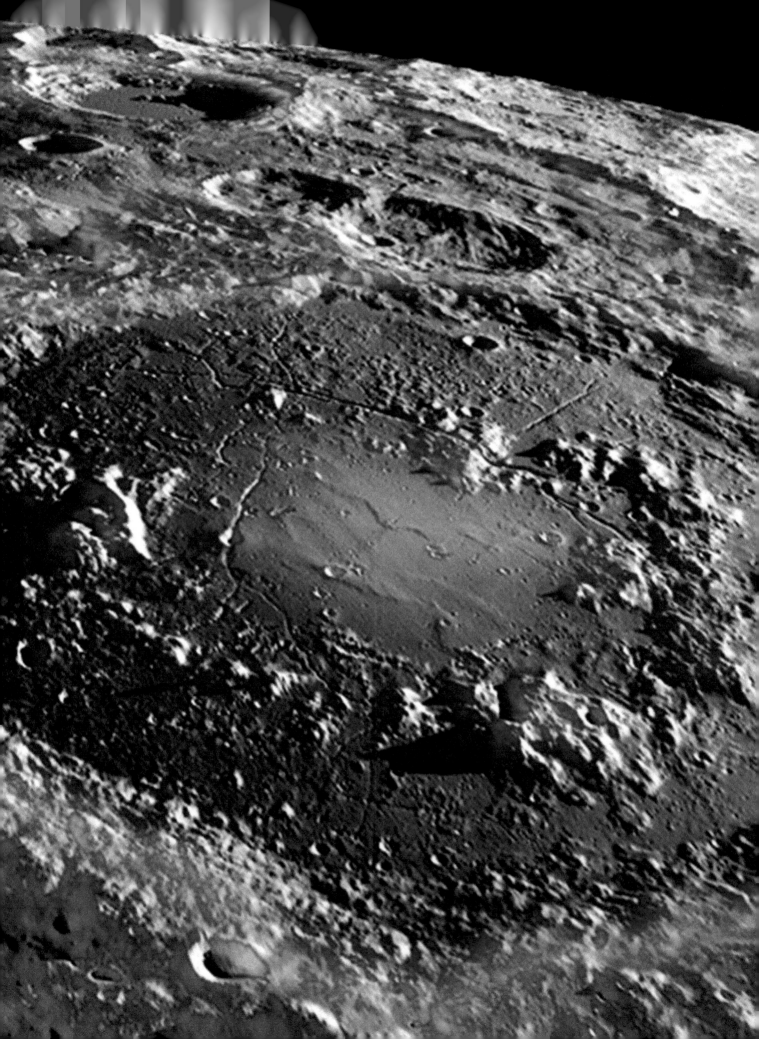

11

Impact Cratering as a Geologic Process

Heavily cratered surfaces emphasize the important geologic role of impacts on almost all planetary bodies. Large impacts (referred to as "hypervelocity" to indicate velocities of tens of kilometers per second) produce micro-, meso-, and macroscale deformations that can influence the structures of planetary crusts. Crater morphologies are described as simple, complex, and multi-ring, and correlate with crater size and inversely with gravity of the target body. Crater formation is envisioned in three stages: contact/compression, excavation, and modification, each characterized by different processes and geologic features. Shock metamorphism has affected all Solar System bodies, producing breccias containing planar deformation features, high-pressure polymorphs, and melts. Craters provide the basis for planetary stratigraphy and chronology. Massive impacts on the Earth have potential consequences for damaging the planetary ecosystem and biosphere.

11.1 Terrestrial Craters: A Little History

Our understanding of crater structures and mechanics depends in large part on the study of terrestrial impact craters. At present, the number of craters recognized on Earth is nearing 200 (an updated compilation and global distribution map can be found in the Earth impact database: www.passc.net/EarthImpactDatabase), although many more craters are presumed to be undetected (Grieve, 1991).

The publication by R. B. Baldwin of *The Face of the Moon* in 1949 provided convincing morphological evidence that lunar craters were formed by impacts. Even though an impact origin for craters on the Moon became generally accepted, prior to the 1960s geologists mistakenly perceived that large meteors striking the Earth were insignificant events that produced no major geologic structures or effects. Early attempts to argue for terrestrial

craters produced by impacts were met with skepticism (see Box 11.1). That view changed once distinctive petrologic criteria for identifying impacts were established; these criteria are shock metamorphic effects described later in this chapter. Soon thereafter, *Apollo* astronauts returned shocked rocks from the Moon's heavily cratered surface, and observations from orbiting spacecraft revealed the ubiquitous role of impact cratering in shaping the surfaces of other planets and satellites. As French (1998) aptly described the situation regarding impacts, "What was once a minor astronomical process has become an important part of the geologic mainstream." Terrestrial impact structures up to hundreds of kilometers in size, such as Sudbury (Canada), Chicxulub (Mexico), and Vredefort (South Africa), have now been documented, and major impact events are now recognized as significant players in the history of terrestrial life.

11.2 Crater Morphologies: Simple and Complex

Whether on Earth or on other worlds, small impact structures occur as bowl-shaped depressions known as **simple craters** (Figure 11.2a). These tend to preserve the approximate diameter of the originally excavated crater, called the **transient cavity**. The transient cavity may then become partly filled with brecciated material that falls back into the cavity. Simple craters, generally up to a few kilometers in diameter, have raised rims and a depth-to-diameter ratio of roughly 0.3, although that ratio depends on the strength of the material and the target planet's gravity.

Larger craters have flat floors, terraced rims, and **central uplifts** (Figure 11.2b). These so-called **complex craters** typically have depth-to-diameter ratios of about 0.1. The central uplifts can be peaks, peaks containing central depressions, or peak rings – a sequence representing the partial collapse of the uplift, which is ultimately

BOX 11.1 **METEOR CRATER CONTROVERSY**

Meteor Crater is a 1.2 km diameter bowl-shaped structure in the Arizona desert (Figure 11.1). Scientists often refer to it as Barringer Crater, in acknowledgment of Daniel Barringer, a mining engineer who championed its origin by impact. Iron meteorites (collectively called the Canyon Diablo meteorite, named for what is now a ghost town) had been recognized in the crater's vicinity as early as 1891. G. K. Gilbert, chief geologist of the US Geological Survey, investigated the crater at the turn of the century and concluded that it formed as a volcanic steam explosion. Ironically, Gilbert would be among the first to argue that similar craters on the Moon formed by impacts.

Barringer's Standard Iron Company staked a mining claim to the crater property in 1903. He assembled numerous observations supporting the crater's impact origin, which were published in 1905. This led to a vehement and protracted argument with a skeptical scientific establishment, and with influential Gilbert in particular. Seeking to bolster his idea, Barringer sought the remains of the impactor within the floor of the crater itself. Although ~30 tons of rusty iron meteorite fragments had been found around the periphery of the crater, Barringer thought that the bulk of the impactor must have been buried beneath the crater, so he drilled (unsuccessfully) into the crater floor for its booty of meteoritic iron–nickel metal. In reality, the bulk of the projectile had vaporized upon impact.

In 1960, a doctoral thesis on Meteor Crater by Eugene Shoemaker finally confirmed Barringer's hypothesis of an impact origin. Stratigraphic and structural similarities to nuclear test craters and the discovery of coesite and stishovite, high-pressure polymorphs of quartz in the target sandstone, were key lines of evidence. More recent work has established the crater's age at ~50,000 years old, and estimated that the impacting meteor was ~50 m across. Today, Meteor Crater, perhaps the best-preserved impact crater on the planet, is a National Natural Landmark and a popular tourist attraction still privately owned by the Barringer family.

Figure 11.1 Meteor Crater, Arizona. The disturbed area on the floor is Barringer's attempt to drill into a buried projectile of meteoritic iron.

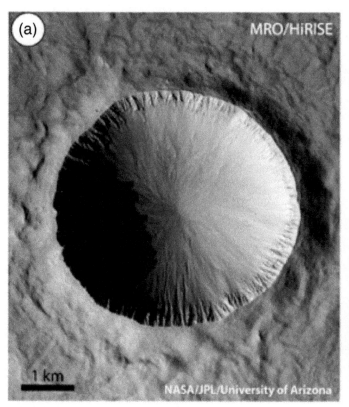 

Figure 11.2 A comparison of (a) simple and (b) complex crater morphologies. Image (a) is from Mars, taken by the camera on *Mars Reconnaissance Orbiter*, and image (b) is from the Moon, obtained by the *Kaguya* spacecraft.

converted into a ring. Because of the greater energy in larger impacts, complex craters are more likely to contain more impact-melted materials.

The most massive impacts produce **multi-ring basins** (Figure 11.3), typically hundreds of kilometers in diameter. These basins appear as gigantic bulls-eyes, with two or more concentric mountain rings and intervening ring-grabens. The transition to multi-ring structures depends on the gravity field of the target body. On the Moon, this transition occurs at crater diameters of ~500 km; the transition on Earth takes place at smaller diameters.

The transition from simple to complex craters also varies inversely with the mass (gravity field) of the target body. The approximate crater diameters for this transition are given in Figure 11.4. Although multi-ring basins are the largest impact features on any planet, it is not known whether their formation is a simple function of crater diameter, or whether special conditions in the target, e.g., a weaker asthenospheric layer at depth, are required (Melosh, 1989).

Some examples of simple and complex craters on various planets and satellites are shown in Figure 11.5. The examples in Figures 11.5a–e indicate that these morphologies are typical of rocky bodies. Craters degrade over time, especially on Earth, where many craters are deeply

Figure 11.3 The Moon's 930 km diameter Orientale basin (~3.8 billion years old) has three ring structures. The color overlay shows the basin's gravitational signature based on data from the *GRAIL* orbiter (Johnson et al., 2016) – red indicates excess mass and blue indicates deficits. Image credit: Ernest Wright, NASA/GSFC Scientific Visualization Studio.

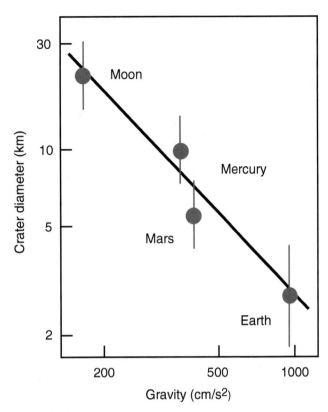

Figure 11.4 The crater diameter at which craters change from simple to complex varies from planet to planet, depending on the gravity field. Icy bodies appear to define a parallel trend displaced from this line to the left.

compression caused by an impacting meteor, the excavation of the crater itself, and its subsequent modification.

11.3.1 Energy and Shock Waves

The energy of an impact event is provided by the kinetic energy of the projectile (also called a **bolide**), and is equal to $\frac{1}{2}mv^2$, where m is the mass of the projectile and v is its velocity. The velocities at which asteroids encounter the Earth are typically 15–25 km/s (Chyba et al., 1994), and comets tend to have even higher encounter velocities. **Hypervelocity impacts** are caused by objects large enough to penetrate the atmosphere without deceleration, striking the ground at these speeds. Because of the high velocities (squared), impacts can be very energetic; a bolide a few meters in diameter can carry the kinetic energy of an atomic bomb, and a kilometer-sized impactor can release more energy in seconds than the entire Earth releases over hundreds of years. Smaller meteors, typically a few meters or less in size, passing through the Earth's atmosphere lose most of their cosmic velocities and, if they do not disintegrate, fall to the ground at no more than a few hundred meters per second. On Venus, only the largest meteors successfully transit the dense atmosphere, so small craters are absent.

The energy of large impacts is released instantaneously as it is converted into **shock waves** in the target that radiate outward from the point of impact at velocities of a few kilometers per second. The passage of a shock wave is marked by an abrupt increase in pressure, temperature, and density. Shock waves travel faster than elastic (seismic) waves, and are often supersonic. They cause flow of the media through which they traverse, which produces irreversible deformation and heating. The shock wave decreases in amplitude with distance from the impact, as its energy is dissipated through an ever-increasing volume. The degree of attenuation depends on the target material; shock waves in porous media decay faster than in solid rocks.

The equations describing shock waves were derived originally by engineer Pierre Hugoniot in 1887. The so-called "**Hugoniot elastic limit**" (HEL) is the maximum stress that a material can experience without permanent deformation. Above this limit rocks undergo shock metamorphism. As much as half of the impact energy is also converted to heat, which can vaporize the projectile and melt large quantities of target rock.

eroded (e.g., Figure 11.5b). Only a few multi-ring basins have been found on Earth, but they occur on all planets. Multi-ring basins on icy worlds, characterized by numerous rings, are even more spectacular (Figure 11.5f).

Other morphological variants of craters reflect different target compositions. For example, rays emanating from some craters (e.g., Figure 11.6a) are streaks of small **secondary craters** and disturbed surface materials. Although rays were once thought to form only on airless bodies like the Moon and Mercury, they also have been produced in Mars' thin atmosphere. They tend to fade with time, so are indicators of the most recent impacts. **Pedestal craters** on Mars (Figure 11.6b) are formed by impacts into targets with subsurface ice or groundwater. The **ejecta** was fluidized and forms lobate aprons surrounding the craters. Glancing impacts sometimes produce elongated craters (Figure 11.6c); even when the crater is not misshapen, oblique impacts can be recognized from asymmetric patterns of ejecta.

11.3 Cratering Mechanics

Let's now consider the violent process that produces impact craters and, in effect, watch in slow motion the

11.3.2 Stages of Crater Formation

The formation of an impact crater can be conveniently separated into three stages. Although crater formation is actually a rapid, continuous event, this subdivision into stages permits examination of the different processes that occur sequentially.

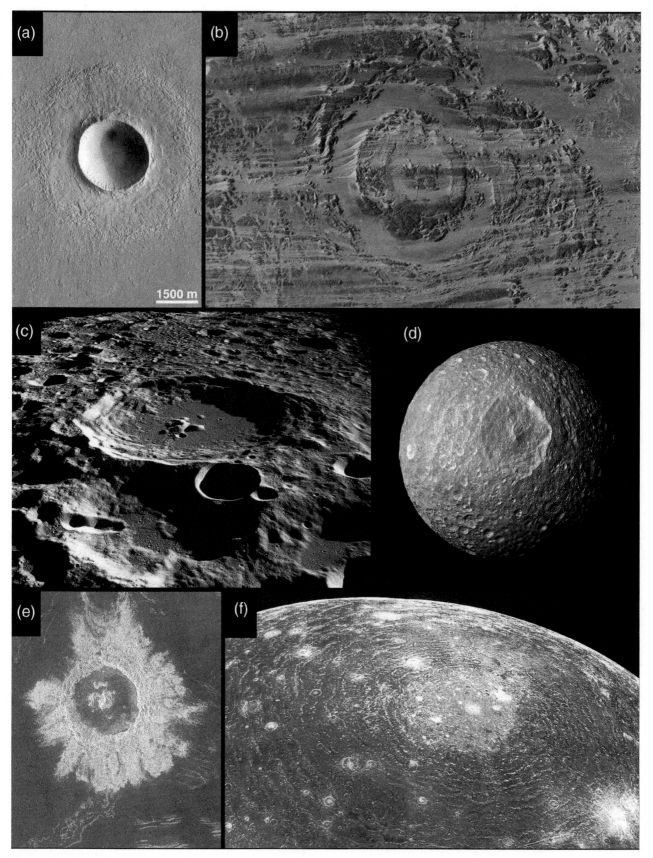

1500 m

Figure 11.5 Examples of simple and complex craters and multi-ring basins. (a) *Mars Global Surveyor* MOC image of an unnamed simple crater with ejecta blanket on Mars. (b) The deeply eroded Aorounga complex crater, Chad. (c) The Daedalus complex crater on the Moon. (d) Complex Herschel crater on Saturn's moon Mimas. (e) Radar image of the Danilova complex crater on Venus. (f) Valhalla multi-ring basin, ~4000 km in diameter, on Jupiter's icy moon Callisto. NASA images.

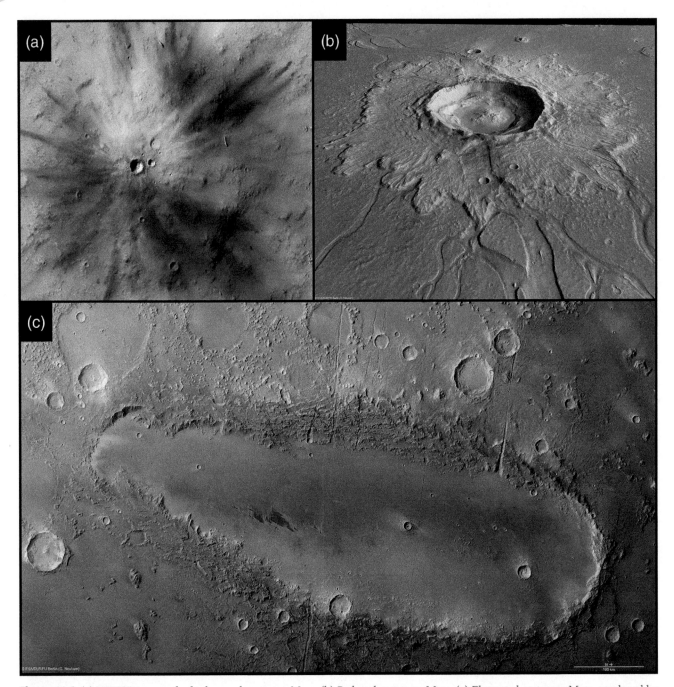

Figure 11.6 (a) HIRISE image of a fresh, rayed crater on Mars. (b) Pedestal crater on Mars. (c) Elongated crater on Mars, produced by an oblique impact. NASA images.

The contact/compression stage begins as the impactor makes contact with the ground. For coherent targets, the projectile penetrates to no more than a few times its own diameter, transmitting its kinetic energy into shock waves that compress the target. One set of shock waves migrates outward into the target rocks, while a complementary shock wave is reflected backward into and through the projectile. This reflected wave causes the projectile to fracture, melt, and/or vaporize. The shock wave passing downward into the target expands, losing energy so that peak shock pressures drop rapidly and exponentially with distance from the impact point. Thus, zones of decreasing shock metamorphic intensity are produced, with fracturing and brecciation occurring farther out. This is illustrated in Figure 11.7a (Melosh, 1989), showing isobars of peak shock pressure on the left and the corresponding effects in target rocks on the right. The contact/compression stage takes no more than a second, even for large impacts.

The excavation stage occurs as the transient cavity is opened up. The shock waves that bend upward (dashed arrows indicate material flow paths in Figure 11.7b) and

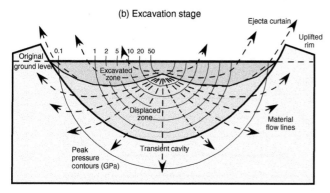

Figure 11.7 (a) Sketch of the contact/compression stage of crater formation. Modified from Melosh (1989). (b) Sketch of the excavation stage. Modified from Grieve (1987).

intersect the ground surface are then reflected downward (rarefaction), accelerating fragments outward at high velocities. This produces a curtain (shaped like an ice cream cone) of ejecta that excavates the depression. The excavated materials are derived from the upper portion of the target (Figure 11.7b). In the case of multi-ring basins, the central uplifted area corresponds to the diameter of the transient cavity. On airless bodies, the ejected debris follows ballistic trajectories and is deposited as a blanket surrounding the crater. The proximal ejecta blanket is continuous, but farther away the ejecta deposit becomes discontinuous. At deeper levels in the target, the flow lines are oriented downward and outward (Figure 11.7b), displacing material but not ejecting it from the transient crater. Once the propagating shock waves reach a distance where they can no longer excavate or displace rock, this stage ends. The excavation stage is still brief, requiring only a few seconds to carve out a 1 km transient cavity and perhaps 100 seconds for a cavity hundreds of kilometers in diameter.

In the modification stage, the transient cavity is altered by gravity-driven geologic processes. In larger craters, isostatic rebound forms central peaks, and collapse and mass movement form terraces in the walls. The

mountainous rings of multi-ring basins also appear during the modification stage. These major structural adjustments require only a few minutes to form. However, the modification stage has no defined end, as craters may continue to be modified much later through erosion and infilling by sediments.

The successive stages during the formation of simple and complex craters are compared in Figure 11.8. The contact/compression and excavation stages are similar, except for different scales and for the more pervasive melting that occurs in larger, complex craters. The modification stage differs in that complex craters form central uplifts and collapse of the crater walls forms terraces. Erosion and post-crater deposition can occur in all these craters.

11.4 Geology of Impact Craters

The mechanics of crater formation can be modeled by numerical simulations (e.g., Artemieva and Pierazzo, 2011). These computer hydrocodes are capable of accounting for three-dimensional crater geometries produced in targets of multiple rock types having varying strengths, as well as fragmentation of the target rocks and interactions of ejecta with the atmosphere. Experiments conducted in the laboratory or on a larger scale through nuclear explosion tests have also provided data used to understand the mechanics of crater formation (Anderson et al., 2003).

In addition, field studies of terrestrial impact craters provide structural and petrologic details that constrain computer simulations and experimental design. These craters provide not only access to shocked materials, but also to the third dimension that can be studied through geologic mapping and drilling. The structural geology of impact craters is complicated. Early deformation results from the rapid loading and unloading of rocks by shock waves during the contact/compression stage, and later deformation occurs during outward ejection of rock during the excavation stage and gravity-driven collapse of the transient cavity during the modification stage (Kenkmann et al., 2014).

11.4.1 Shatter Cones Formed at the Contact/Compression Stage

Besides fracturing and microscopic shock metamorphic effects on minerals, the most notable structures formed at this early stage are **shatter cones**. These objects are defined by fractures that radiate from the top, producing conical shapes (Figure 11.9a). Hypotheses for their origin range from compression as the shock wave passes through the rock to tension as the rock rebounds. Shatter

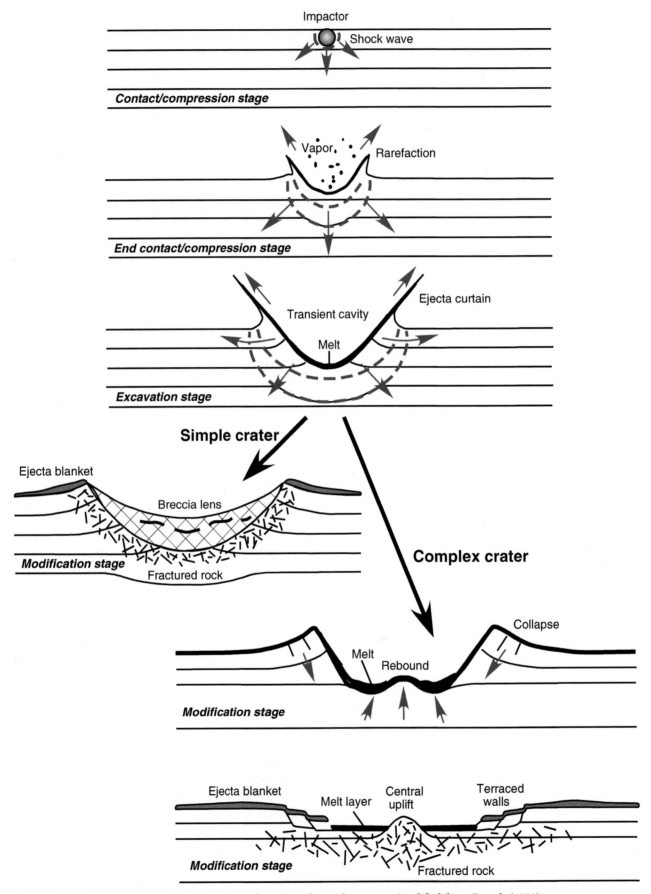

Figure 11.8 Progressive stages of the formation of simple and complex craters. Modified from French (1998).

Figure 11.9 Shocked rocks. (a) Shatter cones. (b) Polymict breccia (Sudbury, Canada). (c) Suevite (Lake Mien, Sweden). (d) Melt dike (Sudbury, Canada).

cones range from microscopic to meter-sized. The axes of shatter cones typically point toward the center of the crater, unless they have been rotated during the modification stage.

11.4.2 Breccias Formed at the Excavation Stage

Breccias deposited from the ejecta curtain partly fill the crater cavity and comprise the surrounding ejecta blanket. During the excavation stage, the material flow lines cross the shock isobars (Figure 11.7b), so breccias formed of excavated and redeposited material can contain fragments that experience various degrees of shock metamorphism. Such breccias (Figure 11.9b) are termed

"polymict." If melted clasts are mixed with a matrix of fragmented material, the breccias (Figure 11.9c) are called "suevite." If the matrix itself is melted, the rock is called "impact melt rock." Within the displaced zone that occurs at deeper levels in the crater, "monomict" breccias (composed of clasts of a single shock grade) are generated through localized fracturing and faulting. The generalized distributions of the various kinds of breccias are shown in Figure 11.10.

11.4.3 Structures Formed at the Excavation Stage

Outward movement of rock in the displaced zone (Figure 11.7b) can produce reverse faults in the crater

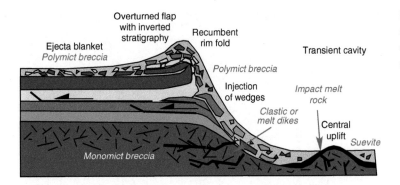

Figure 11.10 Schematic profile of a crater rim showing structural features formed during the excavation stage, and the distribution of various types of breccias. Modified from Kenkmann et al. (2014).

walls. Wedges of intact rock can be injected into the walls along fault surfaces (Figure 11.10). Intrusive veins of melt ("melt dikes") and of brecciated material (clastic dikes) may also be forcibly emplaced into the crater walls (Figure 11.9d). In this way, the crater rim is thickened and uplifted. The elevation of the rim is the sum of the injected blocks and dikes, plus the overlying proximal ejecta blanket. Nearly coherent ejecta on the rim sometimes form an overturned flap with inverted stratigraphy (Figure 11.10), in effect an isoclinal recumbent fold.

Spallation occurs as the shock wave approaches the free ground surface and the tensile strength of the rock is exceeded. This process causes short-term uplift and further enables dike injection into the crater walls. Spalls can also sometimes be accelerated outward to high velocities, and this is thought to account for the launch of meteorites during impacts onto bodies with large gravity fields, such as Mars.

11.4.4 Structures Formed at the Modification Stage

Fold and fault structures in central uplifts formed at the modification stage are exceedingly complex, and can only be unraveled for a few terrestrial craters formed in stratified sedimentary targets. Central uplifts in complex craters typically have anticlines and synclines at their peripheries (Figure 11.11), caused by constriction of mass flowing toward the center. Faulting occurs in the cores of uplifts (Figure 11.11), progressing into intense brecciation. At a certain threshold size, central peaks become gravitationally unstable and collapse to form peak rings.

Gravity-driven collapse of the walls of complex craters is radially symmetric about the crater center. Displacements along normal faults ranging up to kilometers produce terraces in the crater walls.

Multi-ring basins form when the depth of the transient cavity is comparable to the thickness of the lithosphere. For the case of a thin, weak lithosphere and low-viscosity asthenosphere, formation of the basin is followed by multiple oscillations of the cavity and outward propagation of gravity waves, analogous to ripples on a pond. This might explain the unusual multi-ring basin on icy

Callisto (Figure 11.5f). Such gravity waves are damped for more realistic asthenosphere viscosities on rocky bodies, but conjugate systems of normal faults (grabens) will still form around the basin. These faults account for the outer two rings of the Orientale basin (Figure 11.3); the formation of the innermost ring is due to collapse of the central uplift, analogous to the occurrence of peak rings in smaller craters (Johnson et al., 2016).

11.5 Shock Metamorphism

When impacted by meteors, target rocks experience sudden but extreme excursions in pressure and temperature. A shock wave traveling at kilometers per second will transit a mineral grain in microseconds, so the increase and subsequent release of pressure will be almost instantaneous. These shock pressures are far outside the range of normal geologic processes and produce considerable strain in minerals. The shock wave will also deposit energy into the target, causing a rapid increase in temperature, which in turn can result in melting or even vaporization. Collectively, the petrologic effects on impacted rocks are termed **shock metamorphism**.

11.5.1 Changes in Shocked Terrestrial Rocks

Shock metamorphic effects depend on shock pressure (normally a function of the size and speed of the impactor). Low shock pressures of 2–10 GPa cause brittle fracturing of rocks and minerals, producing breccias. Higher shock pressures of >10–45 GPa produce plastic deformation and microstructural changes in minerals (French, 1998).

Shock effects in major rock-forming minerals – quartz, plagioclase, olivine – have been calibrated in shock recovery experiments and can be used to estimate shock pressures in non-porous rocks (Figure 11.12). Crushing of pore spaces leads to local pressures as much as four times the average shock pressure. The shocked minerals develop undulatory extinction, mosaicism, planar fractures (PFs), and planar deformation features (PDFs). Undulatory extinction and mosaicism result from smaller

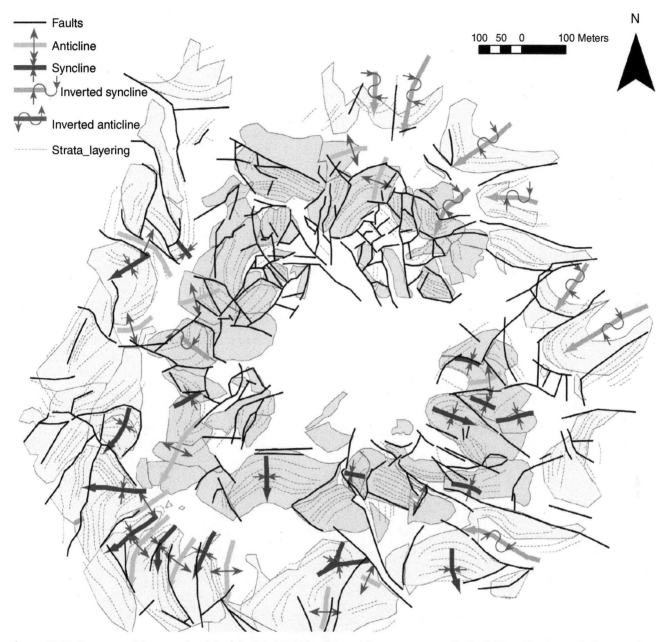

Figure 11.11 Structure of the central uplift of the Jabel Waqf as Suwwan impact structure (Jordan). The oldest, strongly brecciated limestone and marl (purple) form the core, surrounded by chert (blue) that has been folded into synclines and anticlines and segmented into large blocks. The symmetry axis of the uplift trends southwest–northeast and suggests an impact trajectory from the southwest. Modified from Kenkmann et al. (2014).

domains with slightly different optic orientations. The PFs are closely spaced open cracks oriented parallel to crystallographic planes. In contrast, the PDFs are narrow, parallel plates in which the crystalline structure has been transformed into amorphous material. PDFs commonly develop in more than one orientation (Figure 11.13a).

Shock also produces dense, high-pressure polymorphs of some minerals. Graphite can transform to diamond, and quartz can be converted into stishovite or coesite (Figures 11.12 and 11.13b). Hydrous minerals

and carbonates may lose volatiles during shock (Figure 11.13c). Quartz and feldspar can also form "diaplectic" glasses. In this case, the crystalline structures of the minerals are destroyed, producing amorphous materials, but without melting. Diaplectic glass preserves the original grain morphology, and flow structures and vesicles are absent. For example, grains of maskelynite (the name given to diaplectic plagioclase glass) appear as normal plagioclase in plane-polarized light, but are isotropic under cross-polarizers.

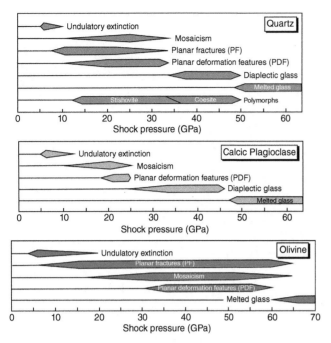

Figure 11.12 The responses of quartz, plagioclase, and olivine to increasing shock pressures. Modified from Stoffler et al. (2018).

At pressures of 50–60 GPa, melting occurs in crystalline rocks. Sedimentary rocks can melt at lower pressures of 15–20 GPa, because more shock wave energy is absorbed by pore spaces and grain interfaces. Shock melting selectively melts certain minerals, unlike normal equilibrium melting which begins at the grain boundaries between different minerals. During shock, each mineral is instantly heated to a temperature that depends on the density and compressibility of that phase. Selective mineral melting thus leads to heterogeneous textures that represent incomplete mixing of incipient melts and contorted, swirling flows (Figure 11.13d). At higher shock pressures above 55 GPa, temperatures increase and mixing becomes more effective, resulting in more homogeneous melts. For large impact melt sheets, it may be difficult to distinguish impact melts from igneous rocks (Orsinski et al., 2018).

The conditions under which various shock metamorphic features occur are summarized in Figure 11.14. Note that the pressure scale in this figure is logarithmic, so as to include the extreme pressures encountered in

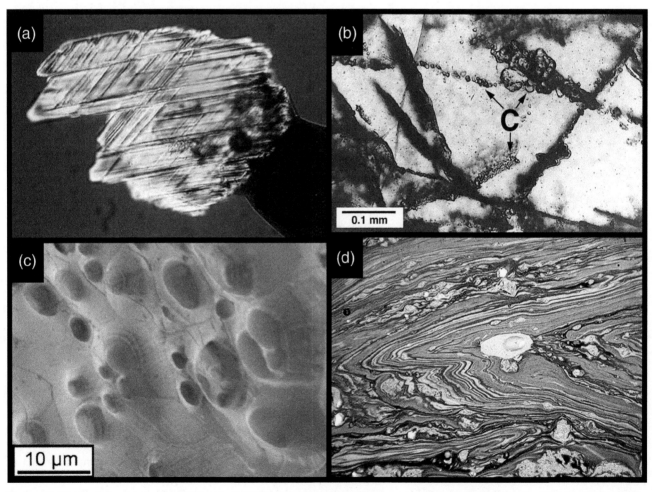

Figure 11.13 Shock features in terrestrial rocks. (a) PDFs in quartz. (b) Shocked quartz sandstone showing polymorphic transformation to coesite (C). (c) Vesicles from volatilization of carbonate. (d) Impact melt, with relict crystals and vesicles.

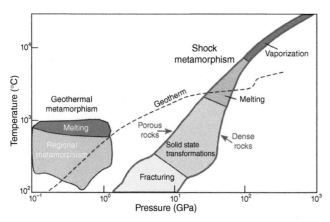

Figure 11.14 Pressure–temperature fields for shock metamorphic effects compared to terrestrial regional metamorphism. The left boundary for shock metamorphism is based on responses of porous rocks, and the right boundary for dense crystalline rocks. Modified from Stöffler et al. (2018).

shocked rocks. These conditions are compared with pressures and temperatures for normal thermal metamorphic facies in the lower left corner.

11.5.2 Shock in Extraterrestrial Rocks

Shock effects in lunar samples and meteorites are pervasive. Breccias are the most common rocks in the *Apollo* sample collection. These include not only regolith breccias but also "fragmental" breccias (Figure 11.15a) that have not been irradiated by cosmic rays and thus represent samples of deeper megaregolith. Both regolith and fragmental breccias are also abundant among chondritic and achondritic meteorites (Figure 11.15b), owing to the collisional histories of asteroids and the impact mechanism that liberates them from their parent bodies.

The shock metamorphic features that characterize rocks in terrestrial craters – PDFs, mosaicism, high-pressure polymorphs, diaplectic glass, and shock melts – also occur in these samples. Noteworthy examples are highly shocked martian meteorites (shergottites). The igneous textures of these basalts are preserved, despite transformation of the plagioclase into maskelynite (Figure 11.15c). They contain veins and pockets of impact melt, into which atmospheric gas was implanted during shock (the composition of this gas constitutes the best evidence that these meteorites are martian rocks). The Tissint shergottite appears to be the most highly shocked sample, having experienced pressures of ~25 GPa and temperatures >2000 °C (Baziotis et al., 2013). Impact melt and high-pressure polymorphs are abundant, including maskelynite (from plagioclase), ringwoodite (from olivine), majorite (from pyroxene), stishovite (from silica), and even more exotic phases.

Stöffler et al. (2018) present a classification system for silicate rocks and soils.

With the exception of maskelynite and stishovite, these high-pressure polymorphs have not been encountered in terrestrial impactites. You may recall from Chapter 6 that ringwoodite and majorite also occur deep within the Earth's mantle. However, they were first discovered as naturally occurring phases, along with bridgmanite, which comprises the bulk of the lower mantle, in shocked chondritic meteorites (Figure 11.15d).

A major difference between shock metamorphism on Earth and on other bodies is that rocks in terrestrial craters record a single shock event, whereas multiple shock events are common elsewhere (Stoffler et al., 2018) and account for regoliths. Shock features on other bodies can only be studied in actual samples. In some instances, however, shock metamorphism does impart a feature that is recognizable in remote sensing spectra.

11.6 Role of Craters in Planetary and Terrestrial Geology

We have already seen in Section 3.1 that craters are the basis for stratigraphy and age dating on the Moon and on other planets and satellites. The ejecta blankets of the most cataclysmic impacts are regional or global time-rock units, and their relative ages are deciphered from superposition. Once crater size–frequency distributions have been calibrated with radiometric age measurements of samples, as on the Moon, absolute ages of these stratigraphic units can be established.

Cataclysmic impacts also have important thermal consequences for the target planet, as described in Section 6.2. However, even modest-sized impacts can produce local heating with geologic effects. If groundwater or ice is present in the target, an impact may set up a hydrothermal system, producing alteration reactions in the subsurface rocks. Metamorphism on Mars is attributed to this kind of hydrothermal activity.

Tektites are blobs of glass that are the distal ejecta from a few large impacts on the Earth. They exhibit a variety of aerodynamic shapes that were molded in flight. Beds containing high quantities of glassy impact spherules also occur far from the putative impact craters and can be used to recognize significant impact events (e.g., Simonson, 1992). The best-known spherule deposits are associated with distal ejecta from the Chicxulub structure (Mexico), which defines the K/T (Cretaceous–Tertiary) boundary (see Section 11.7). Impact spherules appear to be more common in ancient sediments, testifying to a major role for impacts in the early Earth, although they cannot be linked with specific craters.

Figure 11.15 Shock effects in lunar rocks and meteorites. (a) *Apollo 14* fragmental breccia. (b) Beeler (Kansas) ordinary chondrite breccia. (c) Backscattered electron image of the Zagami (Nigeria) shergottite. Despite complete transformation of plagioclase into maskelynite (dark phase) by shock, its igneous texture is preserved. (d) Ringwoodite (blue) replacing olivine in the Taiban (New Mexico) ordinary chondrite.

Some craters also have economic significance or potential (Grieve and Masaitis, 1994). The Sudbury (Canada) impact structure is a significant source of nickel, possibly related to the impactor. Several buried impact craters, discovered during geophysical exploration, are also actual or potential producers of petroleum, as they may become structural traps.

11.7 A Threat to Life and Civilization

Some 66 million years ago at the K/T boundary, a mass extinction wiped out ~70 percent of all species living on Earth. The K/T extinction famously included the disappearance of the last of the great dinosaurs. Alvarez et al. (1980) first noted that the clay layer at the K/T boundary coincided with a spike in the abundance of iridium. Because this element is highly siderophile and was mostly sequestered in the Earth's core, they hypothesized that the iridium spike was the geochemical fingerprint of a chondritic impactor. Other evidence, including shock metamorphic effects and impact spherules, were later found to support this idea. An impacting asteroid large enough to account for the worldwide iridium spike would have been larger than 10 km across. Controversy ensued, and geologists who favored other explanations for the K/T mass extinction clamored for the crater to be identified. Chicxulub, a buried 180 km diameter structure with a peak ring, had already been discovered in the Yucatan region of Mexico by petroleum geophysicists. Once recognized as an impact crater (Hildebrand et al., 1991), it was found to have formed at exactly the right time. Such an impact event must have produced tsunamis and an immense global dust cloud that blocked the Sun's rays for weeks or months, likely ravaging photosynthetic

---

BOX 11.2 **LIVING IN THE FAST LANE**

Figure 11.16, a map of the distribution of small (~1–20 m diameter) object impacts over a 20-year period, reveals just how frequently bolides strike the Earth. These meteors were small members of the population of asteroids with orbits that approach or cross the Earth's orbits. Collectively, these are called near-Earth objects (NEOs). Most NEOs larger than 1 km across have now been discovered, but there are hundreds, and possibly thousands of undiscovered NEOs between 140 m and 1 km diameter (Pierazzo and Artemieva, 2012). A map of the orbits of known potentially hazardous asteroids in this size range is shown in Figure 11.17. The United States, members of the European Union, and other nations fund NEO telescopic surveys that have now discovered more than 15,000 NEOs. The Torino impact hazard scale uses the kinetic energy (in megatons of TNT) of a potential impactor and the probability of impact to assess hazard (Binzel, 2000).

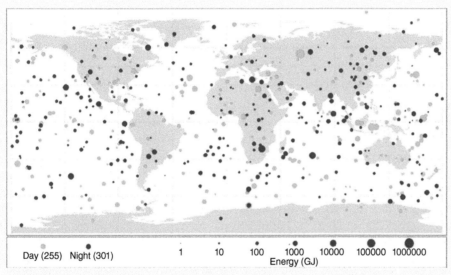

Figure 11.16 Map of 556 daytime and night time impacts from 1994 to 2013. Sizes of dots correspond to optical radiated energies of fireballs, an indirect measure of object masses. NASA map.

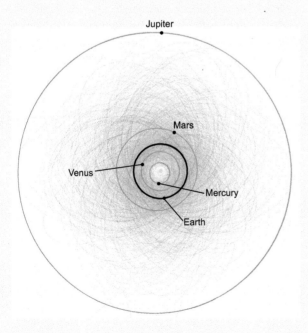

Figure 11.17 Orbits of potentially hazardous asteroids with diameters of 140 m or greater that pass within 7.6 million kilometers of Earth's orbit. The nearly circular orbits of Mercury, Venus, Earth (heavy black circle), Mars, and Jupiter are shown. NASA image.

At the direction of the US Congress, the National Academies conducted an examination of impact hazards. That study, entitled *Defending Planet Earth: Near-Earth-Object Surveys and Hazard Mitigation Strategies*, was published in 2010. It makes sobering reading. Only a few asteroids have been discovered before they impacted the Earth. There are various options to mitigate hazards, once a NEO has been determined to be on a collision course with Earth:

- push or pull tractors, which would slowly change the object's orbit by adjusting its velocity through the application of a small but steady force over decades;
- kinetic impactors, which would instantly change the object's velocity by collision with a massive spacecraft payload;
- nuclear explosions, which would change the object's velocity or possibly disaggregate it;
- civil defense measures, which are likely the only feasible option for advanced warning times shorter than a few years.

None of these options is currently available. For NEOs with diameters greater than a few kilometers, which would inflict horrendous global damage and likely cause a mass extinction event, there is at present no feasible defense.

organisms and producing freezing temperatures. The target rocks at Chicxulub also contained anhydrite layers that would have vaporized, generating sulfur aerosols and acid rain that affected plant life. Another possible consequence is thermal radiation from ejecta reentering Earth's atmosphere after the impact, which heated the surface and ignited global wildfires, accounting for abundant carbon soot found in K/T boundary layers. The wildfires, in turn, pumped greenhouse-warming gases into the atmosphere, further heating the surface.

The idea that large impacts can profoundly affect the terrestrial ecosystem and the biosphere and influence the pattern of subsequent evolution is now a tenant of geology. However, firm links between impacts and other major mass extinctions in the Phanerozoic are not so evident. An extinction event near the Eocene–Oligocene boundary (~35 million years ago) is characterized by impact debris and two large candidate craters: Popigai (Russia) and Chesapeake Bay (USA).

The frequency with which bolides collide with Earth and the ensuing environmental consequences are of more than just scientific interest. The multiple impacts of comet Shoemaker-Levy 9 on Jupiter in 1994 provided a disturbing demonstration that collisions between massive objects are an ongoing process. Any one of these comet fragments would have been devastating to our planet. Future impacts are potential hazards to society, and even to life itself. A globally catastrophic impact has been defined as one that would disrupt agricultural production and lead, directly or indirectly, to the death of more than one-quarter of the world's human population and the destabilization of civilization (Chapman and Morrison, 1994). These authors estimated that collision with an object having a diameter between 600 m and 5 km would be catastrophic by this definition. Such events are estimated to occur on average once every 0.1–1 million years. More recently, the threshold size range for bolides that would be regionally destructive was reduced to 140 m.

Summary

Impact cratering is now viewed as an important geologic process, an integral part of the Earth's evolution. These battle wounds are better preserved on other planets with less active surface modification processes.

Crater morphologies vary with size, and range from simple bowl-shaped depressions at the smallest sizes, complex craters with central uplifts and terraced walls at intermediate sizes, and multi-ring basins with concentric mountain rings and intervening grabens for the largest impact structures. The transition size from simple to complex craters varies inversely with the gravity field of the target planet. The transition to multi-ring basins may require that the excavation cavity reach a weaker asthenospheric zone. Other variants include craters with rays of secondary craters, pedestal craters formed from fluidized ejecta, and misshapen craters with asymmetric ejecta patterns formed by oblique impacts.

The energy for crater formation is equal to one-half of the impactor mass times the square of its velocity. Hypervelocity impacts are large enough that their cosmic velocities are not slowed during atmospheric passage. The formation of craters is conveniently divided into three stages. The contact/compression stage produces shock waves that pass downward into the target and that are reflected back through the bolide. These cause vaporization, melting, or shock metamorphism. The excavation stage produces a transient cavity when ejecta is accelerated outward and is deposited as a blanket around the crater. During the modification stage, a central peak or a peak ring forms in complex craters, and collapse of the walls occurs. Multiple mountain rings are formed in the largest impact structures. Later erosion and sedimentation occur on a longer timeframe.

Shatter cones are distinctive features that form during the contact/compression stage. Dikes of melt and clastic material are intruded into crater walls during the excavation stage. Because material flow lines cross shock isobars, ejecta from the excavation zone forms polymict breccias composed of clasts having different shock grades. In the displaced zone below the transient cavity, monomict breccias are produced. During the modification stage, folds and faults deform and displace material in the central uplift in complex craters, and conjugate systems of normal faults produce multi-ring basins. Shock metamorphic effects are pervasive in terrestrial craters, lunar rocks, and meteorites. These effects include micro-deformation in minerals, formation of high-pressure polymorphs, devolatilization, and melting.

The global ejecta blankets from cataclysmic impacts form the basis for planetary stratigraphic mapping and relative age determination. Effects from cratering include wide dispersion of impact melt droplets (tektites and spherules) and the generation of local hydrothermal systems if the target contains groundwater or ice.

The most massive terrestrial impacts have had devastating consequences for the planetary ecosystem. In at least one documented case (the K/T boundary event), the impact is implicated in the ensuing mass extinction. The disruption that might be caused by future large impacts poses a risk to civilization and even to the survival of our species.

Review Questions

1. What are the morphological differences between simple and complex craters? How do these morphologies scale with crater size and target planet gravity?
2. What are the three stages of crater formation, and what processes characterize each stage?
3. What portions of the target are ejected, melted, or otherwise modified, and how are they distributed in, around, and beneath the final crater?
4. What kinds of materials and structures are diagnostic of impact craters?
5. What are the characteristics of multi-ring basins, and how do they form?
6. How do the pressure and temperature conditions of shock metamorphism compare to thermal metamorphism on Earth?
7. Name and briefly describe shock metamorphic features in rocks and minerals.

SUGGESTIONS FOR FURTHER READING

Impact! – Bolides, Craters, and Catastrophies (2012) *Elements* **8**, 19–60. The seven short papers in this issue describe crater formation, shock metamorphism, catastrophic impacts, and other aspects of cratering in a readily understood manner.

French, B. M. (1998) *Traces of Catastrophe: A Handbook of Shock-Metamorphic Effects in Terrestrial Impact Structures*. Houston, TX: Lunar and Planetary Institute. This excellent introduction to shock metamorphism is freely available online at www.lpi.usra.edu/publications/books/CB-954/CB-954.intro.html.

Kenkmann, T., Poelchau, M. H., and Wulf, G. (2014) Structural geology of impact craters. *Journal of Structural Geology*, **62**, 156–182. This comprehensive review tells you everything you might want to know about what controls crater morphology and structure.

Melosh, H. J. (1989) *Impact Cratering: A Geologic Process.* Oxford: Oxford University Press. This book is considered to be the authoritative work on impact phenomena.

Stoffler, D., Hamann, C., and Metzler, K. (2018) Shock metamorphism of planetary silicate rocks and sediments: proposal for an updated classification system. *Meteoritics & Planetary Science*, **53**, 5–49. This excellent review presents the criteria for assigning a scale for shock metamorphism of rocks and soils.

REFERENCES

Alvarez, L. W., Alvarez, W., Asaro, F., et al. (1980) Extraterrestrial cause for the Cretaceous–Tertiary extinction. *Science*, **208**, 1095–1108.

Anderson, J. L. B., Schultz, P. H., and Heineck, J. T. (2003) Asymmetry of ejecta flow during oblique impacts using three-dimensional particle image velocimetry. *Journal of Geophysical Research*, **108**, 13–21.

Artemieva, N. A., and Pierazzo, E. (2011) The Canyon Diablo impact event: 2. Projectile fate and target melting upon impact. *Meteoritics & Planetary Science* **46**, 805–829.

Baziotis, I. P., Liu, Y., DeCarli, P. S., et al. (2013) The Tissint martian meteorite as evidence for the largest impact excavation. *Nature Communications* **4**, 1404. DOI: 10.1038/ncomms2414.

Binzel, R. P. (2000) The Torino Impact Hazard Scale. *Planetary and Space Science*, **48**, 297–303.

Chapman, C. R., and Morrison, D. (1994) Impact on the Earth by asteroids and comets: assessing the hazard. *Nature*, **367**, 33–40.

Chyba, C. F., Owen, T. C., and Ip, W. H. (1994) Impact delivery of volatiles and organic molecules to Earth. In *Hazards Due to Comets and Asteroids*, ed. Gehrels, T. Tucson, AZ: University of Arizona Press, pp. 9–58.

French, B. M. (1998) *Traces of Catastrophe: A Handbook of Shock-Metamorphic Effects in Terrestrial Impact Structures.* Houston, TX: Lunar and Planetary Institute.

Grieve, R. A. F. (1987) Terrestrial impact structures. *Annual Reviews of Earth and Planetary Science*, **15**, 245–270.

Grieve, R. A. F. (1991) Terrestrial impact: the record in the rocks. *Meteoritics*, **26**, 175–194.

Grieve, R. A. F., and Masaitis, V. L. (1994) The economic potential of terrestrial impact craters. *International Geology Reviews*, **36**, 105–151.

Hildebrand, A. R., Penfield, G. T., Kring, D. A., et al. (1991) Chicxulub Crater: a possible Cretaceous/Tertiary boundary impact crater on the Yucatan Peninsula, Mexico. *Geology*, **19**, 867–871.

Johnson, R. C., Blair, D. M., Collins, G. S., et al. (2016) Formation of the Orientale lunar multi-ring basin. *Science*, **354**, 441–444.

Kenkmann, T., Poelchau, M. H., and Wulf, G. (2014) Structural geology of impact craters. *Journal of Structural Geology*, **62**, 156–182.

Melosh, H. J. (1989) *Impact Cratering: A Geologic Process.* Oxford: Oxford University Press.

Orsinski, G. R., Grieve, R. A. F., Bleacher, J. E., et al. (2018) Igneous rocks formed by hypervelocity impact. *Journal of Volcanology and Geothermal Research*. DOI: 10.1016/j.volgeores2018.01.015.

Pierazzo, E., and Artemieva, N. (2012) Local and global environmental effects of impacts on Earth. *Elements*. DOI: 10.2113/gselements.8.1.55.

Simonson, B. M. (1992) Geological evidence for a strewn field of impact spherules in the early Precambrian Hamersley Basin of Western Australia. *Geological Society of America Bulletin*, **104**, 829–839.

Stöffler, D., Hamann, C., and Metzler, K. (2018) Shock metamorphism of planetary silicate rocks and sediments: proposal for an updated classification system. *Meteoritics & Planetary Science*, **53**, 5–49.

12

Planetary Atmospheres, Oceans, and Ices

Planetary volatiles occur in gas, liquid, and solid forms. In this chapter, we will see that the terrestrial planets have secondary atmospheres formed by outgassing of their interiors. The chemical compositions of the atmospheres of Venus and Mars are dominated by CO_2, but the Earth's atmosphere is distinct because CO_2 is sequestered in the lithosphere and life has added O_2 to the mix. Giant planet atmospheres are mostly hydrogen with some helium. Titan has an atmosphere of N_2 and reducing gases, along with seas of hydrocarbons. Mars has briny groundwater and had lakes and possibly oceans in the distant past. Some moons of the giant planets have subsurface oceans. Noble gases and stable isotopes hold keys to the origin and evolution of volatiles. Differences in temperature and pressure cause atmospheric circulation, controlled by planetary rotation and energy transport. Frozen volatiles are common as polar deposits and sometimes permafrost, but they are especially abundant in the outer Solar System, where they may comprise the crusts of giant planet satellites. Volatile behaviors can be described in terms of geochemical cycles. Greenhouse warming has important implications for planetary climates.

12.1 Planetary Volatile Reservoirs and Dynamics

The reservoirs for volatiles on planets take a number of forms. Gaseous atmospheres may be massive, as on Venus (Figure 12.1a) and the giant planets, or tenuous, as on Mars (Figure 12.1b) and Pluto. Only Earth and Saturn's moon Titan are known to have liquids sloshing about on their surfaces in the present epoch; however, standing or flowing water would have been seen on Mars in its distant past. The dwarf planets Ceres and Pluto, as well as some of the large moons of the giant planets, may host subsurface seas that sometimes leak out as flows or sprays. Ices occur in polar caps on a handful of rocky

bodies, comprise the crusts of icy worlds, and are common subsurface constituents of a few rocky planets, comets, and small bodies in the outer Solar System.

In this chapter we will explore the surprisingly varied chemistry of planetary atmospheres and hydrospheres (this term is not always strictly applicable, since condensed volatiles other than water occur on some bodies), and see how these compositions constrain their origins. We will consider how dynamics can stratify or roil atmospheres, and see how volatiles can be cycled through gaseous, liquid, and solid reservoirs. Describing atmospheres, fluids, and ices is a necessary prelude to understanding them as agents for changing planetary landforms and altering planetary materials, the subjects of the following three chapters.

12.2 Chemistry of Planetary Atmospheres

The chemical compositions of atmospheres have been analyzed in several ways. Some molecules can be identified in the infrared region of remote sensing spectra. Examples include determination that CO_2 is the dominant species in the atmospheres of Venus and Mars, based on telescopic observations from Earth and from orbiting and landed spacecraft, that H_2 comprises the bulk of the atmospheres of the giant planets, as detected by spectrometers on the *Galileo* and *Cassini* spacecraft, and that Pluto's atmosphere is mostly N_2 with small amounts of CH_4 and CO, as observed by spectrometers during the flyby of the *New Horizons* spacecraft.

Direct measurements of the abundances and isotopic compositions of atmospheric gases at the surface of Mars have been made by gas analyzers and mass spectrometers on rovers and landers, and at altitude by the *Trace Gas Orbiter*. The *Galileo* atmospheric probe plunged into Jupiter's atmosphere and directly measured helium and other gases.

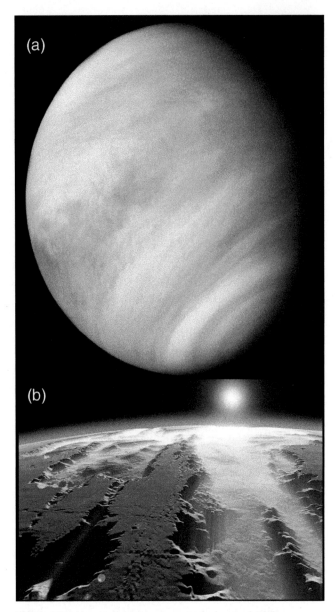

Figure 12.1 Atmospheres on terrestrial planets. (a) The surface of Venus is obscured by its dense atmosphere. (b) The tenuous atmosphere of Mars as seen from orbit. NASA images.

Table 12.1 **Atmosphere compositions of the terrestrial planets**

	Molecule	Abundance (bars)	Fraction of total
Venus	CO_2	86.4	0.96
	N_2	3.2	0.035
	H_2O	0.009	0.0001
	O_2	Trace	Trace
Earth	CO_2	0.000345	0.000345
	N_2	0.7808	0.77
	H_2O	0.01	0.01
	O_2	0.2095	0.21
Mars	CO_2	0.0062	0.95
	N_2	0.00018	0.027
	H_2O	3.9×10^{-7}	0.00006
	O_2	Trace	Trace

markedly different abundances and relative proportions (Table 12.1). Note that the fractional proportions of CO_2 and N_2 in the atmospheres of Venus and Mars are similar, suggesting that atmospheres dominated by these gases should be the norm. The Earth's atmospheric composition is the outlier – its CO_2 abundance is low because most of the carbon has been sequestered in limestone and organic matter, while O_2 from photosynthesis has accumulated in the atmosphere. Its primordial composition was probably similar to the atmospheres of Venus and Mars.

Considerable attention has been focused on the detection of methane in the martian atmosphere. The small amounts detected spectrally vary with location and time. Because methane is rapidly destroyed by ultraviolet radiation, it must be continually produced. It is not known whether the martian CH_4 is of biological or chemical origin. Organisms can release methane as they digest nutrients, and certain mineral reactions, such as serpentinization, release methane.

Mercury and the Moon have very tenuous atmospheres. These are composed essentially of noble gases and other moderately volatile elements like sodium and potassium that have been liberated from the surface by cosmic-ray bombardment.

The atmospheres of the giant planets are basically hydrogen, with relatively small but varying amounts of helium: Expressed as the molar fraction He/H (Lunine, 2004), the proportions are 0.068 for Jupiter and Saturn, 0.076 for Uranus, and 0.1 for Neptune (with considerable uncertainties, except for Jupiter). The lower amounts of helium in the atmospheres of Jupiter and Saturn reflect the rain out of helium at the higher pressures that obtain within these massive planets. Methane and ammonia have also been detected in the atmospheres of all the giant planets. Jupiter's atmosphere has the most complete

12.2.1 Atmospheric Pressures and Molecular Abundances

The masses of the atmospheres of Venus, Earth, and Mars vary significantly, corresponding to surface pressures of 95.6, 1, and 0.0064 bar, respectively. Jupiter has no readily defined surface, and its atmospheric pressure increases to incredible levels before its gases become compressed into solid phases, as described in Section 7.4. Titan's atmospheric pressure at ground level is ~1.5 times that of Earth.

The terrestrial planet atmospheres are composed of the same molecular species – CO_2, N_2, H_2O, and O_2 – but in measurements (Niemann et al., 1998), with small amounts of water and hydrogen sulfide also detected.

Titan, the second-largest moon in the Solar System, has an atmosphere that is mostly N_2, with small amounts of CH_4 and H_2. Because methane condenses at high altitudes, its abundance increases near the surface. A menagerie of trace organic compounds, including ethane, propane, acetylene, and hydrogen cyanide, is also present. The orange color of Titan's atmosphere is thought to result from chemical reactions that produce a smog of tar-like organic particles.

Pluto and Neptune's moon Triton have tenuous atmospheres of N_2, CH_4, and CO_2, produced by sublimation of surface ices. Jupiter's moon Io has a thin atmosphere dominated by SO_2, exhaled from volcanoes. The other Galilean satellites of Jupiter have tenuous atmospheres that contain O_2, formed when cosmic rays sputter surface H_2O ice; the water molecules then dissociate, and loss of hydrogen leaves an oxygen-rich atmosphere.

12.2.2 A Special Role for Noble Gases

Because the terrestrial planets were accreted from solids, excluding gases for the most part, noble gases are very scarce. Owing to their completely filled electron shells, helium, neon, argon, krypton, and xenon do not form chemical compounds. They tend to partition readily into the gas phase, and once in an atmosphere getting rid of them through chemical reactions is next to impossible. Moreover, these atoms (with the exception of helium) are too heavy to escape into space. As a consequence, noble gases are less affected by processes that modify the abundances of other gaseous components in planetary atmospheres.

Despite their simple chemistry, noble gases have complex arrays of isotopes (for example, xenon has nine stable isotopes!) that make them very useful to planetary geoscientists. The isotopic compositions of the noble gases are mixtures of radiogenic and nonradiogenic isotopes. Nuclides like ^{40}Ar are produced slowly by decay of radioactive ^{40}K (half-life of 1.25 Ga) in a planet's interior over time, so its ratio with nonradiogenic ^{36}Ar (which remains unchanged over time) gives an indication of the amount and timing of **outgassing**. Likewise the $^3He/^4He$ ratios of magmas from the Earth's interior differ from the atmosphere, reflecting admixture of primordial ^3He with radiogenic ^3He formed by decay of uranium isotopes in the mantle. Radiogenic ^{129}Xe formed by the rapid decay of now-extinct ^{129}I in the early Solar System, so the ^{129}Xe abundance is a characteristic of the original accreted materials. Cosmogenic nuclides like ^{21}Ne form by interaction with cosmic rays, allowing their use in measuring cosmic-ray **exposure** ages of meteorites in space.

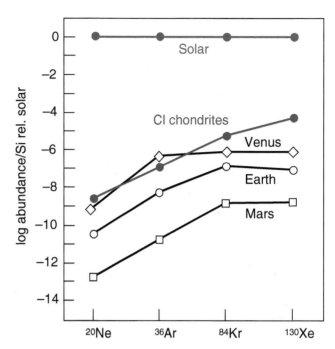

Figure 12.2 Abundances of nonradiogenic noble gases in the atmospheres of the terrestrial planets and in carbonaceous chondrites, relative to solar abundances.

Noble gases are highly depleted in the atmospheres of the terrestrial planets, relative to abundances in the Sun. This is illustrated in Figure 12.2, where the abundances of some nonradiogenic isotopes are plotted. Helium is not included in this diagram because it is not gravitationally bound and thus has been lost from these planets. We can see that the light noble gases are consistently more depleted than heavy elements. In contrast to the terrestrial planets, noble gas abundances in Jupiter's atmosphere are higher than in the Sun.

12.3 Physics of Planetary Atmospheres

Planetary atmospheres are always in motion, driven by differences in temperature and pressure. Let's examine how they vary with altitude, condense to form clouds, and circulate on planetary scales.

12.3.1 Atmospheric Structures

A planet's atmosphere arises from a balancing act between the thermal motions of gas molecules (pressure), which allow them to escape, and the force of gravity, which prevents escape. Pressure decreases with height above the planet's surface, commonly described by **scale height** (the vertical distance over which pressure falls by a factor of $1/e$). The scale heights for Venus, Earth, and Mars are 15, 8, and 16 km, respectively.

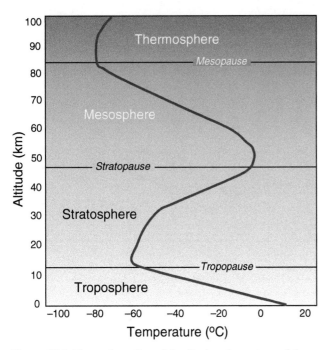

Figure 12.3 Thermal gradient (blue line) and structure of the Earth's atmosphere.

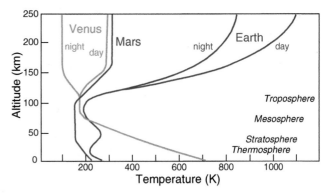

Figure 12.4 Thermal gradients for the atmospheres of Earth, Venus, and Mars. The gradients for Earth and Venus vary for the night and day sides. The structural divisions of the Earth's atmosphere are labeled on the lower right.

Atmospheric temperature is determined by an energy budget, normally heating due to absorption of solar radiation versus the heat reflected back into space by the planet. Thus the temperature depends on solar distance and albedo. If the body generates no internal heat, its **equilibrium temperature** just above the surface is defined by this balance (assuming that it is a black body radiator). The equilibrium temperature for Earth is 33 K lower than the observed value, because the black body assumption is not strictly correct. Comparison of these values provides a convenient way to ascertain whether the planet generates internal heat or the atmosphere experiences heating by the greenhouse effect (described in Section 12.7).

The temperature of atmospheric gas varies with altitude, described by the **lapse rate**, and is governed primarily by the efficiency of energy transport. Let's illustrate this by examining Earth's atmospheric thermal structure (Figure 12.3). Moving upward from the ground surface, temperature decreases throughout the troposphere as thermal energy is transported by convection. Here, infrared radiation is trapped by gas molecules and condensable water forms clouds. Above the troposphere is the stratosphere, where temperature increases with height. Heating results from gases such as ozone (O_3) that trap solar ultraviolet energy, and energy transport occurs by radiation. Above that is the mesosphere, where temperature again decreases because of decreasing ozone production. Above this is the thermosphere, where gases are ionized by solar X-rays and temperature is balanced by conduction. The thermosphere is hot because it is dominated by ions that do not radiate heat efficiently. The atmosphere fades above the thermosphere (called the exosphere, at ~500 km altitude on Earth), where gases escape into space.

Differences in the atmospheric thermal structures of the terrestrial planets are compared in Figure 12.4. The most obvious difference is that Earth's temperature profile has two minima, at the tops of the troposphere and mesosphere, whereas the other planets have only one. In other words, neither Venus nor Mars has a stratosphere. Venus' dense troposphere extends to more than three times the height of Earth's. Pronounced temperature differences in the thermosphere occur on the day and night sides of Venus and Earth. Temperatures in the thermosphere of Venus and Mars are low compared to Earth, because their CO_2 atmospheres cool more efficiently.

The observed temperatures of Jupiter, Saturn, and Neptune are far higher than their equilibrium temperatures, indicating that they emit roughly twice as much energy as they receive from the distant Sun. Uranus emits considerably less excess heat, for unknown reasons. The outer atmospheres of the giant planets have thermal profiles somewhat similar to the terrestrial planets, with temperature maxima at ~100 mbar. The high-pressure transitions and interior structures deeper inside these planets were considered in Section 7.4.

12.3.2 Cloud Formation

Air is said to be "saturated" if the abundance of water vapor (or any other condensable species) is at its maximum partial pressure. **Relative humidity** is the ratio of the partial pressure of the vapor to that in saturated air. The dew point is the temperature at which a certain parcel of air (with a certain relative humidity) begins to condense water. The frost point is similar, except that water ice condenses.

Clouds form when either the dew point or frost point is reached, forming numerous liquid droplets or ice crystals. Clouds on Earth are H_2O droplets, on Venus are H_2SO_4 droplets, and on Mars are CO_2 or H_2O crystals. On the giant planets, clouds of NH_3, H_2S, and CH_4 form. The heights of clouds depend on both temperature and pressure, but are most sensitive to the latter.

Condensation of vapor to form clouds releases heat. Clouds are highly reflective, and thus they reduce the amount of incoming solar energy. They can also block the outgoing heat. These processes combine to alter the atmosphere's thermal structure, although quantification of their effects can be difficult.

12.3.3 Atmospheres in Motion

Atmospheres are constantly in motion, and convection and planetary rotation are keys to understanding circulation patterns. Atmospheric stability depends on the lapse rate, or more specifically the **adiabatic lapse rate** – the lapse rate at which a vertically displaced parcel of air will neither gain nor lose heat with its surroundings. This is expressed as:

$$dT/dz = -\mu g/C_P \tag{12.1}$$

where T is temperature, z is height, μ is molecular weight of the gas, g is gravity, and C_P is heat capacity. For Earth, the adiabatic lapse rate is ~10 °C/km; however, a more realistic moist adiabatic lapse rate is ~5 °C/km – it is lower because of the higher heat capacity of humid air. If a parcel of air is heated near the ground, it will become buoyant in an unstable (superadiabatic) atmosphere, but not in a stable (subadiabatic) atmosphere (Figure 12.5). Unstable atmospheric conditions lead to convection. One clear example is Mars, where ascending hot air carries tiny dust particles from the ground, producing periodic global dust storms (Figure 12.6a).

In addition to vertical motions, horizontal motions (winds) are induced by gradients in atmospheric pressure. Planet rotation causes winds to follow curved rather than straight paths as they blow from a high-pressure region to a low-pressure region. On Earth, this **Coriolis force** (it's really more of an effect than a force) causes winds to be deflected to the right in the northern hemisphere and to the left in the southern hemisphere.

Differential solar heating causes pressure gradients in the atmosphere. The most significant effect is **Hadley cell circulation** (Figure 12.7), seen on all the terrestrial planets except Mercury. A planet's equator typically receives more solar energy than its poles. Hot air near the equator becomes unstable and flows upward and toward the poles, which have lower pressure. There, the air cools and descends, and returns along the surface to the equator. Slowly rotating planets like Venus have one Hadley cell per

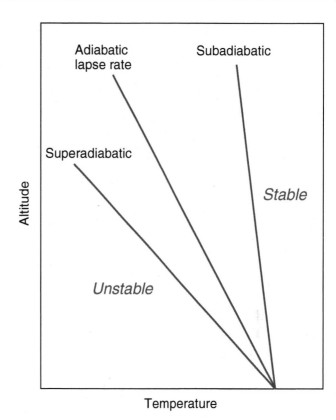

Figure 12.5 Thermal gradients with subadiabatic lapse rates are stable, whereas superadiabatic lapse rates are unstable.

hemisphere. On Earth, which rotates faster, meridional winds are deflected due to the Coriolis effect, causing the circulation pattern to break apart. Consequently, the Earth has three Hadley cells per hemisphere. On the giant planets, which rotate even more rapidly, strong **zonal winds** flow along lines of constant pressure, producing latitude-confined cloud bands (Figure 12.6b). On Earth, such zonal winds (called "jet streams") separate Hadley cells. Stationary **eddies** result from pressure gradients caused by differential heating, in this case because of temperature differences over mountain ranges or at the boundaries between oceans and continents. Eddies also form in the region between two air flows. Eddies produce huge, long-lasting storms like Jupiter's great red spot and Neptune's great dark spot (Figures 12.6c and 12.6d). On Mars, CO_2 flows from one pole to the other annually, as carbon dioxide ice sublimes in summer and recondenses after reaching the winter pole. Such **condensation flows** may also affect N_2 and CH_4 on Pluto.

12.4 Sloshing Oceans, Seas, and Lakes

Planetary scientists tend to use the terms "ocean," "sea," and "lake" to refer to any collected surface liquid, regardless of its composition and only loosely indicating its volume. "Subsurface oceans" refers to significant

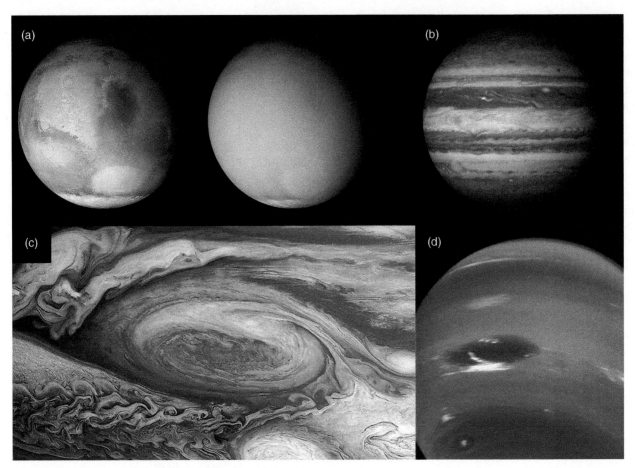

Figure 12.6 Effects of atmospheric instabilities. (a) Two views of Mars taken by *Mars Global Surveyor* about a month apart in 2001, showing the onset of a global dust storm. (b) *Hubble Space Telescope* image of latitude-confined clouds on Jupiter formed by zonal winds. (c) Jupiter's great red spot (*Juno*), and (d) Neptune's great dark spot (*Voyager 2*), storms formed by eddies. NASA images.

amounts of liquid beneath icy crusts, inferred for some smaller bodies in the outer Solar System.

12.4.1 Oceans on Earth and Perhaps Ancient Mars

Earth's oceans amount to 1/1000th of the planet's mass. Seawater has a salinity of ~3.5 percent, due mostly to dissolved sodium and chlorine but also to magnesium, sulfate, calcium, and potassium. The high sodium and chlorine values are due to their long residence times in the ocean; other solutes tend to precipitate more rapidly. CO_2 in the oceans is a critical part of the Earth's carbon cycle. Dissolved phosphorus and nitrogen are important for microorganisms living in the oceans.

Mars' water is mostly long gone. The northern lowlands may once have hosted a vast ocean of water (Baker et al., 1991), but so far the evidence for an ancient ocean is inconclusive (Carr and Head, 2003). Physical observations that may suggest a former ocean are described in Section 14.6. The absence of detectable carbonate

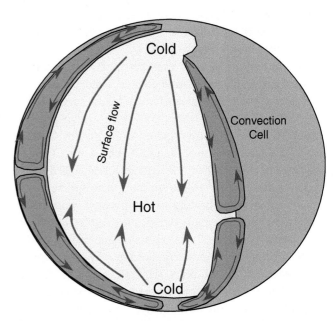

Figure 12.7 Sketch illustrating Hadley cell convection in a non- or slowly rotating body.

deposits in this region suggests that, if there was an ancient ocean, it must have been acidic.

Regardless, ancient Mars clearly hosted smaller bodies of water (Cabrol and Grin, 2010). Sedimentary deposits from a paleolake in Gale crater have been well documented by the Curiosity rover (Grotzinger et al., 2015). Sediments in Meridiani Planum analyzed by the Opportunity rover also indicate large amounts of water, although it was apparently below ground most of the time (Grotzinger et al., 2005). Martian lakes and groundwaters were acidic brines, which stabilized these liquids to low temperatures. Water–rock experiments (e.g., Tosca and McLennan, 2006) suggest that the brines had high concentrations of magnesium, iron, and sulfur, and low contents of sodium and chlorine compared to terrestrial seawater. Evaporation of these brines precipitated magnesium sulfates before halite. Iron sulfates formed from these liquids became oxidized and produced iron oxides.

12.4.2 Titan's Hydrocarbon Lakes

Radar images obtained by the *Cassini* orbiter (Stofan et al., 2007) found that Titan is dotted with lakes of liquid organic compounds (Figure 12.8). The lakes exhibit inflowing channels, coastal features, and drowned topography. The exact blend of hydrocarbons in the lakes is unclear and may vary from one lake to another. The presence of liquid ethane has been confirmed spectrally, and methane is also thought to be an important ingredient. Rain-covered surfaces and hydrocarbon rain showers may also have been imaged.

The light hydrocarbons are stabilized in liquid form by frigid surface temperatures (averaging −179 °C), because distant Titan receives only 1 percent of the solar energy that Earth receives. Benzene snow likely falls and dissolves in the lakes and, once it reaches saturation, it may precipitate as a mud-like sludge on the bottoms.

12.4.3 Subsurface Seas on Other Worlds

Hidden 1.5 km beneath the ice cap at the south pole of Mars may be a lake of liquid water (Orosei et al., 2018). Mapped by bright radar reflections discovered by the Mars Advanced Radar for Subsurface and Ionospheric Sounding (MARSIS) instrument on ESA's *Mars Express* orbiter, the layer exhibits a higher permittivity than rock and ice, consistent with water. High amounts of dissolved salts likely permit the frigid water to remain liquid.

The Jupiter and Saturn satellite systems may be awash in buried oceans. Jupiter's moon Europa, heated by tidal flexing, is thought to have a subsurface ocean. Its magnetic field, detected by the *Galileo* orbiter, provides evidence of a conductive subsurface layer, likely composed of saltwater (Kivelson et al., 2000), which sometimes erupts as geysers and has deposited sulfates on the surface. Ganymede is also believed to have a saltwater ocean sandwiched between layers of ice. In the Saturn system, the *Cassini* spacecraft discovered saltwater-rich plumes spewing from Enceladus and Dione. Analyses of the Enceladus plumes identified simple hydrocarbons and fragments of complex organic molecules, nitrogen, water, and carbon dioxide.

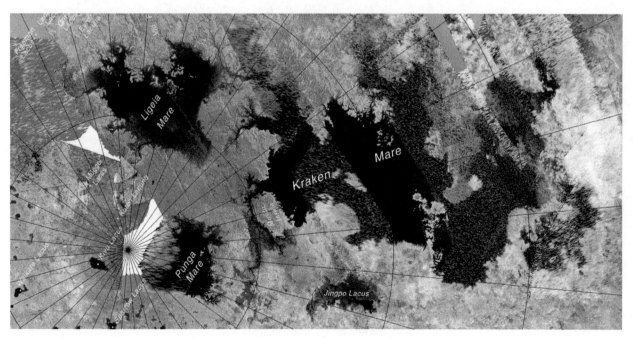

Figure 12.8 Radar image of hydrocarbon lakes on Titan. NASA *Cassini* image.

Pluto's largest impact basin sits directly on the tidal axis linking Pluto and Charon. Its reorientation implies excess mass under the basin, which has prompted the idea that an ocean of liquid water containing methane lurks beneath an icy shell (Nimmo et al., 2016).

12.5 Frozen Volatiles

Planetary ices are much more common than liquids, owing to the limited stability of liquids because of temperature and pressure constraints. Frozen volatile reservoirs can take many forms, as described below.

12.5.1 Surface Ice: Polar Ice Caps

Mars has two polar caps composed mostly of water ice (Carr, 1996). Frozen carbon dioxide cycles between the poles annually as the seasons change and accumulates as a veneer on both, although some CO_2 ice remains on the southern cap continually because of its higher altitude and colder temperatures. The northern cap is larger, 1100 km across, whereas the southern cap has a diameter of 400 km (Barlow, 2008). The total volume of ices in both caps is more than three million cubic kilometers. Spiral troughs (Figure 12.9a) occur on both caps, resulting from winds that spiral because of the Coriolis

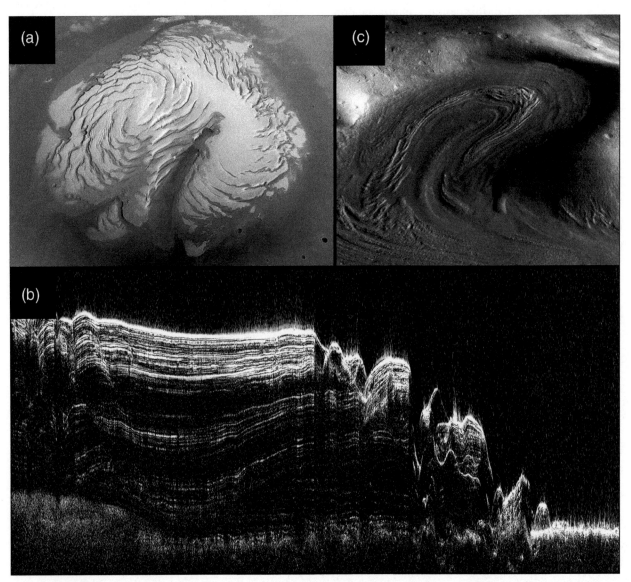

Figure 12.9 (a) *Mars Global Surveyor* image of the northern polar ice cap of Mars. (b) *Mars Reconnaissance Orbiter* SHARAD ground-penetrating radar image of a portion of the northern cap, showing a ~2 km thickness of layered ice deposits extending to the fuzzy basal unit at the bottom (Stuurman et al., 2016). (c) Glacial deposits on Mars. NASA images.

force. The caps consist of layered deposits, formed by seasonal melting and freezing of ice, mixed with varying amounts of dark particulates from dust storms. Images from a ground penetrating radar instrument on the *Mars Reconnaissance Orbiter* reveal ice thicknesses of several kilometers (Figure 12.9b).

Mercury would seem to be an unlikely place to find ice, but the *MESSENGER* spacecraft found water ice deposits in permanently shadowed areas at the north and south poles. Permafrost likewise occurs in shaded craters at the poles of the Moon and dwarf planet Ceres. This water ice was likely brought in by comets, vaporized during impact, and migrated to cold traps at the poles.

12.5.2 Surface Ice: Glaciers

Glaciers, the result of snow self-compaction, are masses of flowing ice. This flow results from sliding over the surface, which may be lubricated by glacial meltwater and by deformation of the ice. Both the ice itself and its meltwater transport voluminous sediment, generated by the massive grinding ice sheet. The results are extensive sedimentary deposits that document the conditions of former glacial periods. Glacial deposits in Canada and the northern United States are evidence of the Pleistocene age Laurentide Ice Sheet that formerly covered half of North America.

Mars also exhibits glaciers and glacial deposits (Figure 12.9c). The largest martian glaciers are located on the sides of the three great Tharsis volcanoes. Their

origin is interpreted as deposition of snow from atmospheric water vapor on the windward sides of these high mountains, where it compacted into ice and flowed downslope. The morphology of the sedimentary deposits shows that the glaciers were cold-based, as might be expected for Mars. The extents of the deposits, which stretch for tens of kilometers in front of the glacier termini, show that Tharsis glaciers have retreated back from their maximum extents. Martian glaciers are also found in impact basins and along tectonic scarps.

Mars glaciers tend to be buried beneath sediment, which limits sublimation of the ice. This covering also prevented the glaciers from being widely detected until the advent of high-resolution images. The unusual grooves on the surface, which show through the sedimentary overburden, mimic flow lines, providing evidence of underlying glacial flow.

12.5.3 Subsurface Ice: Permafrost

Extensive subsurface water ice (up to 70 vol%) has been discovered at mid to high latitudes (mostly poleward of ~60° latitude) on Mars through orbital measurements of hydrogen by the gamma ray spectrometer on *Mars Odyssey* (Feldman et al., 2004). Figure 12.10 shows a global map of water-equivalent hydrogen in the upper few meters of the martian surface; water concentrations near the equator reflect mineral-bound OH and H_2O. **Permafrost** has also been imaged by ground penetrating radar

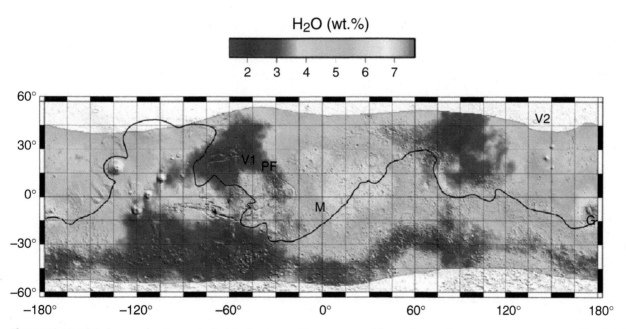

Figure 12.10 Global map of water-equivalent hydrogen on Mars, measured by neutron and gamma-ray spectrometry. Reprinted with permission from Cambridge University Press: *The Martian Surface: Composition, Mineralogy, and Physical Properties*, ed. J.F. Bell III, copyright (2008).

BOX 12.1 **WHAT LIES BENEATH: INFERRING SUBSURFACE ICE ON MARS**

While H_2O on Mars is currently observed as clouds in the atmosphere and polar ice caps on the surface, we can also observe evidence for H_2O in the subsurface. Some of this "seeing" is done using electromagnetic wavelengths shortward of the visible, using gamma rays as described above. However, we can also peer into the subsurface by observing the geomorphic effects that H_2O ice produces on the surface – striking patterns and textures (Figure 12.11) that provide a window into past and present water locations and phases. The expansion of subsurface water when it changes phase from liquid to ice causes the soil around it to expand. This expansion creates surfaces wedges that subsequently become filled with ice or – especially today on Mars, with its very dusty atmosphere – with dust and sand. These wedges often intersect to create extensive polygonal networks. In other locations or under different conditions, this expansion creates subsurface voids, which are preferentially filled with smaller grains of sediment. This infilling effectively pushes larger grains to the surface. This subsurface churning (albeit at a very slow rate) tends to form discrete subsurface convection structures, which move the larger grain sizes upward at regularly spaced intervals. The result is a concentration of rocks at the surface, artfully arranged into a variety of forms, such as circles, nets, or (on slopes) stripes. In the patterned ground at the surface, we see the effect of ice lying beneath.

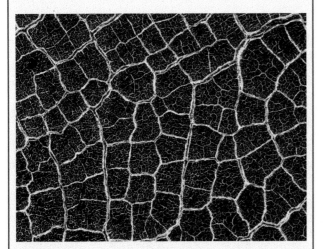

Figure 12.11 HIRISE image of patterned ground on Mars, providing visual evidence of underlying permafrost. NASA image.

(Stuurman et al., 2016). The subsurface ice is protected from sublimation by the overlying regolith. The martian cryosphere, of unknown thickness, may be recharged from below by water circulation from a deep hydrosphere.

12.5.4 Worlds with Icy Crusts

Contrary to terrestrial experience, the vast majority of planetary bodies in our Solar System have icy crusts. Prominent examples are the moons Europa, Enceladus, and Triton, as well as the dwarf planets Ceres and Pluto, and other numerous bodies in the Kuiper belt. The frozen surfaces of these bodies are composed of some combinations of H_2O, N_2, CO_2, CO, and CH_4 ices. Estimates of the thicknesses of ice shells vary from a few kilometers (Enceladus), to 15–25 km (Europa), to many tens of kilometers (Pluto).

The icy crust of Europa is among the best studied, owing to high-resolution spacecraft images. Despite having the smoothest surface at global scale of any solid body in the Solar System, Europa has a structurally complex surface, with multiple generations of ice formation and angular, chaotic blocks resembling icebergs (Figure 12.12a). Criss-crossing cracks record stresses caused by tides in its underlying ocean (Figure 12.12b). Craters are rare, testifying to the youth of the crust. Europa's surface ice is mixed with salts that precipitated from briny liquids.

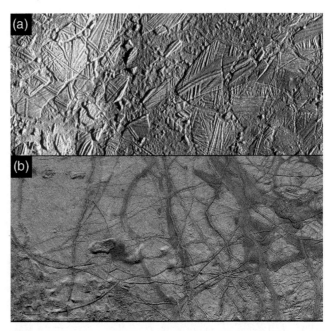

Figure 12.12 Images of Europa's icy crust. (a) Jumbled blocks of ice indicate ongoing tectonic activity. (b) Dark cracks produced by tides in the subsurface ocean. NASA *Galileo* images.

12.6 Origin and Evolution of Planetary Volatiles

Water covers three-quarters of the Earth's surface. Mars, too, once had liquid water, and ice still resides at its poles and in its subsurface. Venus is now bone-dry, but the isotopically heavy hydrogen in its atmosphere suggests loss of a significant amount of water in the past. Even Mercury, baking in the glare of the nearby Sun, has some ice at its poles. If the terrestrial planets formed from differentiated, volatile-depleted planetesimals inside the snow line, how did they obtain sufficient volatiles to form atmospheres and oceans? And how might that explanation differ from the formation of dense atmospheres on the giant planets?

12.6.1 Sources of Volatiles

The giant planets are massive enough to have gravitationally captured solar nebular gas directly, so it is no surprise that their atmospheres are mostly hydrogen and helium (like the Sun). However, their atmospheric compositions are not strictly solar – the *Galileo* probe found that noble gases in Jupiter's atmosphere are enriched by about a factor of three relative to solar abundances, implying that fractionation of gases occurred during accretion. The ratios of deuterium to hydrogen (D/H) are higher in Uranus and Neptune than in the innermost giant planets (Figure 12.13), thought to reflect their accretion of condensed ices as opposed to nebular gases.

Earth and Venus are massive enough to have captured nebula gas if the gas persisted until they grew to their present sizes. However, even if that occurred, any **primary atmospheres** must have been lost, because their atmospheric compositions differ so markedly from solar abundances.

The noble gas abundances are particularly informative about the source of volatiles in planetary atmospheres. Figure 12.2 compares nonradiogenic (unchanging) isotopes of neon, argon, krypton, and xenon in planetary

atmospheres with the abundances in the Sun and in CI carbonaceous chondrites. The strong noble gas depletions, especially of the lighter elements, in the atmospheres of the terrestrial planets actually resemble the pattern measured in chondrites. Consequently, the noble gases (and by extension, water and other volatiles) in the terrestrial planets may have been originally incorporated as trace elements in the accreted solid materials.

Another possible source of volatiles is comets, which contain more ices than do asteroids. Hydrogen isotopes can potentially constrain any contribution to the Earth from accreted comets. The measured D/H in comets is twice as high as that of ocean water (Figure 12.13), suggesting no more than about 10 percent of Earth's water could have come from this source. It is not clear, though, that the few comets in which D/H has been analyzed are representative. However, the D/H ratio of Earth's water is similar to that in carbonaceous chondrites and achondrites (Sarafian et al., 2014).

If the terrestrial planets accreted their volatiles as solid planetesimals, they must have **secondary atmospheres**, that is, they were outgassed during planetary heating and volcanism. Accretion models suggest that the feeding zones for the terrestrial planets grew wider during the later stages of accretion, so that ice-bearing planetesimals from beyond the snow line may have been incorporated. Carbonaceous chondrites, originally containing ice, were processed to form hydrated phyllosilicates (discussed in Section 15.3), and this could have been the originally accreted form of most of the water.

The atmospheres of Titan, Triton, and Pluto probably owe their nitrogen-rich compositions to the original incorporation of ammonia ice that condensed in the frigid outer Solar System. Other components like methane may also have been accreted as ices.

12.6.2 Liquid Condensation

Condensation of water vapor is illustrated by considering our own planet's early oceans. Once atmospheric temperatures cooled below 100 °C and the air became saturated, condensation of water occurred and liquid water collected on the planet's surface. There is some uncertainty about when the first oceans formed. The earliest-recognized sedimentary rocks are ~3.8 Ga old, although oxygen isotopic measurements in ancient zircons have been controversially interpreted to suggest oceans as early as 4.2 Ga. In any case, as extensive continental areas likely had not yet formed by this time, the early ocean was probably shallow and covered much of the Earth's surface.

Condensation of more exotic liquids, such as hydrocarbons on Titan, occurred at much lower temperatures, but the process is basically the same as for water. Titan's

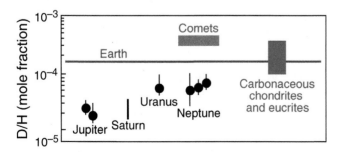

Figure 12.13 Hydrogen isotopic compositions (deuterium/hydrogen ratios) in Earth's seawater (horizontal line), atmospheres of the giant planets, comets, and several classes of meteorites.

low temperature is close to the triple point of methane, so it can exist as gas, liquid, and solid, producing a methane cycle analogous to the water cycle on Earth. Methane and ethane rain falls, then flows across the surface, and collects in lakes.

12.6.3 How Atmospheres Evolve

Outgassing can occur from a global magma ocean or over time by serial volcanism. The timing of outgassing can be estimated from the atmospheric ratio of radiogenic ^{40}Ar to nonradiogenic ^{36}Ar. Early outgassing from a magma ocean would occur before much ^{40}K decayed to ^{40}Ar, producing a lower ^{40}Ar/^{36}Ar ratio. Based on estimates of the original potassium content, the ratio of ^{40}Ar/^{36}Ar suggests that the Earth's atmosphere formed within the first few tens of millions of years.

Atmospheric volatiles can be lost to space, either blown off by very large impacts (**impact erosion**) or gradually lost by **sputtering** (where atoms are accelerated when hit by another fast atom) and **Jeans escape** (where lighter molecules are not gravitationally bound). The latter can produce isotopic changes, as molecules made of light isotopes escape preferentially. The high D/H in the atmospheres of Mars and Venus indicate loss of the light hydrogen isotope (and presumably H_2O) over time.

Photochemical reactions occur when high-energy photons modify the molecular forms of volatiles. Irradiation can also break apart larger molecules into smaller, lighter ones (photolysis), promoting Jeans escape. The penetration depth of photons depends on the optical depth at each particular wavelength of radiation, so most photochemical reactions occur at high altitudes.

Evidence for climate change on Mars, described in detail in Chapter 17, indicates that its atmosphere underwent significant modification. Ancient Mars likely had a denser atmosphere that was somehow lost, and conditions became more oxidizing over time. The details of its atmospheric evolution are unknown and await further exploration.

On Earth, atmospheric volatiles can be recycled into the planet's interior by subduction. This may be a uniquely terrestrial process, since only the Earth is known to have plate tectonics, but it has important consequences. Estimates vary widely, but there may be as much recycled water in the upper mantle as in the oceans, and more in the deep mantle, mostly within nominally anhydrous minerals like olivine and garnet. Atmospheric CO_2 levels have been diminished drastically by its incorporation into the lithosphere, and then subducted back into the mantle (described in Section 12.7.1).

Addition of oxygen produced by photosynthetic organisms has profoundly changed Earth's atmospheric composition. Biogenic oxygen, first produced by cyanobacteria,

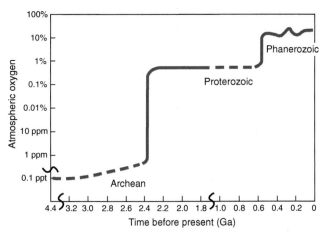

Figure 12.14 The abundance of oxygen in Earth's atmosphere has changed over time. Best-constrained periods are represented by a solid line. Modified from de Pater and Lissauer (2010).

was slow to accumulate because the oceans already had an inventory of reduced solutes, especially Fe^{2+}. Until this was oxidized, producing banded iron formations, this oxygen sink mopped up any free O_2. About 2.4 Ga, atmospheric O_2 rose to ~1 percent (Figure 12.14), marking the "Great Oxidation Event" (Lyons et al., 2014). An alternative explanation for this increase is a change in the composition of continental crust, from mafic to felsic, at that time (Lee et al., 2016); felsic rocks have less iron and thus would have been less effective in removing O_2 from the atmosphere. A later Phanerozoic increase in O_2 at ~600 Ma (Figure 12.14) has been attributed to changes in the amount of carbon in the atmosphere caused by volcanism. This rise prompted the evolution of animals. Another blip at ~300 Ma was influenced by the appearance of abundant land plants.

12.7 Geochemical Cycles and Their Consequences

The behavior of volatiles in planetary systems is commonly described in terms of **geochemical cycles**. Some important cycles include the water and carbon cycles on Earth, the water and sulfur cycles on Mars, and the nitrogen cycle on Titan. Geochemical cycles are difficult to describe quantitatively, so we will focus on the Earth's carbon cycle, since it has received the most attention and has important thermal consequences.

12.7.1 Earth's Carbon Cycle

The Earth's carbon cycle is illustrated in Figure 12.15. CO_2 is added to the atmosphere via volcanic eruptions. It is removed from the atmosphere by weathering of rocks in the presence of water, a chemical reaction that converts

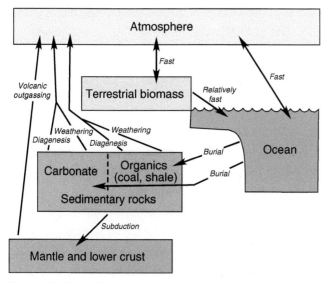

Figure 12.15 Earth's long-term carbon cycle, showing reservoirs and fluxes. Modified from Walther (2009).

CO_2 into bicarbonate ions, which in turn react to form carbonate rock. Dissolved bicarbonate in the oceans precipitates directly to form carbonate or is incorporated into the skeletons of organisms that die and collect on the ocean floor. Some CO_2 is also converted into organic compounds that eventually can form coal and petroleum. In these ways, CO_2 becomes a component of the lithosphere, which can ultimately be recycled back into the mantle by subduction. Mantle rocks are then melted, and ascending magmas carry dissolved CO_2 back to the surface, completing the cycle.

Figure 12.15 is illustrated as a "box model," in which the boxes represent the various reservoirs in which CO_2 is sequestered, with arrows indicating the fluxes of CO_2 transiting between reservoirs at different rates. The most important reservoirs in this model are the ocean $(50 \times 10^{18}\,\mathrm{g\ C})$, carbonate rock $(60{,}000 \times 10^{18}\,\mathrm{g\ C})$, and organic carbon in shales and coals $(15{,}000 \times 10^{18}\,\mathrm{g\ C})$. The rapid fluxes are labeled "fast"; flux rates for volcanic emissions, diagenesis, weathering, and burial are all much slower. This "long-term" model is appropriate for geologic (planetary) timescales, but the carbon cycle would be modeled differently if the duration of observation were shorter. The reservoirs in a "short-term" model of the carbon cycle are the biosphere, oceans, atmosphere, and

soils. These reservoirs exchange carbon on human timescales. Interest in the short-term cycle stems from the fact that it is affecting our climate.

12.7.2 Greenhouse Warming: Now and Then, Here and There

The **greenhouse effect** is the warming of a planetary surface by atmospheric absorption of heat energy. The atmosphere is transparent to ultraviolet/visible incident radiation, which is transformed into infrared energy by interactions on the surface. The emitted infrared radiation is absorbed by bending and stretching of greenhouse gas molecules in the atmosphere, so heat cannot escape. CO_2 is an especially effective greenhouse gas, so anything that increases its abundance can cause global warming.

Increase in atmospheric CO_2 since the Industrial Revolution has resulted in global warming on Earth by altering the atmospheric reservoir in the short-term model of the carbon cycle described above. On planetary timescales, greenhouse warming can also be important, and probably accounts for the lack of global freezing on the early Earth. Four and a half billion years ago, the Sun had a lower luminosity, ~70 percent of today's value. In the modern Sun, lots of hydrogen has been fused into helium, decreasing the Sun's mass (a ^4He atom has 0.7 percent less mass than four ^1H atoms), which in turn increases the rate of fusion and makes it hotter. A faint young Sun would have lowered global temperature by ~20 °C, so some mechanism for surface heating is required to have kept the oceans from being frozen solid. It seems likely that early outgassed CO_2 and H_2O, or possibly CH_4 and NH_3, may have been more abundant and promoted greenhouse warming.

The searing surface temperature of Venus is attributed to a greenhouse effect run amok. A runaway greenhouse effect results from the high pressure and CO_2-rich composition of its atmosphere, combined with its proximity to the Sun. This may account for the loss of Venus' water, since it would have accumulated in the atmosphere as steam, photo-dissociated at high altitudes, and subsequently escaped. Without oceans to dissolve CO_2 and precipitate carbonate, the Earth may have suffered a similar fate.

Summary

The giant planets were massive enough to capture directly and retain nebular gas, producing their hydrogen- and helium-rich primary atmospheres. The secondary atmospheres of the Earth, Mars, and Venus formed by outgassing.

Earth's atmosphere differs from the other terrestrial planets because its CO_2 has been sequestered in carbonate rocks and its O_2 was added by photosynthesis. The tenuous atmospheres of Mercury, Pluto, and various satellites formed by cosmic-ray irradiation or volcanism.

Atmospheric temperature is a balance between absorbed solar radiation and that reflected back into space. Saturation allows volatiles to condense as clouds, and ultimately to collect as oceans, seas, and lakes. Atmospheres are in constant motion, powered by differences in temperature and pressure.

The original sources of planetary volatiles are constrained by noble gas abundances and the isotopic compositions of hydrogen and other elements. Volatiles in the atmospheres and hydrospheres of the terrestrial planets resemble chondritic patterns, indicating that water and other volatiles were accreted as components of planetesimals. Bodies in the outer Solar System, like Titan, incorporated condensed ammonia and methane ices.

The elemental and isotopic compositions of planetary atmospheres have evolved through outgassing and atmospheric loss. The Earth's atmospheric evolution is unique, reflecting the roles of plate tectonics, liquid water, and life.

The behavior of volatiles in planetary systems can be described by geochemical cycles that quantify the partitioning of elements into reservoirs connected by fluxes. Greenhouse warming is a consequence of high concentrations of atmospheric gases that absorb infrared radiation, and has important thermal implications for planets.

Review Questions

1. How and why do the compositions of planetary atmospheres differ from that of Earth?
2. What determines whether an atmosphere is stable or unstable against vertical mixing?
3. How does global atmospheric circulation on different planets with atmospheres compare? What drives this circulation, and what modifies it?
4. How did the Earth originally acquire its volatile elements, and what evidence do we use to constrain the source of volatiles?
5. How are planetary atmospheres formed, and how might they evolve?
6. How does the greenhouse effect work, and why does atmospheric composition affect the surface temperatures on Earth and Venus?

SUGGESTIONS FOR FURTHER READING

Carr, M. H. (1996) *Water on Mars*. New York: Oxford University Press. A comprehensive description of the effects of water on martian geomorphology.

Taylor, F. W. (2010) *Planetary Atmospheres*. Oxford: Oxford University Press. Provides much more detail on energy balance, structures, dynamics, and compositions of planetary atmospheres.

REFERENCES

Baker, V. R., Strom, R. G., Gulick, V. C., et al. (1991) Ancient oceans, ice sheets and the hydrological cycle on Mars. *Nature*, **352**, 589–594.

Barlow, N. G. (2008) *Mars: An Introduction to Its Interior, Surface and Atmosphere*. Cambridge: Cambridge University Press.

Cabrol, N. A., and Grin, E. A., eds. (2010) *Lakes on Mars*. Amsterdam: Elsevier.

Carr, M. H. (1996) *Water on Mars*. New York: Oxford University Press.

Carr, M. H., and Head, J. (2003) Oceans on Mars: an assessment of the observational evidence and possible fate. *Journal of Geophysical Research*, **108**, 5042.

de Pater, I., and Lissauer, J. J. (2010) *Planetary Sciences*, 2nd edition. Cambridge: Cambridge University Press.

Feldman, W. C., Prettyman, T. H., Maurice, S., et al. (2004) The global distribution of near-surface hydrogen on Mars. *Journal of Geophysical Research*, **109**, E09006.

Grotzinger, J. P., Arvidson, R. E., Bell, J., et al. (2005) Stratigraphy and sedimentology of a dry to wet eolian depositional system, Burns formation, Meridiani Planum, Mars. *Earth and Planetary Science Letters*, **240**, 11–72.

Grotzinger, J. P., Gupta, S., Malin, M. C., et al. (2015) Deposition, exhumation, and paleoclimate of an ancient lake deposit, Gale crater, Mars. *Science*, **350**. DOI: 10.1126/science.aac7575.

Kivelson, M. G., Khurana, K. K., Russell, C. T., et al. (2000) Galileo magnetometer measurements: a stronger case for a subsurface ocean at Europa. *Science*, **289**, 1340–1343.

Lee, C-T., Yeung, L. Y., McKenzie, N. R., et al. (2016) Two-step rise of atmospheric oxygen linked to the growth of continents. *Nature Geoscience*. DOI: 10.1038/ngeo2707.

Lunine, J. I. (2004) Giant planets. In *Treatise on Geochemistry*, Vol. 1, ed. Davis, A. M. Oxford: Elsevier, pp. 623–636.

Lyons, T. W., Reinhard, C. T., and Planavsky, N. J. (2014) The rise of oxygen in Earth's early ocean and atmosphere. *Nature*, **506**, 307–315.

Niemann, H. B., Atreya, S. K., Carignan, G. R., et al. (1998) The composition of the Jovian atmosphere as determined by the Galileo probe mass spectrometer. *Journal of Geophysical Research*, **103**(E10), 22831–22845.

Nimmo, F., Hamilton, D. P., McKinnon, W. B., et al. (2016) Reorientation of Sputnik Planitia implies a subsurface ocean on Pluto. *Nature*, **540**, 94–96.

Orosei, R., Lauro, S. E., Pettinelli, E., et al. (2018) Radar evidence of subglacial liquid water on Mars. *Science*, **361**, 490–493.

Sarafian, A. R., Nielsen, S. G., Marschall, H. R., et al. (2014) Early accretion of water in the inner solar system from a carbonaceous chondrite-like source. *Science*, **346**, 623–626.

Stofan, E. R., Elachi, C., Lunine, J. I., et al. (2007) The lakes of Titan. *Nature*, **445**, 61–64.

Stuurman, C. M., Osinski, G. R., Holt, J. W., et al. (2016) SHARAD detection and characterization of subsurface water ice deposits in Utopia Planitia, Mars. *Geophysical Research Letters*, **43**, 9484–9491.

Tosca, N. J., and McLennan, S. M. (2006) Chemical divides and evaporate assemblages on Mars. *Earth & Planetary Science Letters*, **241**, 21–31.

Walther, J. V. (2009) *Essentials of Geochemistry*, 2nd edition. Sudbury, MA: Jones and Bartlett Publishers.

13

Planetary Aeolian Processes and Landforms

We describe the near-surface wind profile, its relation to environmental conditions, and how it can be quantified. The freestream wind speed can be converted to a friction wind speed, which relates to the flow at the atmosphere–surface interface and thus to the entrainment of sediment. The minimum wind speed for entrainment of aeolian sediment depends on gravity and grain size, so that threshold wind speed differs for varying planetary conditions. The difference in transport mechanism for grains leads to different depositional morphologies, which provide clues to the wind speed, wind direction, and sediment availability. Erosional landforms likewise provide information on near-surface atmospheric processes and surface sediments, as well as bedrock lithologies. The study of aeolian landforms thus informs our understanding of the atmosphere, surface geology, and sedimentology on other planets.

13.1 Bringing the Atmosphere Down to the Surface (and Why We Care)

In Chapter 12 we learned about atmospheric flow and the planetary boundary layer. But what happens when atmospheric flow comes into contact with the surface of a planet? Contextually, this question is both strange and obvious. It's strange because the large majority of planetary objects in the Solar System do *not* have both an atmosphere and a surface. For example, the giant planets, which constitute the greatest mass in our planetary system after the Sun, have atmospheres but no discrete surfaces. Conversely, the millions of asteroids, which are the most numerous planetary objects, have surfaces but no atmospheres. So for these massive and common populations, respectively, the question makes no sense. At the same time, the answer is obvious to anyone who resides on Earth. We are all familiar with wind-driven processes and the resultant landforms, the most familiar of which

perhaps are sand dunes. Dunes and other wind-derived landforms are termed "**aeolian,**" after Aeolus, the Greek god of the winds, and they have been found on virtually every planetary body that has a solid surface and a gaseous envelope (Figure 13.1). This gaseous envelope can take the form of a dense and stable atmosphere, as on Venus, or of a very tenuous, even transitory gaseous covering (sometimes called an "**exosphere**"), as on comets and other smaller bodies.

If aeolian processes require this rather uncommon confluence of both a solid surface and a gaseous envelop, why do we care about them? In part, we care because we live on Earth, where aeolian processes occur everywhere, influencing our everyday lives and even human history. In fact, scientific research dedicated to aeolian processes was initiated by Ralph Bagnold, a British engineer who conducted military countermeasures in North African deserts in World War II. Following this insightful field work, Bagnold spent decades conducting innovative wind tunnel research into sand movement and dune formation, thus laying the foundation for understanding aeolian processes. Despite their requirement for a somewhat limited confluence of conditions, aeolian processes are prevalent on our nearest planetary neighbors. Aeolian landforms have been documented on Mars, where aeolian sand transport is pervasive, and to a lesser extent on Venus, for which current data are less extensive and of lower resolution than for Mars. In addition, aeolian processes occur in the outer Solar System, such as on Titan and Triton, the largest satellites of Saturn and Neptune, respectively. They even occur on comets and on Pluto, which means they likely occur on the millions of other **Kuiper belt** objects in the outer Solar System.

Beyond their prevalence on terrestrial bodies, what information can aeolian landforms give us? Wind-driven landforms tell us both about the atmosphere (or exosphere) of the body and about the surface, as well as

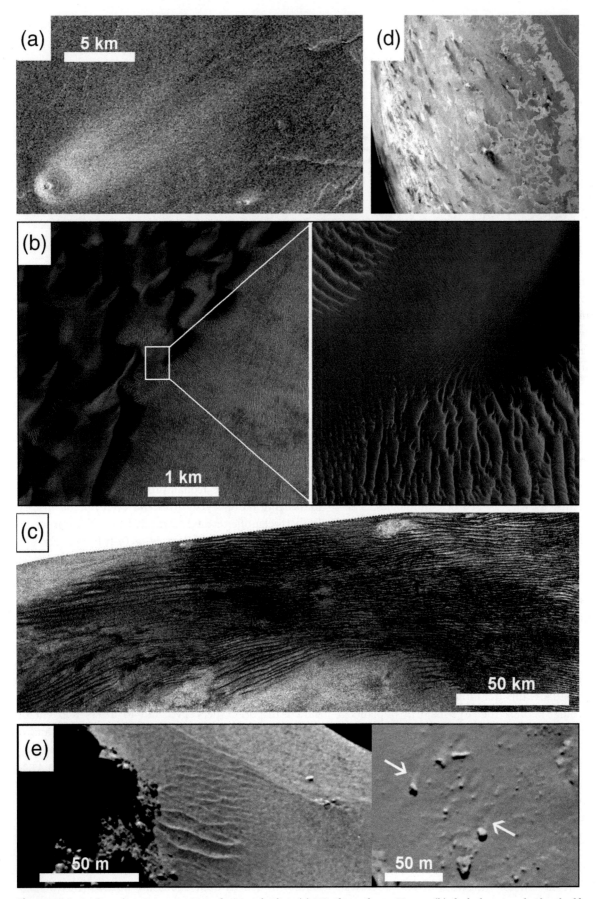

Figure 13.1 Aeolian deposits on various planetary bodies: (a) Wind streaks on Venus; (b) dark dunes and other bedforms on Mars; (c) synthetic aperture radar image of dark linear dunes on Titan; (d, upper right) dark wind streaks on Triton; (e) dune-like forms (left) and possible wind streaks (right) on Comet Churyumov-Gersimenko. NASA images.

BOX 13.1 USING WIND TUNNELS FOR AEOLIAN EXPERIMENTS

Following Bagnold's pioneering use of wind tunnels for investigating sand motion, wind tunnels have been used extensively to better understand aeolian processes and sedimentation. Most wind tunnels are terrestrial in nature, facilities that control ambient air to precise conditions to investigate processes that occur on Earth. Some terrestrial wind tunnels are portable, which enables geologists to run semi-controlled experiments under natural conditions in the field. Other wind tunnels are found in laboratories, where conditions, although more artificial, may be more precisely controlled. Examples of terrestrial laboratory wind tunnels are found at the Arizona State University in the USA and in the Trent University in Peterborough, Canada.

Planetary wind tunnels have also been constructed for investigating extraterrestrial aeolian processes. In the Planetary Aeolian Laboratory at the NASA Ames Research Center, facilities exist for simulating low-pressure atmospheric conditions, such as are found at the surface on Mars (the Mars Surface Wind Tunnel, or MARSWIT). Another high-pressure wind tunnel exists, originally constructed to simulate conditions on Venus, now refurbished for simulating surface atmospheric conditions on Titan (the Titan Wind Tunnel, or TWT – Figure 13.2).

Wind tunnels permit simulation of specific conditions relevant to different planetary processes. For example, one of the most basic aeolian processes is **entrainment** – getting stationary grains into motion. Entrainment on planetary bodies with thin atmospheres occurs largely through impact of upwind grains, whereas under thick atmospheres, entrainment occurs through fluid flow. To numerically model this difference, however, requires experimental data with which to test and tune the model. Conducting entrainment experiments at different pressures in wind tunnels provides such data.

Wind tunnels can also be used to investigate erosional aeolian processes. For thinner (low-density, lower-pressure) atmospheres, the wind must be moving more quickly to generate the force necessary to move grains, compared to thicker (higher-density, higher-pressure) atmospheres. Thus, aeolian **abrasion** should be more effective under thin atmospheres. Wind tunnel experiments provide data to quantify this difference in abrasion effectiveness with atmospheric density, and thus support interpretations of **ventifacts** and **yardangs** (discussed in Sections 13.6.2 and 13.6.1, respectively).

Simply put, aerodynamic forces are dominant in aeolian processes under high-density atmospheres, whereas ballistic (gravity-driven) interactions are dominant under low-density atmospheres. Having planetary wind tunnels that enable simulation of both of these extraterrestrial conditions has left us an impressive legacy in understanding the effects of different conditions on aeolian transport. As we expand our awareness of the pervasiveness of aeolian processes, laboratory simulation of planetary aeolian conditions will continue to provide fundamental information for interpreting these extraterrestrial planetary landforms (Figure 13.1).

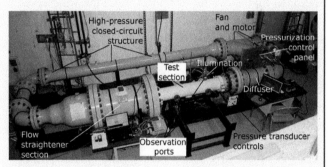

Figure 13.2 The Titan Wind Tunnel, with components of the structure labeled. For experiments, the test section is rolled toward the viewer, the test plate is inserted and covered with sediment, and the test section is rolled back into place. After sealing, the wind tunnel is pressurized. Lastly, wind is generated with the fan at the back right of the image, and observations of sand motion are made through the observation ports.

about the existence, characteristics, and cycling of both volatiles and sediments. In this way, the study of aeolian geology provides critical clues to the development and evolution of planetary surfaces. To access this information, however, we first have to understand how the wind interacts with and shapes these surfaces.

13.2 The Near-Surface Wind Profile

Imagine wind flowing across a land surface. Personal experience tells us that the wind speed decreases with proximity to the surface, as a result of the land surface exerting friction on the near-surface wind. The region of decreasing wind speed is referred to as the **boundary layer**.

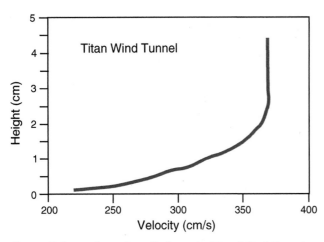

Figure 13.3 Wind speed profile from the Titan Wind Tunnel, with a typical logarithmic shape showing an increase in wind speed with distance from the surface.

This friction so retards the flow of the gas molecules closest to the surface that, at some theoretical height above the surface – termed "roughness height" and symbolized as z_0 – the wind flow ceases. For a relatively smooth surface, the lowest portion is denoted the laminar sublayer, where gas molecules move in layers (laminae) without turbulence. In contrast, the flow of gas molecules higher above the surface experiences less friction. This reduced friction effect means less slowing of the wind speed, so that above the boundary layer the wind speed no longer decreases with height. The result is a characteristic wind profile of logarithmically increasing speed with height (Figure 13.3).

The wind speed that can be experienced or measured by instruments is referred to as the **freestream wind speed**. Although measurement of the freestream wind speed is quite useful (think of weather forecasts), its variation with distance above the surface makes its application to surface processes somewhat ambiguous. Thus, a more independent measure of wind speed at the surface is desirable. For this purpose, the aeolian community uses the **friction wind speed** – think of this as a shear speed, a quantity that represents the strength of the gas flow at the interface with the surface. It is derived from (and directly proportional to) the freestream wind speed, but is independent of height. The quantitative relationship between freestream wind speed (u) and friction wind speed (u^*) is given by the "law of the wall," namely:

$$u_z = \frac{u^*}{\kappa} \ln \frac{z}{z_0} \qquad (13.1)$$

where z is the height above the surface at which the freestream wind speed is measured and κ is the von Karmán constant (~0.4). In theory, the equation can be inverted to solve for friction wind speed (u^*) from the

freestream wind speed measured at a single height, although in practice deriving z_0 requires measurements of the freestream wind velocity at two heights.

13.3 The Physics of Particle Entrainment

The wind speed required for entrainment of particles depends on the interplay of a number of forces, as well as the size of the particles.

13.3.1 Force (Torque) Balance: The Conditions for Entrainment

This near-surface wind exerts a force on the surface. For a granular surface, the wind forces acting on each grain include both a drag force (F_D) that results from the flow of gas over the grain surface, and a lift force (F_L) that results from the greater wind velocity over the top than the bottom of the grain. At the same time, the grain experiences gravity (F_g) as well as an interparticle force (F_{ip}) that results from water absorption, electrostatic attraction, and van der Waals forces. Gravity – or actually weight – increases for more massive (e.g., larger) grains. The interparticle forces increase with decreasing grain size (i.e., smaller particles stick together better), because smaller grains have a larger surface area-to-mass ratio, and the interparticle forces are a function of the surface area. Even dry, clay sticks together better than sand!

A balance of these forces provides the conditions under which the movement – or entrainment – of the grain by the wind would occur. In the natural world, grains are almost always perched on other grains, as in a sand dune, so that a grain at rest lies on a non-smooth surface. Thus, the forces acting on the grain must be expressed at torques (force multiplied by a lever arm distance) around a pivot point (Figure 13.4).

The individual torques can be expressed as:

$$\begin{aligned}
r_D F_D &\sim K_D \rho_a D^3 u^{*2} \\
r_g F_L &\sim K_L \rho_a D^3 u^{*2} \mathrm{Re}^* \\
r_g F_g &\sim K_g (\rho - \rho_a) g D^4 \\
r_{ip} F_{ip} &\sim K_{ip} D^2 \; (\text{assumes } F_{ip} \propto D)
\end{aligned} \qquad (13.2)$$

in which $K_{subscript}$ denotes some empirically derived constant, ρ is the grain density, ρ_a is the atmospheric or gas density, and D is the grain diameter. Re^* denotes the **Reynolds number** calculated with the friction wind speed (u^*), or $u^* D / v$, where v is the kinematic viscosity. The variable "r" with different subscripts denotes the distance of the lever arm for each of the corresponding forces.

Entrainment occurs when the wind-derived torques just barely exceed the torques holding the grain in place, that is, when

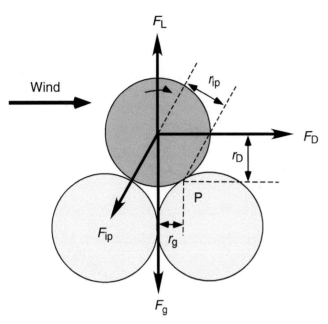

Figure 13.4 Force (torque) balance diagram, showing the various forces and their lever arms acting on a stationary grain before entrainment.

$$r_D F_D + r_g F_L \approx r_g F_g + r_{ip} F_{ip} \qquad (13.3)$$

Solving this torque balance for u_* gives the minimum, or threshold, friction wind speed at which entrainment occurs. The equation for this threshold wind speed is:

$$u_{ft}^* = A \sqrt{\frac{\rho - \rho_a}{\rho} gD} \qquad (13.4)$$

where u_{ft}^* is the threshold (minimum) friction wind speed for entrainment occurring only under the flow of the gas or "fluid." (As described in the next section, entrainment can also occur through the impact of other grains.) The variable A has a lot of hidden complexity because the interparticle forces are more complicated than a simple dependence on D. The equations for A contain (D^n) where n is ~2. Because of this exponent, the threshold wind speed curve as a function of D has a U-shape (Figure 13.5a). The minimum value (the bottom of the U) is termed the optimum diameter, or the grain size that is most easily moved by the wind.

13.3.2 Entrainment by Fluid and by Impact

The above derivation for the threshold wind speed considers only the flow of the gas. As gas is a fluid, this threshold speed is referred to as the fluid threshold speed. However, in the natural world, grains are also moved by the impact of other grains. The impacts of these other grains impart force to the bed, and thus affect the simple force balance shown in Figure 13.4.

The effects of impact vary with the density of the planetary atmosphere. For thinner atmospheric densities, as on Earth or Mars, the wind speeds have to be greater in order to impart the minimum necessary force to the grain to cause entrainment (Figure 13.6a). Thus, the impact of these fast-moving grains adds significantly to the force imparted by the wind to the grains on the surface, and this "splash" effect contributes to the movement of the grains. As a result, the force from the wind does not have to be as great to entrain the grains on the surface. Under these conditions, the impact threshold wind speed is less than the fluid threshold wind speed. On Earth, the impact threshold is roughly 90 percent of the fluid threshold. On Mars, with its even less-dense atmosphere, the differential is even greater: the impact threshold is roughly 10 percent of the fluid threshold. This significant differential means that grains may be entrained by a passing gust of wind, but because they then impact the grains around them, they may produce a cascading effect that – even at low wind speeds – results in greater movement than would be expected based on the wind speed alone. This effect is likely a significant factor in the global dust storms observed on Mars.

In contrast, on planetary bodies with dense atmospheres, such as Venus or Titan, the wind speed necessary to entrain the grains can be considerably slower. Because the grains are moving more slowly, they do not impart much additional force when they impact the grains on the surface. Thus, under denser atmospheric conditions, the wind speed for impact threshold is no less, and may be even numerically greater, than for the fluid threshold. Under these conditions, the wind speed for fluid threshold has to be maintained for continued transport (Figure 13.6b).

13.4 Aeolian Transport of Sediment

The movement of sediment particles occurs by several mechanisms. Before we explore these processes, we must consider how particle size affects grain transport.

13.4.1 Terminal Velocities for Sand versus Dust

Once grains of sediment have been entrained into motion through the atmosphere, they tend to fall. This simple and rather obvious statement actually provides an important means for distinguishing types of sediment, because different sediment sizes fall at different rates. Grains that fall more quickly do not travel as far (although they have more of an abrasive effect on the land surface over which they travel). Grains that fall more slowly can travel farther, and in some cases, much farther.

The terminal rate (or velocity) at which grains fall can be calculated using a force balance approach, in this case,

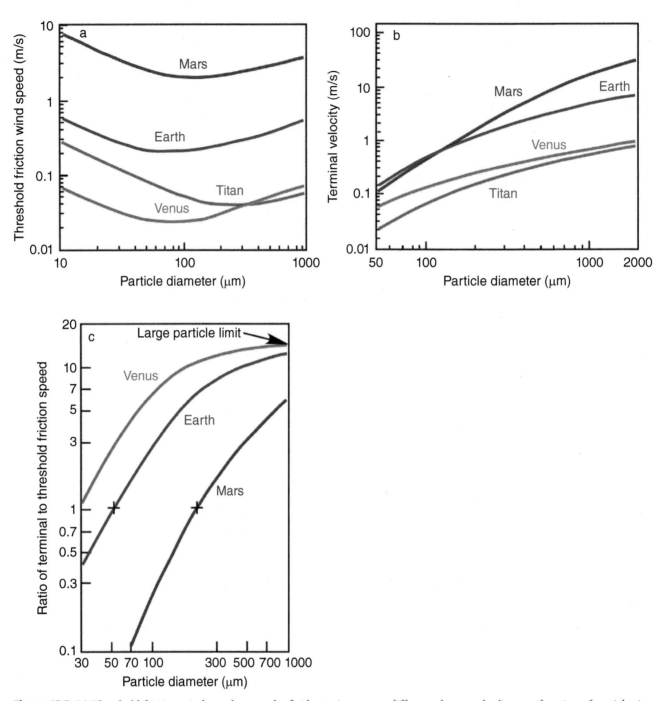

Figure 13.5 (a) Threshold friction wind speed curves for fluid entrainment on different planetary bodies as a function of particle size, derived from Equation 13.4. (b) Terminal velocities on different planetary bodies as a function of particle size, derived from Equation 13.6. (c) Ratio of terminal to threshold friction wind speeds. Grain sizes with values above 1 indicate transport through the atmosphere by suspension, whereas grain sizes with values above 1 indicate transport over the surface.

between the weight of a falling body and the drag exerted on it by the deflection of the atmosphere. Solving this force balance for velocity gives the equation

$$u_t = \sqrt{\frac{4\,(\rho - \rho_a)}{3\,\rho_a}\frac{gD}{C_D}} \qquad (13.5)$$

where u_t is the terminal velocity and C_D is a drag coefficient, largely a function of shape, which equates to ~0.4 for a sphere. This equation shows, as expected, that bodies fall faster if they are larger, denser, or under greater gravitational acceleration, and that they fall more slowly if the atmosphere is denser and/or the drag for the body shape is

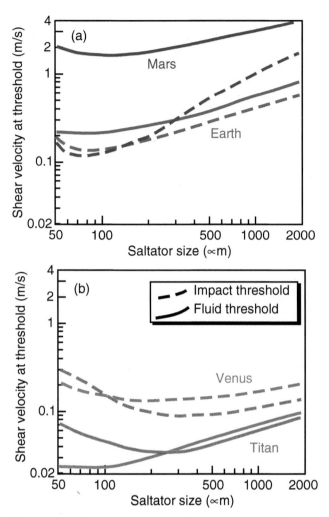

Figure 13.6 Threshold friction wind speed curves for fluid and impact entrainment on a variety of planetary bodies. (a) Planets with less dense atmospheres (Earth and Mars) have impact thresholds that are less than their fluid threshold, such that the wind speed may decrease (significantly for Mars) and the sand grains will remain in motion. (b) Planets with more dense atmospheres (Venus and Titan) have impact thresholds that are greater than their fluid thresholds. Modified from Kok et al. (2012).

greater. This approximation holds best for grains in the turbulent region, where the atmosphere creates drag due to deflection around the falling body. This condition implies that the falling body is relatively large, and under this condition, the atmospheric viscosity is not a factor (as shown by its absence from Equation 13.5).

For very small particles, the atmospheric viscosity (instead of deflection of the atmosphere around the body) provides the drag. In this case, the terminal velocity is approximated using **Stokes' Law**, which is

$$u_t = \frac{(\rho - \rho_a)gD^2}{18\mu} \tag{13.6}$$

For small particles, this equation shows that the terminal velocity is inversely proportional to the viscosity of the gas and directly proportional to the square of the diameter (or surface area) of the particle. The square of a small number – such as the size of a $\sim 10^{-6}$ m cloud droplet – is very small, showing that the terminal velocity of small grains is very slow. Together, these equations can be combined to yield a continuous curve for terminal velocity (Figure 13.5b).

Armed with this understanding of how grain size affects grain transport, we can now distinguish between major classes of grains. Larger sediment readily falls back to Earth. Here, we define "readily" as the condition in which the threshold friction velocity of the wind exceeds the terminal velocity of the grain, or where

$$u_t / u_{ft,it}^* < 1 \tag{13.7}$$

in which u_t is terminal velocity of the grain under the specific gravity and atmospheric viscosity conditions, and $u_{ft,it}^*$ is threshold friction velocity for fluid or impact entrainment, governed largely by the atmospheric density. These larger grains move in more frequent contact with the surface, produce specific bedforms when they are deposited, and because they interact with the surface during transport, they cause more erosion. Conversely, if the terminal velocity is small compared to the wind speed that entrained the grains, or

$$u_t / u_{ft,it}^* > 1 \tag{13.8}$$

then the grains tend to stay in suspension (Figure 13.4c). Suspended grains travel farther and, because they fall more uniformly out of suspension, then tend to blanket landscapes without forming discrete bedforms.

13.4.2 Transport Mechanisms

Based on this understanding of how grain size affects transport, we can now be more specific about the different processes by which sediment is transported by the wind (Figure 13.7). Grains that move along the surface are transported by multiple, interrelated processes. The least energetic process is **creep**, in which grains are pushed, slid, or rolled over the surface by the force of the wind. With increasing wind speed, grains begin to bounce or hop over the surface in a process called **saltation** (from the Latin for jump). Saltating particles, when they impact, can cause other particles to be entrained in low hops for short distances, a process called **reptation** (related to the word reptile, which is from the Latin for crawl).

Smaller grains move in suspension, with infrequent or no interaction with the surface. Although small grains require relatively high wind speeds to be entrained

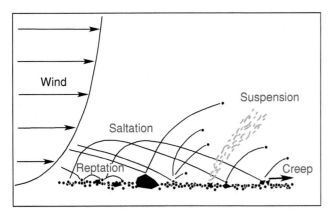

Figure 13.7 Diagram of sediment transport modes.

(remember the discussion of Figure 13.5), they require less energy than larger grains for transport because of their lesser mass (weight). Thus, dust is transported in suspension over very long distances. Dust from the Sahara Desert in Africa (Figure 13.8a) has been found in South America, and dust storms on Mars can become near-global in extent (Figure 13.8b). Dust can also be moved in dust devils, localized vortices that can exert sufficient lift to entrain dust (Figure 13.8c). Dust devil tracks on Mars are common in orbital images, showing both the frequency of dust devils and the prevalence of surface dust (Figure 13.8d). Dust devils in action have even been imaged by Mars rovers!

13.5 Aeolian Deposition and Planetary Landforms

When the wind flow slows to a speed below that necessary to keep grains in transport, the grains are deposited onto the surface. However, the mode of transport that the grains *were* experiencing before deposition strongly governs the resultant depositional landforms. The reason for the reduction in wind speed also influences the morphology and location of the deposition.

13.5.1 Depositional Landforms for Sand

As distinctive landforms that are often visible from orbit, sand dunes can provide valuable clues to the near-surface atmospheric processes (e.g., wind speed, direction), sediment locations and the transport pathways by which they arrive at those locations, and surrounding surface characteristics. Dunes, which build up through the progressive accumulation of sand, form different morphologies based on the conditions under which the sand was deposited.

This accumulation of sand commonly occurs in the lee of flow obstacles, such as rock outcrops. Dunes can also

form on relatively flat terrain, where surface roughness or variability in wind flow may result in an initial sand deposition, which acts as a flow obstacle, causing more sand deposition in its lee and building the dune through positive feedback. The transport of sand over a dune begins with saltation and, to a lesser extent, reptation/creep up the stoss (or upwind) side of the dune (Figure 13.9). As the sand builds up at the dune crest, this localized accumulation becomes unstable and the sand avalanches down the lee (or downwind) side of the dune. The morphologic result is an asymmetric deposit with a more shallowly sloping upwind side (on Earth, common upwind slope values are 10–15°) and a lee side that dips more steeply, closer to the **angle of repose** (on Earth, common downwind slope values are 30–35°). The avalanching of sand down the lee side also engenders an internal stratigraphy to the dunes, specifically, downwind tilting strata also near the angle of repose. The lee slope does not attain an angle of repose due to deposition by other mechanisms, such as airfall, in which grains are lofted over the crest and settle out onto the lee face.

This net movement of sand from the upwind to the downwind side of a dune results in downwind migration of the dune. This dune movement may not be equal at all points in the dune field or even on a single dune, and the variation in sand transport results in variation in dune morphology (Figure 13.10). In general, smaller amounts of sand move more quickly than larger amounts of sand. This tendency means that for a single dune that is initially transverse to the wind, the ends (or horns) of the dune advance more quickly than the main body of sand. The bending of these dune tips downwind results in curved morphology in plan view, recognizable as a barchan dune. Under wind regimes that are more variable in direction, other morphologies, such as dome or star dune morphologies, result. Linear (or longitudinal) dunes occur where wind directions are bimodal, transporting sand first from one side of the dune, then from the other.

Thus, dune morphology and orientation (Figure 13.11) are fundamentally controlled by variability in wind direction and by sand availability (Figure 13.10). Sand availability is a combination of both the presence of sand-sized sediment and the readiness of that sediment for transport. For example, moist sand is less readily transportable than dry sand, so that sand availability may vary by season or even time of day. Where sand availability is high, dunes are larger and/or more closely spaced, even forming continuous ridges of dunes. If sand availability is low, the resultant dunes tend to be smaller, fewer in number, and/or more dispersed. While variability in the direction of winds above threshold influences dune morphology, the direction (or directions) of those winds

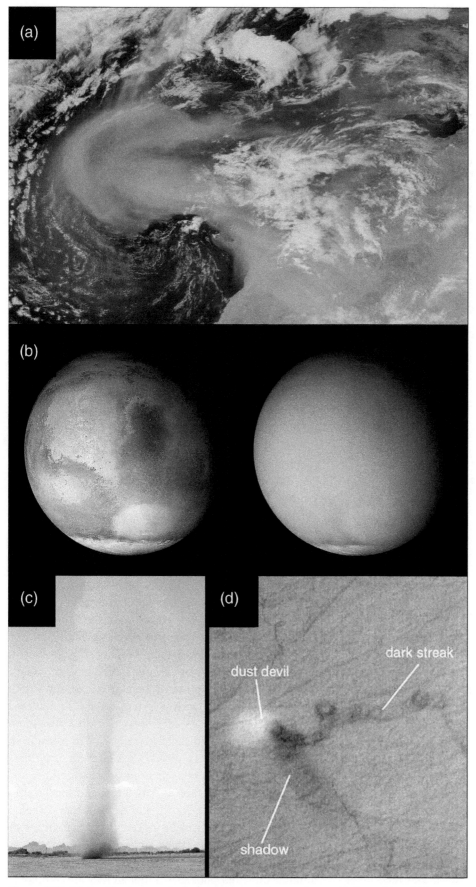

Figure 13.8 (a) Satellite image of a dust storm on the West Coast of Africa. (b) Hubble Space Telescope images of Mars showing a relatively dust-free atmosphere (left) and a global dust storm (right). (c) Ground image of a dust devil in Arizona. (d) Satellite image of dust devil and its track on Mars. NASA images.

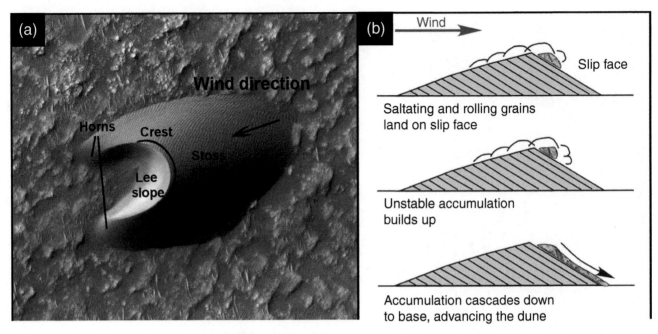

Figure 13.9 (a) Nomenclature for the parts of a dune on Mars (HiRISE image). (b) The granular processes of dune formation and the resulting internal stratigraphy.

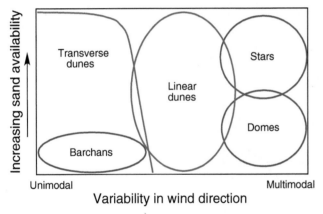

Figure 13.10 Plot of sand availability versus wind speed, showing stability fields for dune morphologies.

govern(s) the orientation of the resultant dunes. As the names imply, transverse dunes are transverse or perpendicular to the direction of the prevailing winds above threshold, whereas linear dunes are oriented between the two directions of the bimodal winds. In addition to wind direction and variability and sand supply, dune morphology and orientation are also influenced by anchoring or impedance. This effect of impedance means that certain dune types are particularly useful wind direction indicators.

Sand dunes, both individually and as fields, can be quite extensive in size. Dunes grow through the overtaking of larger dunes by smaller dunes and their amalgamation. Individual sand dunes reach ~200 m in height and occupy ~15–20 percent of the Sahara Desert, which is roughly the same size as the continental USA. Of course, this estimate means that ~75 percent of the Sahara is *not* covered by sand, but by areas of stone pavement or bedrock. Bedrock outcrops provide a ready source for sand on Earth, which is recycled through formation and erosion of sandstone and other sedimentary lithologies. On Titan, linear (or longitudinal) dunes cover most of the low latitudes, with individual dunes attaining lengths of tens of kilometers and heights similar to those in the Sahara on Earth. The origin or primary source of sand on Titan is not known – possibly atmospheric aerosols accreted into sand-sized particles – but the large size and extent of the dunes indicate that sand availability must be (or have been) high. Sand and dunes are globally distributed on Mars, and are concentrated within craters, around the layered deposits at the North Pole, and within the Valles Marineris canyon system, as well as scattered on intercrater plains. As on Titan, the source of sand on Mars is not known. Volcaniclastic deposits or ejecta from impacts into lava plains, possibly dating from early Mars when impacts were more numerous, are possibilities.

Sand deposits may form at both larger and smaller sizes than dunes. At the larger size, sand deposits may form flat or topographically low plains of sediments that are larger than sand size. Because these large grain sizes are not as readily transported by the wind, sand sheets exhibit no or minimal dune morphologies, although they are commonly found on the margins of dune fields. At the smaller size,

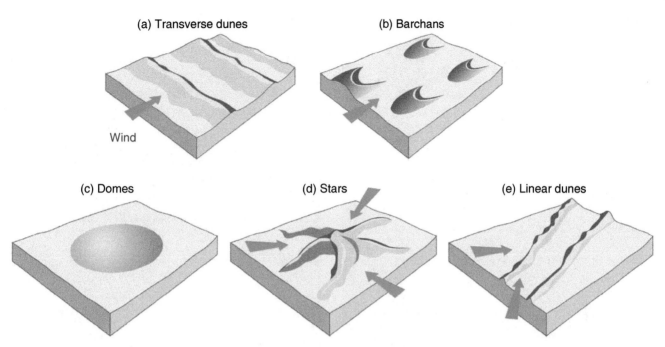

Figure 13.11 Dune morphologies are strongly controlled by wind direction and sand supply. Adapted from McKee (1979).

sand deposits form ripples as the products of reptation or creep. Mars exhibits some aeolian deposits that are larger than ripples on Earth but smaller than dunes. These "transverse aeolian ridges" may be specific to the martian conditions of a thin atmosphere and low gravity.

13.5.2 Depositional Landforms for Dust

When fine-grained dust settles out of suspension, it tends to blanket the landscape. Thus, in contrast to sand dunes, dust deposits do not form discrete landforms but rather a mantle that covers pre-existing landforms. Because it rarely develops landforms, this mantle can be difficult to discern remotely, but it can be inferred from a variety of clues. Tracks, where a thin layer of dust has been removed during the passage of a dust devil or a rover, are evidence of a thin dust mantle on Mars (Figure 13.8d). Dust deposits may be quite thick, as on Earth where they form deposits of loess (fine glacially scoured sediment) that are hundreds of meters thick. The Yellow River in China, which drains formerly glaciated regions, gets its name from the enormous volumes of loess or glacial dust that it transports.

On Mars, dust deposits are also areally extensive, although the thicknesses of these deposits are not well known. Although requiring higher wind speeds to entrain, dust is nevertheless extensively mobilized on Mars. Regional or global dust storms occur every few years (the frequency varies with orbital configuration).

Because dust responds differently to heating than does sand or rock, thermal inertia maps can be used to detect surficial dust; they show that dust is distributed throughout the equatorial and low-latitude zones on Mars.

Dusty mantles can also include ice. The mid-latitudes of Mars show land surfaces that are currently undergoing disaggregation. This disaggregating terrain is interpreted as an ice-rich dust mantle that was emplaced during the last glacial (high-obliquity) period on Mars, but that is currently undergoing disaggregation or breakdown due to the current interglacial (low-obliquity) period. As this breakdown releases fine grains, dust loading in the martian atmosphere may increase.

13.6 Planetary Erosional Landforms

In addition to creating landforms by depositing material, aeolian sediments can also create landforms by removing material.

13.6.1 Yardangs

Yardangs (from the Turkic word meaning "steep bank") are erosional landforms composed of sedimentary rock that has been abraded by wind-borne sand into a streamlined form (Figure 13.12a). Yardangs are extensive on Mars (Figure 13.12b), perhaps because the high threshold

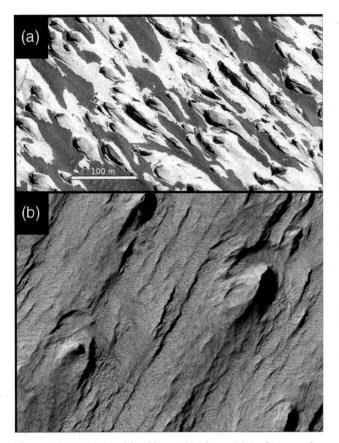

Figure 13.12 Erosional landforms. Yardangs (a) in the central Andes mountains, Chile (Google Earth), and (b) in Tithonium Chasma, Mars (NASA image). The scale bar in (a) applies also to (b).

Figure 13.13 Ventifacts (a) in Death Valley, California (released under terms of the GNU FDL), and (b) near the Spirit rover landing site in Gusev crater, Mars (NASA image).

wind speed, which gives wind-borne sand a very high kinetic energy, is more erosive than under denser atmospheres (e.g., on Earth). In addition, extensive sedimentary layers provide the necessary bedrock for yardang formation. Yardangs have also been hypothesized to exist on Titan and Pluto.

13.6.2 Ventifacts
Whereas yardangs are landscape in scale, ventifacts occur at sub-outcrop scale. Ventifacts result from abrasion of crystalline rock, although less defined examples can also form in sedimentary lithologies. Thus, like yardangs, they serve as current or paleo-wind indicators. Abrasion of crystalline boulders results in faces (or facets) with polished surfaces, pits, subparallel grooves, and/or flutes where the facets join. Such ventifacts have been observed on Earth (Figure 13.13a) and by rovers on Mars (Figure 13.13b), where the facets' dip slopes and groove direction give evidence of the direction of wind

with speeds above threshold and an upwind source of available sand.

13.7 Combined or Ambiguous Planetary Landforms

Beyond known depositional or erosional landforms, the wind also creates landforms whose mechanism of formation is ambiguous.

13.7.1 Stone Pavements
Stone pavements are composed of extensive surficial sheets of coarse sediment, generally grains that are pebble-size or larger. Two models of formation are proposed for pavements. In one, the pavement is a lag deposit left behind by aeolian winnowing of more easily

Figure 13.14 Stone pavements (a) in Australia (Mark Marathon, https://commons.wikimedia.org), and (b) in Gale Crater, Mars (NASA image).

transportable (e.g., sand-sized) grains. In this model, the land surface is deflated (lowered) during the formation of the pavement. In the other, dust and fine silt accumulate on a pebble- or cobble-strewn land surface but, because of their smaller sizes, sift downward between the coarser particles. In this model, the land surface was largely pre-existing but has been inflated (raised) by the addition of fine-grained sediments. Stone pavements on Earth and Mars are shown in Figure 13.14. The Mars pavement is strewn with hematite concretions ("blueberries") eroded out of the rock during deflation.

13.7.2 Wind Streaks

Elongated areas of different albedo than the surrounding landscape can be wind streaks. These aeolian features occur downwind of obstacles and may result either from deposition of sediment due to a reduction of wind speed in the obstacle lee or from the removal of sediment due to increased turbulence. Thus, wind streaks on Mars are commonly bright due to deposition of dust, but may also be dark where a dust covering has likely been removed by movement of sand by turbulent eddies. Wind streaks that appear bright in radar images of Venus are evidence of larger grain sizes in the lee of the obstacle, whereas wind streaks that appear darker are smoother, perhaps being covered with smaller (e.g., silt) grain sizes (Figure 13.1a). In other situations, wind streaks may form downwind of a sediment source, for example on Triton (Figure 13.1d), where they are interpreted as deposits from geyser-like eruptions of gas, dust, and ice from the polar cap.

Summary

While atmospheric phenomena themselves, such as clouds, provide evidence of atmospheric conditions (Chapter 12), aeolian landforms give valuable clues to atmospheric processes occurring at the surface, including wind direction, strength, and variability. These processes can fundamentally affect the geology of planetary surfaces, redistributing sediments and changing surface albedo. These changes, in turn, cause variations in surface heating, which feed back into atmospheric processes, and provide records of sediment transport and redistribution. Dune morphologies provide evidence both of winds above threshold for extended periods of time and of sand availability in the upwind direction, whereas the orientation of the dunes provides evidence for the direction of these winds and the location of the sand source. Wind streaks provide similar information about the wind, whereas dust is more strongly linked to climate. Yardangs and ventifacts provide evidence not only of wind and sediment supply, but also of the bedrock lithology. Thus, the study of aeolian landforms is an important source of information for understanding planetary geologic processes.

Having considered how flowing gases influence planetary geology, we will next address the same question for flowing liquids.

Review Questions

1. What conditions are necessary for aeolian landscapes? Does your answer to this question differ between erosional and depositional landscapes? For different planets?
2. What information do aeolian landscapes provide? Does your answer to this question differ between erosional and depositional landscapes? For different planets?
3. How can aeolian landscapes be used to understand sediment transport paths?
4. How can aeolian landscapes be used to constrain atmospheric models?
5. Compare and contrast aeolian processes on Earth, Mars, Venus, and Titan.

SUGGESTIONS FOR FURTHER READING

Bagnold, R. A. (1941) *The Physics of Blown Sand and Desert Dunes*. London: Methuen and Co. Originally published by Methuen and Co, reprinted by Dover Publications. A classic in terrestrial aeolian science, no worse for age and inspiring in its technological creativity and the resultant insights.

Greeley, R., and Iversen, J. (1985) *Wind as a Geologic Process*. Cambridge: Cambridge University Press, An older but still useful "first view" of planetary aeolian processes.

Lorenz, R. A., and Zimbelman, J. R. (2014) *Dune Worlds: How Windblown Sand Shapes Planetary Landscapes*. New York: Springer-Praxis. A recent peripatetic review of sand movement by wind, how it can be studied, and what it can tell us.

McKee, E. D., ed. (1979). *A Study of Global Sand Seas*. Washington, DC: USGS. https://pubs.er.usgs.gov/publication/pp1052. A nice survey of global-scale deposits on Earth.

REFERENCES

Bridges, N. T., Greeley, R., Haldemann, A. F. C., et al. (1999) Ventifacts at the Pathfinder landing site, *Journal of Geophysical Research*, **104**, 8595–8615.

Burr, D. M., Bridges, N. T., Marshall, J. R., et al. (2015) Higher-than-predicted saltation thresholds on Titan. *Nature*, **517**, 60–66.

Ewing, R. C., Hayes, A. G., and Lucas, A. (2015) Sand dune patterns on Titan controlled by long-term climate cycles. *Nature Geoscience*, **18**, 15–19.

Greeley, R., Bender, K. C., Saunders, S., et al. (1997) Aeolian processes and features on Venus. In *Venus 2: Geology, Geophysics, Atmosphere, and Solar Wind Environment*, eds. Bougher, S. W., Hunten, D. M., and Phillips, R. J. Tucson, AZ: University of Arizona Press, pp. 13–62.

Greeley, R., Whelley, P. L., Neakrase, L. D. V., et al. (2008), Columbia Hills, Mars: aeolian features seen from the ground and orbit. *Journal of Geophysical Research*, **113**, E06S06.

Kok, J. F., Parteli, E. J. R., Michaels, T. I., et al. (2012), The physics of wind-blown sand and dust, *Reports on Progress in Physics*, **75**(10), 72. DOI: 10.1088/0034-4885/75/10/106901.

McKee, E. D., ed. (1979). *A Study of Global Sand Seas*. Washington, DC: USGS.

14

Planetary Fluvial and Lacustrine Landforms:
Products of Liquid Flow

We describe the flow of liquids – water in the inner Solar System, hydrocarbons on Saturn's moon Titan – and its effects on planetary surfaces. Liquids fall onto, flow through, and emerge from planetary landscapes. The resultant entrainment, transport, and deposition of sediment are observed in a variety of forms, which can be ascribed to the variety of surficial and subsurface flow conditions. As the area within the highest topographic elevations surrounding a river network, the drainage basin provides a natural hydrologic unit for defining and discussing these various processes. In cratered landscapes on Mars and Titan, drainage divides are often obscured by impact craters and by atmospheric degradation, although in younger terrains the crater rims themselves often demarcate the drainage divides. River networks exhibit morphologies based on surface and subsurface controls on the flow. Whereas networks on Earth are primarily dendritic (branching in a tree-like fashion), the majority of network morphologies on Titan are rectangular, suggesting tectonic influence. Deposits from channelized flow provide data on the flow conditions and sediment load. Fans on Mars and Titan provide evidence of subaerial deposition. Deltaic deposits on those bodies, along with possible shorelines and inferred tsunami deposits around the northern lowlands of Mars, imply deposition and erosion in lakes, seas, and perhaps even a vast martian ocean. Fluvial, alluvial, and lacustrine landforms thereby provide insights into climate, surface, and sedimentologic processes on planetary bodies.

14.1 Volatile Landscapes

If the confluence of requirements is unusual for aeolian landscapes (Section 13.1), that description is even more apt for landscapes shaped by flowing volatiles. Earth is, once again, a poor example in this regard. Earth is 70 percent covered by vast oceans that have played a controlling role in the evolution of our planet. In addition to their overwhelming physiochemical effects on the composition of terrestrial rocks, they are both source and sink for the hydrologic cycle, which leaves unmistakable geomorphic signatures on land surfaces. No other body in our Solar System has an active hydrologic cycle in which temperature and pressure conditions permit a vast volatile reservoir to change among the three different phases of matter. So of what use for planetary geology is studying our planet's hydrologic landforms?

Exactly because water is so constrained to specific temperature–pressure conditions, hydrologic landforms provide distinct clues to changes in planetary conditions over time. Thus, the existence of ancient river (**fluvial**), lake (**lacustrine**), and even ocean (**marine**) deposits provides unmistakable evidence that Mars has not always been the cold desert that we see today. And because water is requisite for life as we know it, correctly interpreting hydrologic landforms is integral to astrobiological exploration (Chapter 16). Other volatiles besides water can participate in cycling, as on Saturn's largest satellite, Titan. And any cycling of volatiles also involves movement of sediments, which produces fundamental geologic changes in the distributions and compositions of planetary materials. Thus, for the forensic science of geology, we study volatile landforms for critical clues in the discernment both of planetary evolution and the potential for life.

Earth provides the foundation for this study, so its hydrologic processes and landforms are the basis for this chapter. Hydrologic processes also apply to Mars, whereas the operative volatiles on Titan are **hydrocarbons**, compounds like methane that – although

Table 14.1 Parameters relevant to fluvial processes on Earth, Mars, and Titan

	Earth	Mars	Titan
Sediment σ = particle density	Quartz $\sigma = 2650\,\text{kg/m}^3$	Basalt $\sigma = 2900\,\text{kg/m}^3$	H_2O ice $\sigma = 992\,\text{kg/m}^3$ Organics $\sigma = 1500\,\text{kg/m}^3$
Fluid ρ = fluid density η = dynamic viscosity	Water $\rho = 1000\,\text{kg/m}^3$ $\eta = 1 \times 10^{-3}\,\text{Pa·s}$	Water $\rho = 1000\,\text{kg/m}^3$ $\eta = 1 \times 10^{-3}\,\text{Pa s}$	CH_4/N_2 $\rho = 450\,\text{kg/m}^3$ $\eta = 2 \times 10^{-4}\,\text{Pa·s}$
Gravity	$9.8\,\text{m/s}^2$	$3.7\,\text{m/s}^2$	$1.4\,\text{m/s}^2$

gaseous at terrestrial temperatures and pressures – are liquid under Titan's conditions (Table 14.1).

14.2 Liquid: Falling Down, Soaking In, Flowing Over, Flowing Through, Coming Out

Liquids and landscapes interact through various means. This interaction starts with liquid falling onto the landscapes, soaking in – to a greater or lesser degree – and eventually leaving the fluvial landscape unit, also called the drainage basin or watershed. These processes leave geomorphic signatures that can provide evidence for fluvial flow.

14.2.1 How Liquids Interact with Landscapes

As a starting point for our discussion of hydrologic (or volatile) cycling, we'll begin with rainfall. The first effect of rain on a land surface is rainsplash, in which the impact of the droplets loosens or even ejects surface material. In cohesive sediments or where the rain was limited in duration, the imprints of the raindrop impacts can be seen gigayears after the fact. Where the surface material is less cohesive or the rain drops bigger or longer-lasting, the loose material is transported preferentially downslope, which smooths the surface. Rain on early Mars, during its heyday of a thicker atmosphere, may have been relatively ineffective on the surface, whereas, as the atmosphere thinned over time (Chapter 12), rainsplash would have become more geomorphologically effective. During this transition period, precipitation likely occurred preferentially as snow, until the current aridity shut down rain entirely. On Titan, rain drops might grow several times the size of those on Earth, although the slow descent under Titan's low gravity and thick atmosphere likely limits the effect of rainsplash. Whereas rain occurred only on ancient Mars, present-day rain has been inferred remotely for Titan from the formation and dissipation of clouds.

The next process, at least on a porous surface, is infiltration. The infiltration rate decreases with time (even for a constant rainfall rate) as near-surface pores become filled with loosened sediments and/or filled with water. Liquid that doesn't infiltrate into the subsurface generates surface runoff, either because the rainfall rate exceeds the infiltration rate (of a low-porosity bedrock surface, for example) or "saturation excess overland flow" where pore space has become saturated with water (Figure 14.1). Runoff promotes the redistribution or erosion of surface material. In fact, one piece of evidence for present-day rain on Titan was a change in surface albedo associated with clouds and interpreted as removal of sediments by runoff. Surfaces with high infiltration rates are relatively immune to erosion compared to surfaces with low infiltration rates. Thus, land surfaces with low infiltration rates tend to form networks as **overland flow** coalesces downslope into channels, which promotes deeper flow and therefore enhanced landscape incision. This positive feedback mechanism leads to the formation of channel networks, which move water to streams quickly. On Mars and Titan, fluvial networks, composed of river channels within wider river valleys, provide persuasive evidence for rainfall in the past and present, respectively, of those planetary bodies.

Besides overland flow on the surface, liquid may flow through the porous subsurface (Figure 14.2). Shallow **subsurface flow** tends to move water to streams relatively

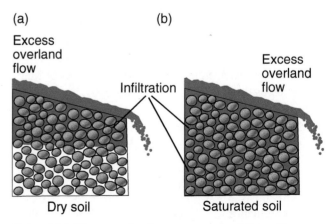

Figure 14.1 Diagrams showing excess overland flow with infiltration into dry soil (a) and soil saturation (b).

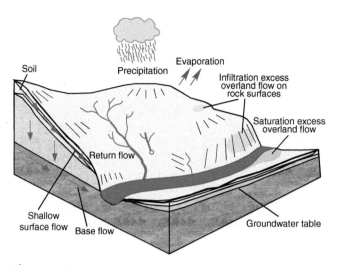

Figure 14.2 Diagram of the drainage basin including subsurface flow and surface flow. Modified from Charlton (2015).

rapidly because the path length is short and the topography provides a downslope gravitational pull. Where this shallow groundwater flow intersects the land surface before encountering a stream (termed "return flow"), it flows out of the hillside. This re-emergence of liquid carries sediment with it, thereby undermining the overlying material. The result is localized collapse of the hillside, which produces steep-sided, rounded ("amphitheater-shaped") depressions or cliff-faces that tend to incise backward into the land surface with each new rainfall. This erosional process that forms amphitheater-headed valleys through headward migration in response to fluid discharge is termed "**sapping.**" It occurs on Earth, occurred on ancient Mars, and is hypothesized to occur today on Titan, where the liquid would be methane and other hydrocarbons rather than groundwater. The stubby, amphitheater-headed channels formed by sapping are morphologically distinct from the channels formed by the coalescence of overland flow, whose initiation point may be hard to determine and whose width increases with distance downstream (Figure 14.3, Table 14.2). Nonetheless, determining an exclusive or uniquely sapping genesis for a channel or channel network is challenging, as any rainfall needed to supply the shallow subsurface groundwater flow likely also resulted in at least some overland flow.

Deep subsurface flow, or base flow, is the continual slow delivery of stored groundwater to a stream. In contrast to overland or shallow subsurface flow, which are responses to discrete rainfall events, base flow keeps streams flowing between rain events and, for a while at least, even during droughts. In large, deep aquifers, stored groundwater may be thousands of years old, as is, for example, the water in the Ogallala Aquifer that supplies drinking water to cities in central North America.

14.2.2 The Drainage Basin as the Fundamental Unit in Hydrology

On the basis of the above discussion, we can now define the fundamental unit in hydrology: the **drainage basin**, also referred to as the watershed. Because liquid flows downhill, even in the subsurface (Figure 14.2), the topography of the land surface controls liquid flow. On Earth, the drainage basin, circumscribed by the highest topographic elevations surrounding an enclosed area, is that portion of the surface within which all surface water from rain, snow, or ice flows inward to lower elevations. Through branched networks of river channels, this centripetal flow converges at the basin exit, beyond which the waters join another body of water, such as another river, a lake, a sea, or eventually an ocean.

Because of its defined area and single outlet, the drainage basin provides a convenient means to evaluate the liquid budget of a landscape. Fundamentally, terrestrial water budgets are built on the principle of conservation of mass for a closed system, so that water entering the system (precipitation) minus water leaving the system (evapotranspiration and stream flow) equals the change in mass stored in the system (groundwater). This equation requires that the drainage basin be a closed system. If it is not, for example when shallow subsurface flow follows dipping strata that outcrop in an adjacent drainage basin, adjustments to the equation are required.

Drainage divides must have existed during the warmer, wetter climate of early Mars (Section 12.4.1), but have been largely erased by the pervasive pounding by impactors. The degraded or breached rims of older martian craters – and almost all craters on Titan – illustrate the concomitant atmospheric effects. In younger landscapes, where craters have experienced less degradation, it is the crater rims themselves that demarcate the drainage divides. Drainage divides, separating the flow of liquid hydrocarbons, must also exist on Titan, as evidenced by the numerous distinct fluvial networks observed there. However, the relative low-resolution and incomplete surface data make the delineation of Titan drainage divides challenging.

14.3 Processes that Channelize the Flow of Liquid

Within the fluvial landscape, liquids coalesce into channels. This channelized flow leads to movement of sediment, which can be understood through simple physical relationships. The processes include both sediment deposition into bedforms and erosion. The physics-based nature of these processes means that the same hydraulic equations can be used to understand how these processes work on different planetary bodies.

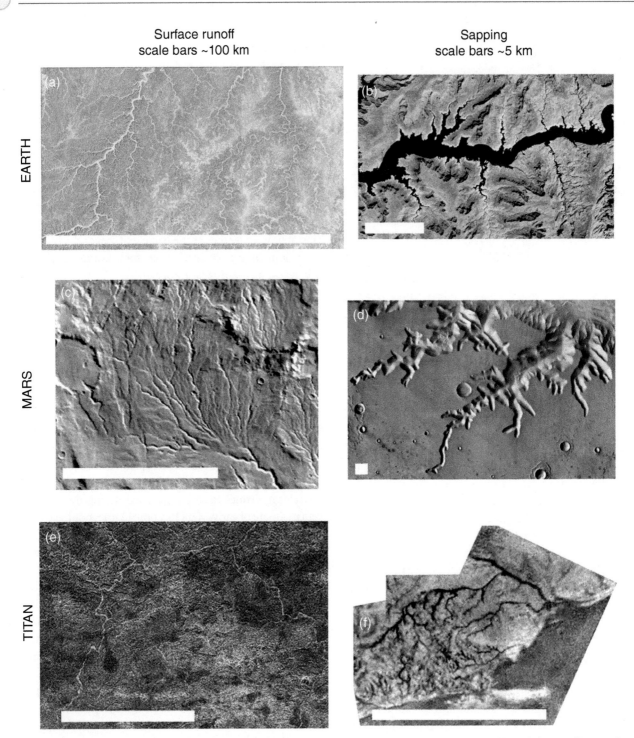

Figure 14.3 Channels formed by surface runoff and by sapping on Earth, Mars, and Titan. Earth: (a) surface runoff networks in South Yemen, (b) sapping channels on the Colorado Plateau. Mars: (c) surface runoff networks of Warrego Vallis, (d) sapping network on the Valles Marineris Pleteau. Titan: (e) (radar images) radar-bright surface runoff network on southwestern Xanadu, (f) sapping networks near the Huygens landing site. NASA images.

14.3.1 Flow Velocity Profile

Like gas (Section 13.2), liquid flowing over a surface experiences friction at that surface that decreases with height. Thus, like air, liquid in a channel exhibits a logarithmic profile with greatest flow speeds near the top and slower speeds, decreasing to zero, at the channel bed (Figure 14.4a). Unlike free airflow, channelized flowing liquid also experiences friction along the surfaces

Table 14.2 Characteristics of channels formed by surface versus subsurface liquid

	Valley growth mode	Drainage pattern characteristics
Surface runoff	Erosion by runoff of meteoric liquid	Fills available surface area Scale-invariant Channels: V-shaped cross-sections
Groundwater sapping	Headward undermining, collapse, and rock material removal at groundwater spring sites	Does not fill available surface area (undissected interfluves) Tends to have short, stubby tributaries Alcove-shaped heads Channels: U-shaped cross-sections

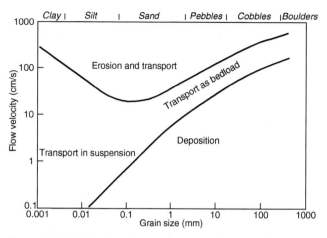

Figure 14.5 Hjulström diagram showing sediment behavior as a function of grain size for a given flow velocity.

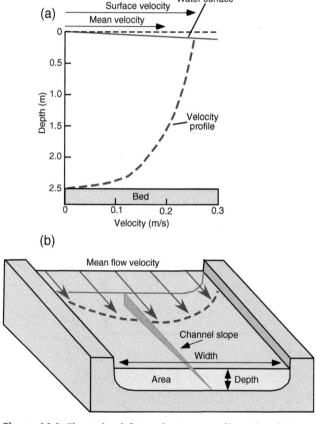

Figure 14.4 Channelized flow velocity in profile and in three dimensions.

that constitute the channel margins or banks. Thus, the flow velocity in plan view exhibits a curved profile with the maximum in the center of the channel banks (Figure 14.4b).

On Earth, calculations of discharge, or the volumetric flow rate (in dimensions of length³/time such as m³/s), entail measuring the velocity at multiple locations across the channel and at multiple depths to find the geometric

average value, and then multiplying that value by the cross-sectional area of the channel:

$$Q = V \times W \times D \qquad (14.1)$$

where Q is the discharge, V is an average velocity, W is the width of the channel, and D is representative depth.

14.3.2 Entrainment

In a river with a bed of loose transportable sediment, the immediate effect of this flow is sediment entrainment into transport. In his doctoral thesis published in 1935, Filip Hjulström published curves (different curves for different depths) that delineate the flow velocities necessary for transport (Figure 14.5). A **Hjulström diagram** illustrates that, as we might expect, transport is a function of grain size and flow velocity: Larger (heavier) grains require stronger (faster) flow for transport. At the same time, the diagram also shows that this direct correlation between grain size and flow speed, seen on the right side of the curve, does not describe the entire curve. The left side of the curve shows that the smallest grains – clay – also require high flow speeds to be entrained or eroded from the bed. So we have the same U-shaped curve that we saw previously (Section 13.2) for aeolian transport and for the same reasons: Small grains have high surface-to-volume (or mass) ratios and so have strong interparticle forces. We know these forces from experience – from walking through mud and having it stick to our shoes!

The flow velocity on the Hjulström diagram is the physical flow speed that can be measured in a stream, as explained above. However, as we learned for aeolian processes, a useful quantity when considering sediment transport is the friction or shear speed, which represents the strength of the flow exerted at the surface of the bed.

As for the wind, the shear speed for a river can be quantified using the "law of the wall" or by solving for friction wind speed, u_*, in Equation 13.1. In fluvial environments, however, the strength of the flow at the bed is usually described as a shear stress τ, given as

$$\tau = \rho g z \sin \alpha \tag{14.2}$$

where ρ is the density of the liquid, g is the gravitational acceleration of the body, z is the depth of the flowing liquid, and α is the slope of the channel. Like the Hjulström curve, this shear stress can be calculated as a function of grain size, as was done by A. F. Shields in 1936. The **Shields curve** (Figure 14.6a) is based on the equation for the shear stress, but made nondimensional using other boundary conditions for the flow. The equation for the dimensionless shear stress, τ^*, as a function of grain size, is:

$$\tau^* = \tau / \left[\left(\rho_g - \rho \right) g D \right] \tag{14.3}$$

where, as for aeolian equations, the * denotes a shear quantity, ρ_g and ρ denote grain and liquid densities, respectively, and g and D denote gravity and grain diameter, respectively.

The inclusion of gravity and material properties (densities) in the Shields curve gives it the very helpful characteristic of being translatable to non-Earth conditions. Thus, the Shields curve can be used to calculate the entrainment conditions for Mars and Titan (Figure 14.6b). From this comparison, we can see that flow velocities can be lower on Mars and Titan than on Earth and still entrain sediment. As a result, sediment

deposition is likely much more areally distributed on Mars and Titan than on Earth. This comparison implies that fluvial and lacustrine deposits, such as deltas (see Section 14.5.2), might take much longer to build up into a recognizable form on these bodies compared to on Earth.

14.3.3 Transport Mechanisms of Fluvial Sediment: Three Regimes

Once sediment is entrained, how does it move? The "easiest" mode of motion – the one requiring the least energy or flow speed of the liquid – is motion as bedload. As the name tells us, bedload moves in contact with the channel bed, by sliding, rolling, or bouncing. In most rivers on Earth, bedload typically includes larger size fractions such as pebbles or cobbles, although in very slow flows or very fast flows, it may include smaller or larger grain sizes, respectively. At smaller grain sizes and/or faster flows relative to bedload conditions, sediment moves as suspended load, suspended within the water column by turbulence, but concentrated near the bed. A final mode of fluvial sediment motion is termed washload, in which the sediment is uniformly distributed with height above the bed and remains in transport even in the slowest flows.

Although these different modes of transport are distinguished conceptually, they are gradational. We can imagine – and video shows – that some rolling bedload grains may occasionally bounce high enough into the water column to mimic suspended load. Likewise, some suspended grains may be small enough so as to remain suspended even in very slow flows, and so grade into washload. However, for the purposes of estimating

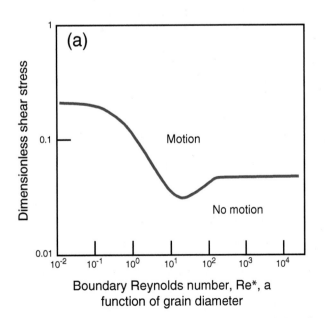

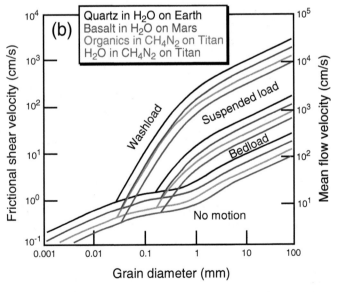

Figure 14.6 (a) Idealized Shields curve. (b) Plots of modes of transport as a function of (friction) velocity for various planetary conditions. Modified from Burr et al. (2006).

sediment discharge by rivers, hydrologists have quantified the distinction among the modes of motion using a ratio of flow velocities, denoted by k, as:

$$k = w_s/u^* \qquad (14.4)$$

where w_s is the settling velocity of the particle (its rate of fall through the liquid in the absence of any turbulence) and u^* is the shear speed, related to the shear stress as $u^* = (\tau/\rho)^{1/2}$, and a proxy for the turbulence that suspends the particles. Various studies set the value of k for the bedload-to-suspended load transition at 1–1.79 and for the suspended load-to-washload at 0.05–0.13. By selecting a single value to represent the gradation from one mode of transport to another, we can plot the conditions under which each mode of transport occurs on each planetary body. Such a plot shows that it requires less energy, or flow speed, to transport sediment in the various modes on Titan than on Mars, and less energy on Mars that on Earth (Figure 14.6b). This difference is a result of the lower gravity on Mars and Titan than on Earth, and the lower grain densities on Titan.

14.3.4 Fluvial Bedforms

When the flow velocity falls, perhaps temporarily or in certain regions of the channel, deposition occurs. Obviously, perhaps, this process is vital for geologists! It provides us with evidence – the sedimentary rock record – by which we can intuit the extent, magnitude, and characteristics of the flow that created them.

Although many processes can affect these in-channel deposits, their shapes depend primarily on the size of the sediment and the speed of the flow. Bedform stability diagrams provide a visual representation of the bedforms that develop with particular grain size and flow conditions. These conditions can then be adjusted arithmetically for the lower gravity and the different sediment and liquid densities that occur on other planets. A bed stability diagram for Earth and Titan (Figure 14.7) shows, for example, that dunes on Titan form at a lower bed shear stress – a lower flow velocity – than dunes on Earth. Such diagrams enable researchers to make inferences from the observation of planetary bedforms about the liquid flow and sediment that created them. Although our current images from the surface of Titan are not yet adequate to show any bedforms (if they exist – recall that sediment dispersal is probably quite broad on Titan), images of cataclysmic outflow channels on Mars show some examples of flood-formed dunes. Modeling of these dunes, developed for modeling terrestrial dunes in the post-glacial outflow channels on Earth, provide constraints on the flow velocity that formed them. And flow velocity, in conjunction with measures of channel width

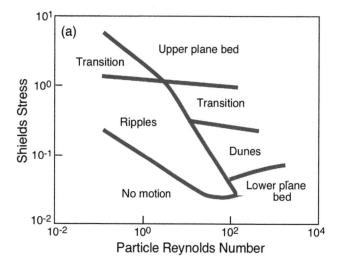

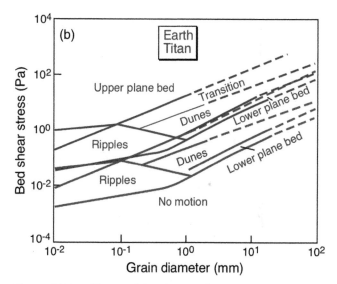

Figure 14.7 Bedform stability diagram for Earth and Titan. Adapted from Burr et al. (2013a).

and depth (see Equation 14.1), provides an estimate of flood discharge.

14.3.5 Fluvial Erosion

In addition to the entrainment and transport of loose sediment, channelized flow erodes bedrock. This erosion may occur through one of three mechanisms. At the lowest energies, the bedload and suspended load being carried by the flow abrades the bedrock. This abrasion is more significant in faster flows and flows carrying more sediment. Calculations suggest that, because of the lower gravity on Mars, water flows more slowly than on Earth under the same discharge and slope and so may abrade less than on Earth. The same is likely true on Titan, where the lower gravity and slower flow cause abrasion to proceed at a slower rate. At higher energies, plucking can occur. During this process, blocks of bedrock are slid or even pulled into the flow by the turbulence of the flow. Plucking has been inferred, based on the size and shaped of debris, for cataclysmic flooding over jointed basalts on Earth. The observation of jointed basalts on Mars and the inference of even larger-than-terrestrial floods in Mars' history strongly suggest that plucking was an important erosive process. Under very turbulent flows with extensive bubble formation, the implosion of the bubbles against the bedrock chips away at the rock. This process, called cavitation, is observed on submarine propellers, and might contribute to bedrock erosion on Mars where the low atmospheric pressure may enhance the effect of bubble implosion.

14.4 Channelized Flow of Liquid: Landscape Results

The sedimentary processes of channelized liquid flow are seen across landscapes, both at local scales as channels and at regional scales as drainage networks. Such networks can be seen on Earth and Titan that have active three-phase volatile cycles, and in ancient terrains on Mars, evidence of a volatile cycle billions of years ago.

14.4.1 Fluvial Channels

Channelized flow over loose sediment produces **alluvial channels**. Alluvial channels have a variety of morphologies (Figure 14.8), which are a product of the history of the landscape, the discharge of the river, and the amount and nature of sediment transport. Thus, channel morphology may change as discharge and/or sediment supply changes. Channels with high bedload and irregular discharge tend to produce wide, shallow, braided rivers, where the bedload forms temporary bars or islands that are remobilized downstream during flooding. On Earth, this morphology is often displayed, for example, in coastal regions where rivers exit mountains that were formerly glaciated and thus have extensive glacial debris

for fluvial transport. Such deposits have not been identified on Mars, perhaps because the lower martian gravity enables transport of most material as suspended load instead of bedload or because the bedload that does exist is broadly dispersed during deposition under low martian gravity. However, braided outflow channels have been identified on Titan, so low gravity alone may not explain their seeming absence on Mars.

Channels with high suspended load and more regular discharge tend to form meandering channels, as displayed on Earth by the lower Mississippi, Amazon, and Nile rivers. These rivers have relatively deep and narrow cross-sections, with stabilized banks. This terrestrial bank stabilization was greatly enhanced ~0.5 billion years ago by the rise of vascular (rooted) plants, before which time, the terrestrial rock record shows less evidence for meandering rivers. However, the discovery of inferred meandering river deposits on Mars (e.g., Figure 14.9) indicates that meandering rivers are not uniquely associated with plants and that other mechanisms can provide the necessary bank stability. The best current hypothesis for the source of cohesion for non-vegetated meandering rivers on Mars is cohesive clays, although permafrost and geochemical cements are also possibilities. Meandering rivers have also been claimed for Titan, but the image resolution is not yet sufficient to distinguish the smaller river channel from the river valley, so that claim awaits testing with higher-resolution images.

On Earth, combinations of these braided and meandering morphologies are possible, providing evidence of a mixed type of sediment load and/or of a mixed fluvial

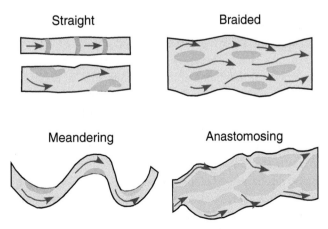

Figure 14.8 Diagram showing plan view morphologies exhibited by alluvial channels.

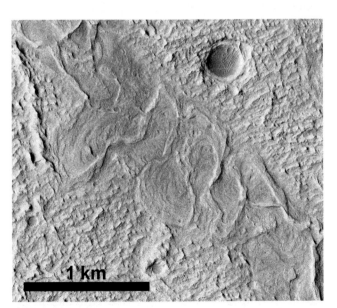

Figure 14.9 Fluvial deposits (oriented upper left to lower right across image) left by ancient meandering rivers on Mars. NASA image.

history. However, the identification of such combined morphologies on Mars or Titan awaits better imaging data.

Channelized flow over bedrock produces channels whose behavior is controlled by the rate of erosion. Because this erosion – by abrasion or, in jointed rock, by plucking – is generally slower than entrainment of grains in alluvial rivers, the rate of migration and incision is often slow, although tectonic uplift or a large supply of sediment accelerates it. Because of the lack of sedimentary banks and floodplains, the morphology of rivers in bedrock is generally simpler than that of rivers in alluvium, and may include straight segments, particularly when flowing over bedrock with fractures or fissures.

14.4.2 Channel Drainage Networks

As channels intersect and grow, they form drainage networks that reflect the controls on overland flow, such as the rate and timing of water delivery to the surface and the land surface over or through which the water flows. Thus, quantitatively characterizing networks provides both relative and absolute information on a variety of drainage controls. Networks can be characterized by the number of branches or links in the network. The number of links in the network reflects the amount of runoff and, because more porous terrain generates less runoff, provides insight into the behavior of rain on the surrounding land surface.

Drainage networks can also be characterized by their planview morphology or pattern (Figure 14.10). Each pattern forms under a different range of circumstances and so has different implications for the controls on overland flow (Table 14.3). The canonical and perhaps most common drainage pattern on Earth is dendritic, in which streams intersect at moderate angles, reflective of a moderate surface slope without significant pre-existing weaknesses. Drainages with lower-angle junctions include parallel networks, which form when the land surface slope is steeper or when elongate landforms, such as longitudinal dunes (Section 13.5) affect the direction of flow. Drainages with higher-angle junctions include rectangular networks, which form over a landscape with near-right-angle joints or faults, and trellis networks, which form most commonly over dipping or folded rocks, as in the Valley and Ridge terrain of the Appalachian Mountains.

Because the broad river valleys that form drainage networks may be visible even in low-resolution data, drainage network morphology can often be characterized when channel morphology cannot. Thus, drainage networks provide early insights into regional land surface features, history, and controls on runoff. For example, the characterization of drainage networks on Titan revealed that over half of the drainages are rectangular in morphology, indicating that – beneath the mantle of atmospherically

derived aerosols, aeolian sediments, and other surface veneers – the crust of Titan is tectonically fractured.

Table 14.3 **Planview drainage patterns and the geologic implications of each**

Drainage pattern	Geologic implications
Dendritic	Uniformly resistant rocks, gentle/moderate regional slope at time of drainage inception
Parallel	Moderate to steep regional slopes and/or parallel, elongate landforms
Rectangular	Joints and/or faults at near-right angles
Trellis	Dipping and/or folded sedimentary, volcanic, or low-grade metasedimentary rocks with contrasting resistance to erosion

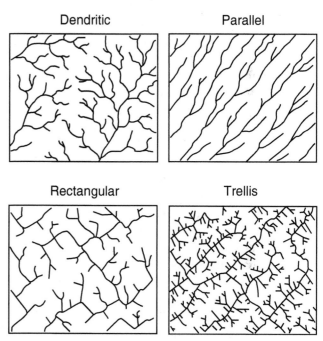

Figure 14.10 Drainage network patterns. Table 14.3 lists the implications of each pattern. Modified from Burr et al. (2013b).

14.5 Deposition from Channelized Flow

The landscape results of channelized liquid flow are not only in the channels and networks that transport the flowing liquid and sediment, but also in depositional features, as seen on Earth, Mars, and Titan.

14.5.1 Subaerial Deposition: Fans and Bajadas

The channelized flow and its sediment may be deposited either in a body of water or on land, i.e., subaerially. Landscape-scale subaerial landforms deposited by channelized running water include both individual fans and coalesced fans or bajadas.

BOX 14.1 CATACLYSMIC FLOODING

In contrast to river valleys, which result from continuous or at least frequent flow, **outflow channels** are produced by sudden and massive deluges of liquid. Whereas river valleys form tributary networks, outflow channels form giant scars across the landscape of interconnected gorges that are hundreds of meters deep, tens of kilometers wide, and orders of magnitude greater in length. On Earth, outflow channels are associated with glaciation, which provides a means for storing the massive amounts of water that are necessary. This storage occurs commonly in pro- or subglacial lakes as a product of the melting of the basal ice of the glacier. Floating or bursting of the ice dam releases this stored water as a sudden flood. The largest outflow channels on Earth are in the Altai Mountains of Siberia, Russia, and in the Channeled Scabland in eastern Washington State, USA, both of which resulted from the damming and sudden release of glacial meltwater. Because glaciers often form on volcanoes, whose heat then melts these ice caps, outflow channels on Earth often debouch from intracaldera lakes, such as in Iceland or New Zealand.

On Mars, outflow channels apparently originate from three different sources or types of water storage (Figure 14.11), whose relative importance has changed over geologic time. The oldest (Noachian) outflow channels on Mars date from the era of rivers and resulted from overflow of **crater basin lakes** fed by runoff. An example of a crater basin overflow flood channel is Ma'adim Vallis, which originates full width from the Eridania Basin. The medial ridges that cover the floor of the Vallis suggest that, at least in some locations, the flow occupied the entire valley, characteristic of a flood, instead of a smaller inset channel. In the Hesperian period, outflow channels originated from the Valles Marineris canyons (Figure 14.11). These largest of the martian flood channels originate at **chaos terrain**, areas of large jumbled blocks inferred to have formed during catastrophic breakout of confined groundwater. The trigger for breakout is suggested to be thinning of the confining cryosphere, perhaps by fluvial activity, or drainage of canyon lakes following collapse of sedimentary or ice barriers. The resultant floods may have fed a vast northern ocean. The youngest outflow channels, which date to the Amazonian Period, originate at volcano-tectonic fissures (Figure 14.11), fractures from which lava also originated. The exact age of these youngest water floods is open to question because the lava embays the channels, but this embaying lava has been dated to just a few million years in age – just last week in geologic time! And the floods could have occurred around the same time. As with the Hesperian-aged channels, the storage of water for these Amazonian floods apparently occurred in the subsurface, but the mechanism for storing such massive water volumes in the subsurface and releasing them suddenly enough to form outflow channels is not yet understood. A plot of the change in flood-generating mechanisms on Mars (Figure 14.11) shows that floodwater storage moved from the surface to the subsurface over geologic time.

Thus, geologically extreme aqueous flooding on Earth and Mars is a function of the ability of water to assume the solid phase. The operative volatile on Titan, which is methane, cannot as easily assume the solid phase under current conditions. On Mars, the largest floods result from subsurface water storage, followed by some mechanism of tectonic or fissure-fed release, or from water storage in crater lakes. Fissures have not been detected on Titan, and craters are sparse, perhaps due to screening by the episodically thick atmosphere. However, precipitation has been inferred on Titan. Thus, extreme precipitation events, which have produced smaller but more frequent floods on Earth, are likely to be the most important flood generators on Titan in the recent past.

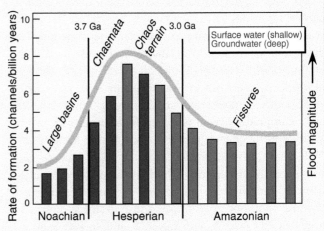

Figure 14.11 Causes of outflow channels over geologic time on Mars. Adapted from Burr (2010).

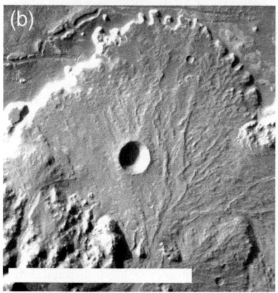

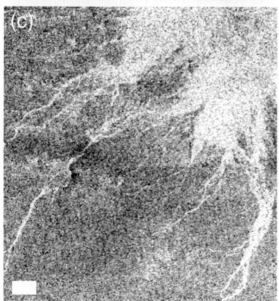

Alluvial fans (Figure 14.12) are sedimentary landforms that develop at the mouths of upland canyons that contain significant transportable sediment. During high-discharge precipitation events, the channelized flow emerges from the canyon, where it changes from confined to unconfined flow, resulting in an abruptly decreased flow depth. The decrease in flow depth, in turn, results in an abrupt decrease in shear stress (see Equation 14.2), causing deposition of the sediment load (alluvium) at the canyon mouth. Over time, repeated events result in the natural construction of a wedge- or fan-shaped deposit between the canyon mouth and the floor of adjacent plains. In addition to a long-term hydrological cycle, fans also generally require the flow events to be intermittent in time, in order to allow alluvium to build up in the catchment. Lastly, to produce the necessary change in flow confinement, alluvial fans require relief or a sudden change in elevation, such as occurs at a mountain front on Earth (Figure 14.12a) or inside impact crater rims on Mars (Figure 14.12b). On Titan, features hypothesized to be alluvial fans (Figure 14.12c) are observed in the mid- to high latitudes, although the poor imaging data makes their identification somewhat uncertain.

Fans can be formed either through debris flow deposits, which are poorly sorted, with grains ranging from mud to boulders in size, or sheetflood deposits, which consist largely of sand-sized grains. Besides differences in grain size and sorting, the two types of deposits have measurable differences in topography, with debris flow fans exhibiting leveed channels and lobes and sheetflood fans exhibiting smoother surfaces and termini. These differences have been used to distinguish debris flow and sheetflood fans on both Mars and Titan.

A **bajada** develops where multiple alluvial fans coalesce. The lateral spreading is often a function of repeated side-to-side shifting of the contributing streams. Like fans, bajadas can form either in arid climates where flash floods produce repeated sediment depositional events, or in wetter climates where streams deposit sediment more continuously. These broad deposits occur along mountain fronts on Earth and have been documented within impact craters on Mars.

Figure 14.12 Images of alluvial fans. (a) Alluvial fan, Taklimakan Desert, China. The false-color blue on the left side of the fan indicates water currently flowing. (b) Alluvial fan on Mars in Holden Crater, with the part of the inner crater rim visible in the lower left. (c) Elivagar Flumina on Titan, interpreted as a compilation of alluvial fans, which are fed by flow sourced from an impact crater off the image to the left. In all images, flow is toward the top of the page and scale bars are ~10 km. NASA images.

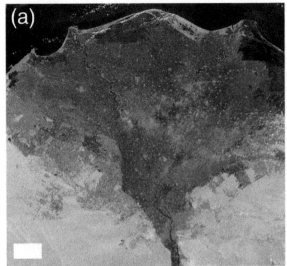

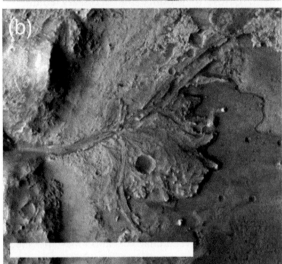

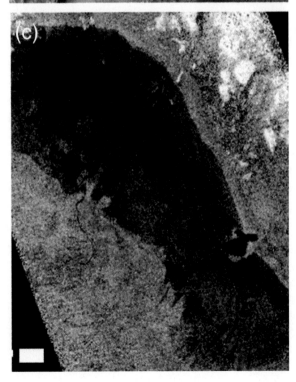

14.5.2 Subaqueous Deposition: Deltas

Deltas (Figure 14.13) derive their moniker from the triangular-shaped deposit of land at the mouth of the Nile River, which the Greeks understood correctly to look from the air like a capital "D" or "delta." Study since that time has shown that deposits of sediment at the mouth of rivers emptying into lakes, seas, or ocean have a range of shapes, which reflect the dominant control. River-dominated deltas, such as the Mississippi River delta, form a digitate shape (sometimes referred to as a "bird's foot" shape) as the river switches course over time. Tidally dominated deltas, such as the Ganges/Brahmaputra delta in the Bay of Bengal or the Fly River delta in Papua New Guinea, show more disaggregated form with lobes or islands perpendicular to the shoreline and separated by tidal flats or channels. Wave-dominated deltas, of which the Nile delta is the classic example, tend to have smooth shorelines where wave action has removed sediment (Figure 14.13a). A fourth kind of delta, named after the geomorphologist G. K. Gilbert, has a distinctive, coarse-grained, sedimentology, such as results from steep mountain streams.

Deltas on Mars also span a range of morphologies. Bird's foot (Figure 14.13b) and Gilbert deltas have been hypothesized, along with other more exotic varieties. Consistent with Mars' small moons, no tidally dominated deltas have been suggested. The observation of deltas on Mars is perhaps a bit surprising. The low gravity on Mars would cause sediment to disperse farther in the basin than it would on Earth, so to be recognizable as deltas the fluvial transport and sediment deposition would take considerably longer on Mars. An increased timescale for delta formation should be even more true for Titan, where both lower gravity and lower sediment densities would result in even broader sediment dispersal. Broad dispersal on Titan might account for the relative dearth of deltas observed to date, although a few suspects have been noted (Figure 14.13c).

14.6 Large Bodies of Standing Liquids

The end result of channelized flowing liquid and sediment is a standing body of liquid. Such standing liquids are seen on Earth and Titan, and have been inferred for

Figure 14.13 Examples of the three types of delta. (a) Nile River delta, a wave-dominated delta on Earth. (b) Jezero Crater delta on Mars, a bird's food delta (colors indicate different types of clay) formed by river channel switching. (c) Possible deltas on Titan in Ontario Lacus (two on the left shoreline, one on the lower right shoreline) inferred to have resulted from channel switching and/or wave modification. In all images, scale bars are ~10 km. NASA images.

Mars. Their existence provides singular evidence for volatile collection and cycling.

14.6.1 Marine and Lacustrine Morphologies on Mars

In addition to the geochemical evidence for oceans and lakes in our Solar System (discussed in Section 12.3), we can also point to geomorphic evidence for standing bodies of water. The low elevation and flat topography suggest that the northern lowlands of Mars once hosted an ancient ocean of water, as illustrated in Figure 14.14a (Baker et al., 1991). Valley networks terminate in the lowlands, as rivers do into the ocean basins on Earth, and a number of deltaic features occur at nearly the same elevation around the lowland margins. Putative remnants of ancient shorelines have been recognized; their elevations, which vary considerably, are inconsistent with standing liquid in equilibrium but may be consistent with geophysical changes on Mars "after the fact." Tsunami "run-up" deposits have also been inferred around the lowland margins.

Mars clearly hosted smaller bodies of water (Cabrol and Grin, 2010). The geomorphic evidence for lakes includes deltaic deposits in impact crater basins that are closed except for an inflow channel. Other crater basins were open (or overflow) lakes, having both inflow and outflow channels. Overflow of some of these crater lake basins produced considerable floods (see Box 14.1). Sedimentology investigated by the Curiosity rover in Gale crater suggests subsurface standing water (Section 12.3).

14.6.2 Hydrocarbon Lakes and Seas on Titan

Cassini orbiter radar images found that Titan is dotted with standing bodies (Figure 14.14b) of liquid organic compounds (Stofan et al., 2007). The lakes give no radar return, as appropriate for smooth surfaces, and Ontario Lacus, in the southern polar region, appears to host a delta. One lake has a measured depth of ~180 m, but many more have depths of just a few meters; empty basins, interpreted as dried lakes, have also been detected. These lakes and seas are clustered in the north polar region, apparently a result of orbital control on Titan's volatile cycle. An estimate of the combined volumes of all

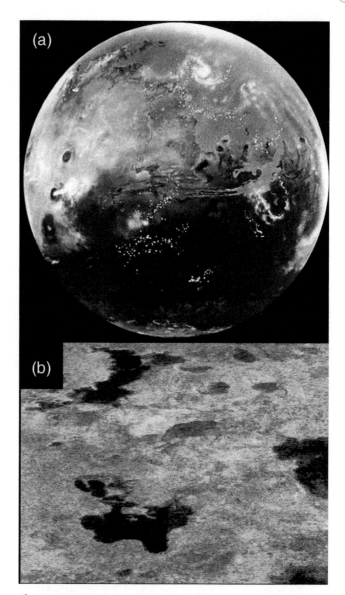

Figure 14.14 Oceans and lakes on other worlds. (a) Artist's rendition of a Noachian ocean in the northern plains of Mars. (b) Colorized radar image of hydrocarbon lakes on Titan. Image is 205 km across. NASA images.

the extant lakes is 300 times that of Earth's proven oil reserves. A Titan "boat" – a spacecraft mission that would float on a Titan sea, taking data of wind, waves, and composition – is under discussion.

Summary

Liquids on other bodies, though unusual in our Solar System, are diagnostic of atmospheric conditions and hugely influential modifiers of planetary surfaces. Rivers, lakes, and their geomorphic results provide critical clues to understanding planetary geologic processes. River discharges provide constraints on atmospheric models, whereas river networks and associated depositional landforms constrain factors that control the flow and serve as data on the material transport over the surface. Shorelines, deltas, and other lacustrine byproducts point to the locations and extents of standing bodies of liquid. Because water is necessary for life as we know it on Earth, "follow the water" has been a mantra for exploration of Mars with a focus on discovering past habitable environments. On Titan, hydrocarbon liquids provide a medium for facilitating organic processes that might echo prebiotic processes on Earth. Thus, the study of fluvial and lacustrine landforms is a significant source of information relevant to the fields of planetary geology, sedimentology, and astrobiology. Although this chapter has focused on remotely detected evidence of fluvial processes, *in situ* data – such as geochemical analyses or observation of rounded cobbles – support the remote inference of fluvial sediment transport on these two bodies. Continued exploration of such extraterrestrial data, along with improved understanding of fluvial deposits on our own planet, will further deepen our understanding of fluvial processes elsewhere.

Review Questions

1. What conditions are necessary for fluvial landscapes? Does your answer to this question differ between erosional and depositional landscapes? For different planets?
2. What information do fluvial landscapes provide? How does the type of information that can be derived from fluvial landscapes vary with resolution?
3. How can fluvial landscapes be used to constrain atmospheric models?
4. What is surprising about the appearance of meandering fluvial deposits on Mars?
5. Compare and contrast fluvial processes on Earth, Mars, and Titan.
6. Compare and contrast lakes, seas, and oceans on Earth, Mars, and Titan (e.g., composition, size, locations).
7. Why might we expect lacustrine deposits on Earth, Mars, and Titan to be similar? Why might we expect them to be different?

SUGGESTIONS FOR FURTHER READING

Baker, V. R., Hamilton, C., Burr, D. M., et al. (2015) Fluvial geomorphology of Earth-like planetary surfaces: a review. *Geomorphology*, **245**, 149–182. A nice overview of the effects of channelized surface liquid flow in the Solar System.

Burr, D. M. (2010) Palaeoflood-generating mechanisms on Earth, Mars, and Titan. *Global and Planetary Change*, **70**, 5–13. A synthesis of liquid floods on the three bodies in the Solar System with present or past volatile cycling.

Burr, D. M., Perron, J. T., Lamb, M. P., et al. (2013) Fluvial features on Titan: insights from morphology and modeling. *Geological Society of America Bulletin*, **125**, 299–321. A description of the morphologic effects of channelized liquid flow on Titan and how they could (or could not) be modeled using terrestrial flow modeling.

Hynek, B. M., and Phillips, R. J. (2003) New data reveal mature, integrated drainage systems on Mars indicative of past precipitation. *Geology*, **31**, 757–760. A short, early publication on the discovery of fluvial channels within fluvial valleys on Mars.

REFERENCES

Baker, V. R., Strom, R. G., Gulick, V. C., et al. (1991) Ancient oceans, ice sheets and the hydrological cycle on Mars. *Nature*, **352**, 589–594.

Burr, D. M. (2010) Palaeoflood-generating mechanisms on Earth, Mars, and Titan. *Global and Planetary Change*, **70**, 5–13.

Burr, D. M., Emery, J. P., Lorenz, R. D., et al. (2006) Sediment transport by liquid overland flow: application to Titan. *Icarus*, **181**, 235–242.

Burr, D. M., Perron, J. T., Lamb, M. P., et al. (2013a) Fluvial features on Titan: insights from morphology and modeling. *Geological Society of America Bulletin*, **125**, 299–321.

Burr, D. M., Drummond, S. A., Cartwright, R., Black, B. A., and Perron, J. T. (2013b) Morphology of fluvial networks on Titan: evidence for structural control. *Icarus*, **226**, 742–759.

Cabrol, N. A., and Grin, E. A., eds. (2010) *Lakes on Mars*. Amsterdam: Elsevier.

Charlton, R. (2015) *Fundamentals of Fluvial Geomorphology*. London: Routledge.

Stofan, E. R., Elachi, C., Lunine, J. I., et al. (2007) The lakes of Titan. *Nature*, **445**, 61–64.

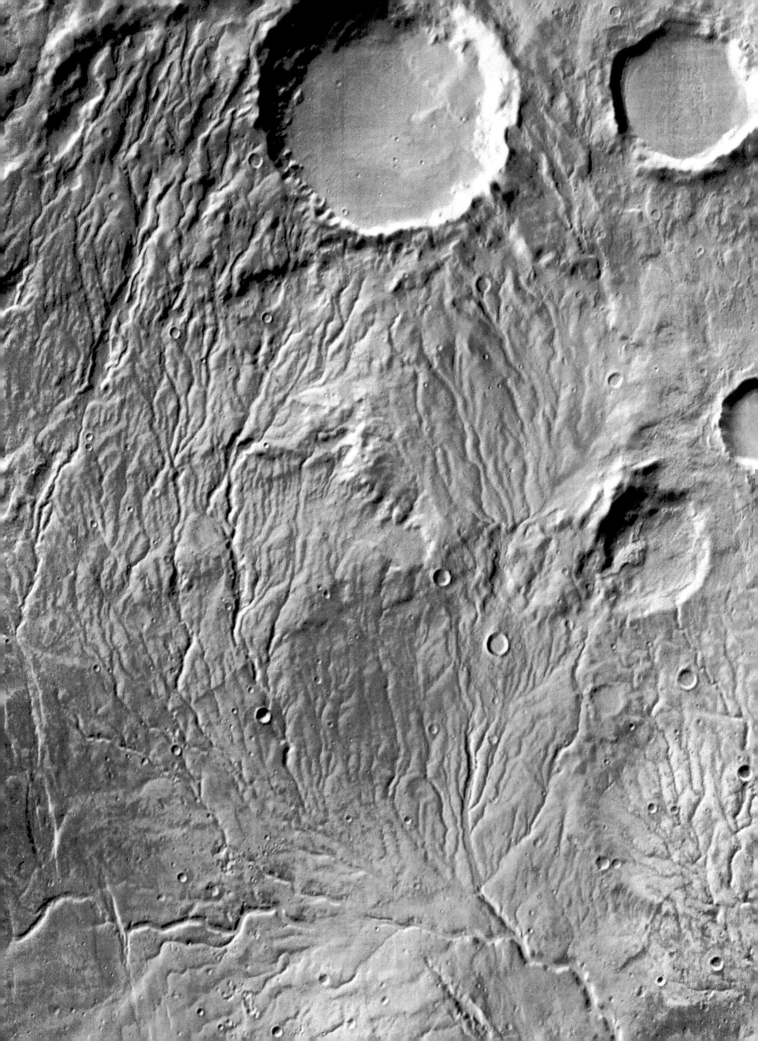

15

Physical and Chemical Changes: Weathering, Sedimentology, Metamorphism, and Mass Wasting

Physical weathering of rocks on bodies other than the Earth occurs mostly through impact fragmentation, producing regoliths. The lunar regolith is finer-grained and contains more agglutinates than asteroidal regoliths, indicating its greater maturity. Mars exhibits both physical and chemical weathering, and its sedimentary deposits superficially resemble those on Earth. However, its basalt-derived sediments differ from those formed from felsic protoliths on Earth, and evaporation of its aqueous fluids is dominated by sulfates, distinct from terrestrial evaporites that are mostly carbonates and halides. On the surfaces of airless bodies, recondensation of vapor produced by micrometeorite impacts accounts for spectral changes, known as space weathering. In the interiors of carbonaceous chondrite asteroids, isochemical reactions of rocks with cold aqueous fluids produced by melting of ice have altered their mineralogy. Thermal metamorphism of dry chondritic asteroids has modified all but near-surface rocks. Hydrothermal metamorphism on Mars, likely associated with large impacts, has produced low-grade mineral assemblages in metabasalts and serpentinites. Conditions at Venus' surface are severe enough to cause thermal metamorphism, and reactions with rocks may control the composition of the atmosphere. Because all bodies have gravity, some sloping topography, and some unconsolidated materials, mass wasting is among the most common processes modifying planetary surfaces.

15.1 Petrologic Changes and the Rock Cycle

Igneous, sedimentary, and metamorphic rocks comprise the Earth's crust and interior. The processes that account for this petrologic variety can be summarized by the familiar rock cycle (Figure 15.1). Melting produces magmas that crystallize to form igneous rocks. Rocks on the surface are weathered and eroded, producing unconsolidated sediments that are transported, deposited, and subsequently lithified during diagenesis to form sedimentary rocks. Deeply buried sedimentary and igneous rocks are transformed by heat and pressure into metamorphic rocks. Plate tectonic processes recycle these rocks back into the mantle, where melting occurs, regenerating the rock cycle.

Many of the rocks we encounter on other planets or collect as meteorites have also been modified from their original igneous (or primitive, in the case of chondrites) states. Fragmentation by impacts is the dominant form of physical weathering for most extraterrestrial bodies. Impact-comminuted sediments cover the surfaces of other bodies and may become solidified to form sedimentary rocks. Chemical weathering is less common on other planets, because of the requirement for liquid water. Thermal metamorphism occurs within the deep interiors of other bodies, but metamorphic rocks are rarely exposed because of the lack of tectonic uplift mechanisms. However, shock metamorphism by meteor impacts (discussed in Section 11.5) is pervasive in extraterrestrial rocks; although this is not commonly illustrated in the rock cycle, it is an important part of the cycles on other planets (Figure 15.1). The rock cycle is truncated on planets other than our own, because the lack of plate subduction does not allow recycling of rocks back into the deep interior, where melting occurs. In this chapter, we will explore the petrologic processes that modify the rocky materials that comprise extraterrestrial bodies.

All bodies, large and small, have some topographic relief. Gravity acting on unstable slopes results in downslope

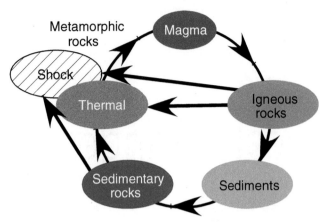

Figure 15.1 The Earth's rock cycle, with the addition of shock metamorphic rocks that are pervasive on other planetary bodies.

movements. We will also consider how surfaces are modified by this physical process.

15.2 Regoliths: Physical Weathering

On Earth, various physical weathering processes break down surface rocks into sediments. On other planets, however, the most effective physical weathering process, by far, is comminution by meteor impacts. As noted in Section 8.5.4, thermal stresses on airless bodies can also cause fracturing of surface rocks. Fragmentation produces an unconsolidated surface layer, called **regolith**. Soil scientists sometimes distinguish regolith from "soil" – by their definition, soil must contain an organic component, a distinction lost on most planetary scientists who use the terms interchangeably.

Surface regoliths on other bodies are generally fine-grained, and grade downward into larger transported blocks, and even deeper into fractured crust (Figure 15.2); collectively, the disturbed outer stratigraphy is the megaregolith. Micrometeorite impacts account for most of the pulverization of rocks, and larger meteor impacts slowly churn and mix ("garden") the regolith.

15.2.1 The Lunar Regolith

Our knowledge of planetary regoliths is based mostly on *Apollo* astronauts' field observations and on laboratory studies of regolith samples returned from the Moon (McKay et al., 1991; Lucey et al., 2006). The thickness of the lunar regolith, as estimated from penetrating craters, ranges from 4–5 m in maria regions to 10–15 m in older highlands regions. Regolith thickness correlates with the density of impact craters and the age of the underlying rocks, indicating its progressive development

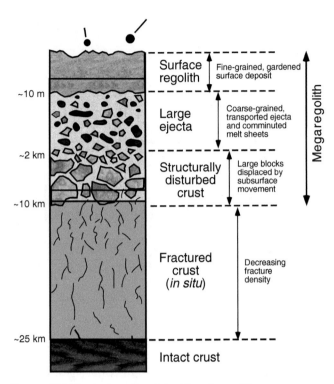

Figure 15.2 Vertical profile of the Moon's regolith.

over time. At most lunar sites where core samples or scoops of soil were taken (Figure 15.3a), the regolith is remarkably homogeneous and becomes highly compacted below depths of ~10 cm. The needs of future lunar explorers have focused considerable interest on the geotechnical properties of the regolith.

The surface regolith is extremely fine-grained (Figure 15.3b), with particle sizes varying down to silt (typically 60–80 μm). Over time, crushed particles can become aggregated by impact-produced melts, forming glass-welded clusters of clastic grains called **agglutinates** (Figure 15.3c). The textural "maturity" of the regolith is determined by the balance between two opposing impact processes, one destructive and the other constructive with respect to particle size.

Bulk chemical analyses of regolith at the various sites can be understood in terms of mixing only a few components (highlands anorthosite, mare basalt, and KREEP; see Box 10.1). The regolith is a mixture of mineral grains, rock fragments, and agglutinates (Figure 15.3d), the abundance of the latter increasing with soil maturity. Lunar surface soils contain implanted hydrogen from the solar wind and have been irradiated by cosmic rays, which produced particle tracks and cosmogenic isotopes in target mineral grains.

The *Apollo* sample collections also contain numerous regolith breccias (Figure 15.3e,f) – lithified soils that constitute the Moon's only sedimentary rocks. The

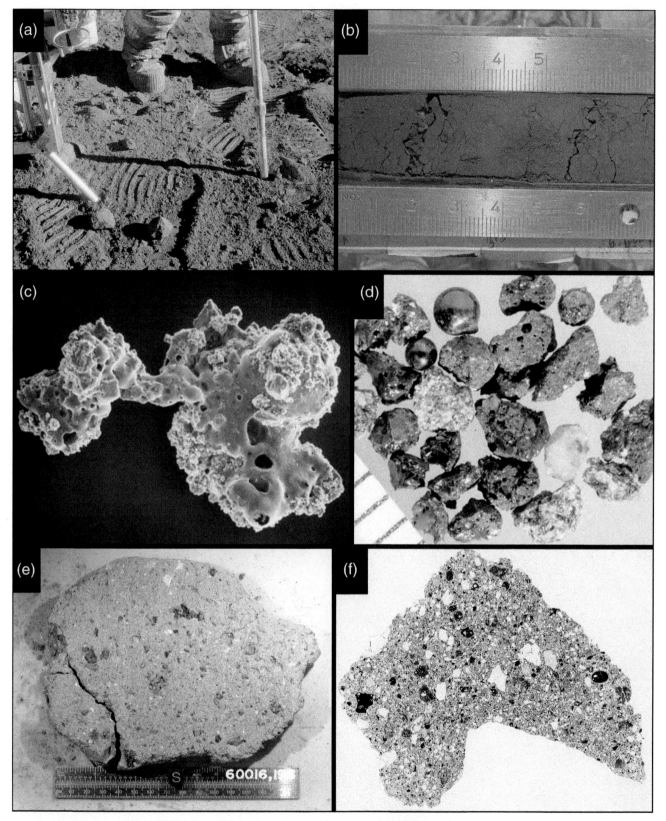

Figure 15.3 Lunar regolith. (a) Astronaut collecting a regolith core sample at the *Apollo 12* landing site. (b) Portion of an *Apollo 11* regolith core sample. (c) Agglutinate, composed of glass-bonded clastic particles. (d) *Apollo 17* soil particles consist of mineral grains (white grain is plagioclase), rock fragments including volcanic beads (spheres), and agglutinates (vesicular particles). (e) *Apollo 16* regolith breccia. (f) Thin section of regolith breccia shown in (e). NASA images.

induration of loose regolith into coherent rock results from compaction and cementation by impact melts.

15.2.2 Asteroid Regoliths

Regoliths have been imaged on the surfaces of asteroids and other small bodies like the moons of Mars. These tend to be blocky and thinner than regoliths on larger bodies (Figure 15.4a), which retain more impact fragments and have longer lifetimes against collisional disruption. Small asteroids may be covered by just a meter or so of regolith, but the regolith thickness on Vesta, the second-most massive asteroid, is estimated at greater than 1 km in places, as revealed in craters and landslides (Figure 15.4b). Models suggest that the irregular shapes and complex gravitation fields of asteroids lead to unequal global distributions of regoliths, including infilling of valleys, a prediction borne out by the discovery of ponds of dust on asteroid surfaces.

Petrologic studies of meteorite regolith breccias indicate that they are less fragmented and contain fewer agglutinates than lunar soils. In other words, asteroid regoliths are less mature. Howardites (Figure 15.4c), representing the regolith of asteroid Vesta, show significant differences in the mixing ratios of different igneous components. Depending on the size of impacts, these breccias can sample otherwise unobtainable rocks; for example, a few howardites contain small clasts of mantle rocks (Hahn et al., 2018). Howardites also contain tiny fragments of exogenic carbonaceous chondrite, the source of localized concentrations of hydrogen (in the form of hydrated phyllosilicates) discovered by the *Dawn* spacecraft.

Chondritic and achondritic breccias are common, and some of them must represent regoliths. However, not all breccias reside on the surface. The lack of agglutinates makes distinguishing regolith breccias from other fragmental rocks challenging, but cosmic ray tracks and implanted solar wind are diagnostic.

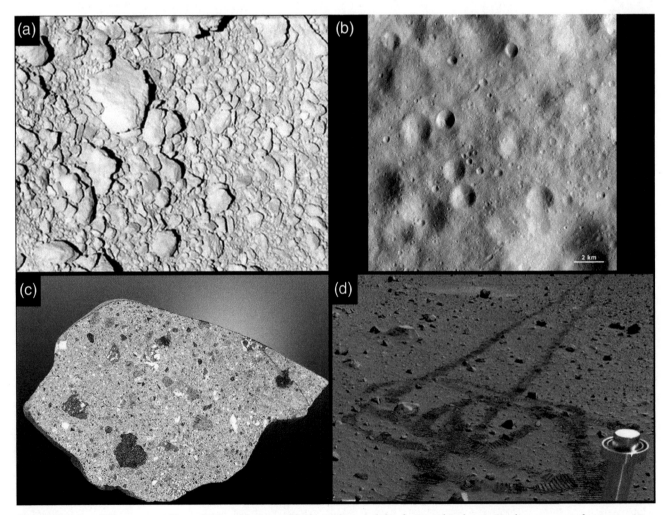

Figure 15.4 Regoliths on asteroids and Mars. (a) Coarse blocks in the regolith of asteroid Itokowa. *Hyabusa* spacecraft, image ~25 m across. (b) *Dawn* image of thick regolith on asteroid Vesta. (c) Dag 844 howardite, a sample of the vestan regolith. (d) Mars regolith, showing tracks of the Spirit rover. NASA and JAXA images.

BOX 15.1 SPACE WEATHERING

The effects of **space weathering** on the visible/near-infrared spectra of the Moon are lowering of albedo (darkening), increasing reflectance with increasing wavelength (reddening), and reduction in the depth of absorption bands (flattening). Lunar igneous rocks do not show these spectral effects, so space weathering is understood to be a characteristic of the surface regolith. Relatively young craters on the Moon expose bright materials that darken and redden over time through space weathering.

These spectral changes are attributed to tiny inclusions of nanophase iron metal, which occur in agglutinates and on the surfaces of soil particles (Figure 15.5). These minute blebs of metal are formed when iron-bearing minerals are vaporized during micrometeorite impacts, and the iron is subsequently recondensed from vapor in its native form.

Even thin atmospheres can screen out micrometeorites, so we would not expect space weathering on Venus or Mars. Curiously, though, Mercury exhibits no spectral effects, despite absence of an atmosphere. Mercury's surface rocks are nearly devoid of iron, as judged from its spectrum. Although energetic micrometeorite impacts should produce more melt and vapor than on the Moon, there is no source for condensable iron.

Many asteroids, particularly the abundant S-types, show darkening and spectral reddening, relative to the spectra of ordinary chondrites thought to be derived from them. For many years, this inability to match meteorite and asteroid spectra was a conundrum, but space weathering on asteroid surfaces now provides an explanation. As further evidence, a gradation between the spectra of chondrites and near-Earth S-type asteroids has been documented, and recent craters on the surface of asteroid Gaspra show no space weathering.

Despite its compositional similarity to the Moon, asteroid Vesta does not show the spectral effects of space weathering. Impact velocities in the asteroid belt are slower and consequently produce less vaporization, perhaps accounting for this difference. Also, the lower gravity on asteroids suggests more extensive overturn of regoliths, so there should be less exposure to micrometeorite impacts. Indeed, examination of howardites from Vesta shows virtually no nanophase iron, although other shock effects are common.

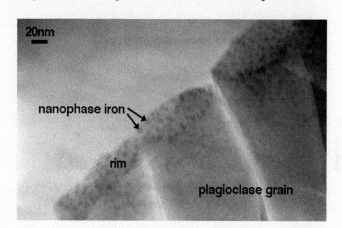

Figure 15.5 Transmission electron microscope (TEM) image of a space-weathered rim of glass containing nanophase iron on grains of lunar plagioclase. Image from NASA.

15.2.3 The Martian Regolith

Besides impacts, the martian surface has been affected by sedimentary processes that are familiar from terrestrial experience. Aqueous and aeolian processes are described in Chapters 13 and 14, and chemical weathering is considered in Section 15.3. Here we focus on physical weathering.

Since Mars' crust is dominated by basaltic volcanism, it should not be surprising that its regolith consists largely of basaltic minerals, in contrast to Earth, where sediments are mostly derived from felsic rocks (McLennan and Grotzinger, 2008). Soils, dominated by sand-sized particles, on modern Mars are transported and sometimes swept into dunes by winds. Most soils are covered by fine dust, which is revealed by rover tracks (Figure 15.4d). Basaltic detritus comprises two-thirds of these soils, with the remainder being chemically altered materials (McSween et al., 2010). The physical properties (grain sizes and shapes) of excavated soils, as determined from rover microscopic images (McGlynn et al., 2011), match those of crushed rock and are consistent with derivation by impacts. In contrast, the physical sorting of surface soils indicates reworking by aeolian activity.

The only martian sedimentary rock occurring as a meteorite is NWA 7034 (and a few meteorites paired with it that broke up during its plunge through the atmosphere). This rock is an ancient regolith breccia (Figure 15.6), composed of basaltic fragments and

Figure 15.6. Cut surface of the NWA 7034 martian meteorite, illustrating its brecciated texture. Photograph courtesy of Carl Agee. Reprinted with permission.

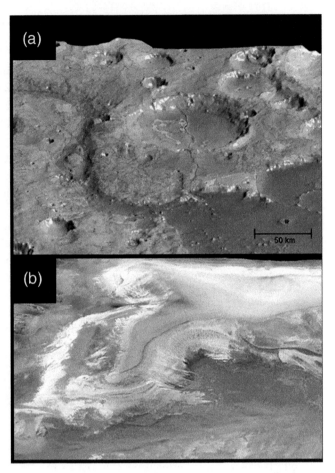

Figure 15.7 Aqueously altered rocks on Mars. (a) Phyllosilicates (blue) around Mawrth Vallis. (b) Layered sulfate deposits (brown) in Candor Chasma. Both images are colorized to show minerals mapped by *Mars Express* OMEGA. ESA and NASA images.

impact-melted clasts. Less coherent sedimentary rocks on Mars are unlikely to survive impact ejection.

15.3 Chemical Weathering and Aqueous Alteration

The Earth has the only planetary surface where liquid water is common and persistent and where chemical weathering is pervasive. Consequently, reactions of minerals with aqueous fluids are generally more common underground than on planetary surfaces. Water flowed or ponded on the surface of ancient Mars, allowing some chemical weathering, but any aqueous reactions on other planets or asteroids occurred in the subsurface by interaction with groundwaters. For asteroids, we do not use the term "chemical weathering" and instead talk about "aqueous alteration."

15.3.1 Chemical Weathering on Mars

Because Mars had surface water, at least in its distant past (see Box 15.2), chemical weathering occurred. This process likely produced the amorphous material, clay minerals (Figure 15.7a), silica, hematite, and other phases that

comprise portions of Noachian rocks. Examples of martian rocks that have experienced chemical weathering (Grotzinger et al., 2005, 2014) include sandstones, siltstones, and shales encountered by the Curiosity rover in paleolake deposits in Gale crater. The igneous detritus in these rocks is associated with amorphous material, and sometimes silica and clays, and is typically cemented by evaporative salts (sulfates, chlorides, and carbonates) or hematite.

The smectite clays on Mars come in two varieties: dioctahedral $(Mg, Fe)^{2+}$ clays are nontronites, and trioctahedral $(Al, Fe)^{3+}$ clays are saponites. Nontronite does not form in cold, limited-water environments, so these clays likely formed by hydrothermal processes in the subsurface. Saponites likely formed during weathering on the surface during intermittent warm and wet periods (Bishop et al., 2018). Ehlmann et al. (2011a) mapped the global distributions of clays formed by these competing processes.

In addition to clastic rocks, Mars has chemical sediments, mostly layered sulfate evaporates (Gendrin et al.,

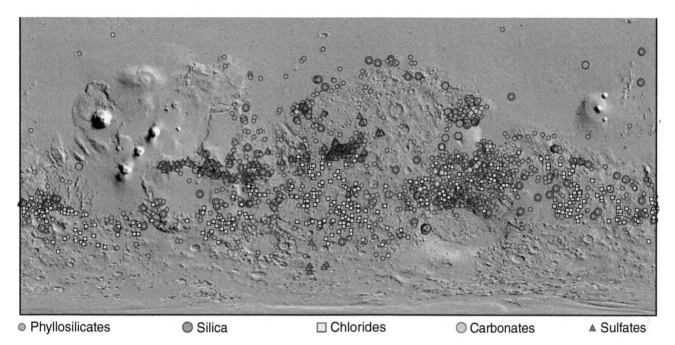

● Phyllosilicates   ● Silica   □ Chlorides   ○ Carbonates   ▲ Sulfates

Figure 15.8 Global distribution of the major classes of alteration minerals on Mars. Modified from Ehlmann and Edwards (2014).

2005) identified in orbital spectra (Figure 15.7b). Chlorides (Osterloo et al., 2008) and small amounts of carbonates also occur. These ancient rocks demonstrate that chemical weathering occurred when water was more prevalent early in Mars' history; in contrast, Mars during the last several billion years has been dominated by physical weathering and aeolian processes.

The global distribution of the major classes of alteration minerals on Mars (Ehlmann and Edwards, 2014) is shown in Figure 15.8. These minerals are detected wherever Noachian crust is exposed.

We can summarize the effects of chemical weathering on Mars using the molar A-CNK-FM diagram (Figure 15.9) where $A = Al_2O_3$, $CNK = CaO + Na_2O + K_2O$, and

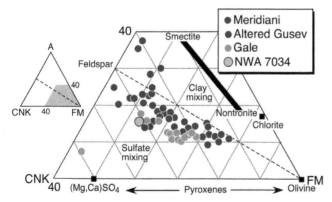

Figure 15.9 A–CNK–FM diagram illustrating the compositions of martian sedimentary rocks analyzed by rovers and the NWA 7034 meteorite. Reproduced with permission: Harry Y. McSween (2015) Petrology on Mars, *American Mineralogist*, v. 100, i. 11–12, p. 2380–2395.

$FM = FeO + MgO$. Rover-analyzed sedimentary rocks from Meridiani, Gusev, and Gale are chemically similar to basalts at the same locations, pointing to their volcanic provenance and to limited chemical change. Although mineralogic evidence for chemical weathering is widespread on Mars, bulk geochemical data do not show the leaching of water-soluble elements characteristic of open system alteration. The martian sedimentary rocks form a nearly linear array on this diagram, interpreted to reflect either some dissolution of olivine during acidic chemical weathering (Horowitz and McLennan, 2007) and/or physical sorting of olivine and other minerals during transport (McGlynn et al., 2012). The martian trend is distinct from terrestrial basaltic sediments, which migrate toward the A–FM join as CNK-rich soluble minerals are removed during weathering under neutral to slightly basic conditions.

Aqueous weathering processes are also recorded in many martian basaltic meteorites. Small amounts of clays, salts, and carbonates, as well as stable isotope data, indicate limited alteration of these igneous rocks (Leshin and Vicenzi, 2006).

15.3.2 Asteroids: Cosmic or Cosmuck?

Carbonaceous chondrites have experienced aqueous alteration (unlike other chondrite classes which experienced thermal metamorphism, as described below). Their parent bodies originally accreted mixtures of anhydrous rock with H_2O ice, which melted when heated by ^{26}Al decay. The presence of ice moderated large temperature increases, so for the most part these aqueous fluids were

BOX 15.2 **A WEATHERING CHRONOLOGY FOR MARS**

The observation that clay minerals were widespread in Mars' Noachian terrains has prompted the popular hypothesis that early Mars was warmer and wetter than later periods. This was followed in the Hesperian by precipitation of spatially restricted sulfate deposits, resulting from the evaporation of discharged groundwaters. Thus, early Mars is commonly thought to have been dominated by chemical weathering. In the Amazonian the martian surface became desiccated, characterized by iron oxides and dominated by physical weathering. A chronology for Mars based on these abundant weathering products (bottom of Figure 15.10) was developed by Bibring et al. (2006).

Although some crater lakes certainly existed for a time, e.g., in Gale crater (Grotzinger et al., 2014), the existence or persistence of widespread surface water on early Mars is now questioned. Substantial Noachian clay formation is recognized to have occurred in the subsurface by reaction with hydrothermal groundwaters (Ehlmann et al., 2011a). If those clays did not form on the surface by chemical weathering, they cannot be used as evidence of climate change. Instead, cold, arid conditions with only transient surface water may have characterized early (and later) Mars, limiting the amount of chemical weathering. This accords better with the difficulty for models to produce enough greenhouse warming and to explain what happened to a thick early atmosphere.

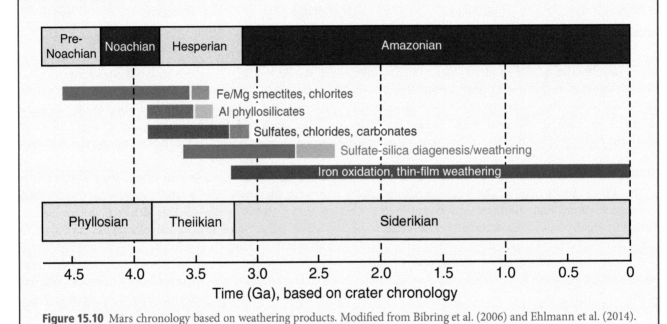

Figure 15.10 Mars chronology based on weathering products. Modified from Bibring et al. (2006) and Ehlmann et al. (2014).

fairly cold, generally 0 to 100 °C. The original chondritic minerals – mostly olivine, low-calcium pyroxene, and metal – were highly reactive and formed a variety of secondary minerals. The earliest-formed minerals were cronstedtite (Fe-serpentine) and tochilonite (interlayered $Mg(OH)_2$ and FeS) (Figure 15.11a). As alteration progressed, these phases were converted to antigorite (Mg-serpentine), clay minerals (smectites), magnetite, sulfates, and carbonates (Brearley, 2006; Howard et al., 2011). In chondrites, all these alteration minerals are fine-grained, requiring analysis by electron microscopy (Figure 15.11b). Organic matter is also an important constituent of these meteorites, and its association with phyllosilicates suggests

reactions to form more complex molecules must have accompanied the alteration of minerals.

Although aqueous alteration has also been suggested to occur by reaction with nebula gas prior to accretion, textural and stable isotopic data generally favor alteration within chondrite parent bodies. The biggest surprise is that aqueous alteration in carbonaceous chondrites was nearly isochemical. We have already seen that CI chondrites have retained their solar-like element abundances, despite having the appearance and mineralogy of congealed mud puddles. In contrast, terrestrial alteration by fluids typically causes significant chemical changes (metasomatism).

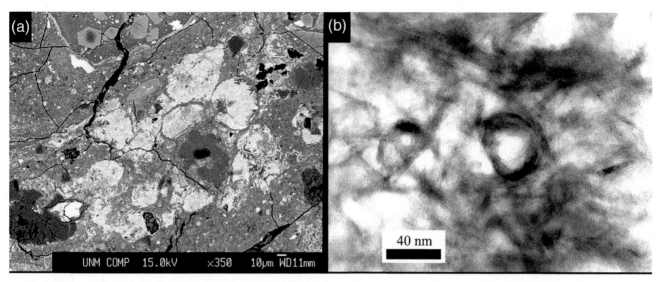

Figure 15.11 Aqueously altered CM carbonaceous chondrites. (a) Cronstedtite and tochilinite (light-colored clots in backscattered electron image). (b) Hollow serpentine tubes (TEM image). Images courtesy of Adrian Brearley. Reprinted with permission.

Dwarf planet Ceres is a carbonaceous chondrite-like body that has experienced pervasive alteration. Ceres' spectrum is similar to carbonaceous chondrites, but it has some distinct absorption bands. The spectrum, as measured by the *Dawn* spacecraft (Figure 15.12), consists of Mg-serpentine, ammonia-bearing clay, magnetite, and carbonate. Ammonia, perhaps produced by heating organic matter or incorporated as nitrogen-bearing ices, apparently exchanged with alkalis in clay minerals, and alteration produced more abundant carbonates than are found in carbonaceous chondrites. Unlike other carbonaceous asteroids, Ceres is differentiated, with an altered crust containing as much as 40 percent ice, and a coating of altered mineral grains representing a lag deposit formed when ice sublimated.

Several other massive carbonaceous asteroids have similar spectra to Ceres, implying that larger bodies suffered more extensive aqueous alteration than smaller ones.

15.4 Sedimentary Petrology on Other Worlds

The dominant sedimentary rocks on extraterrestrial rocky bodies, excepting Mars, are breccias physically weathered by impacts. Whether on bodies of asteroidal, lunar, or planetary size, impact processes have pulverized target rocks, scattered and deposited the fragments as ejecta, and cemented the buried clasts with small amounts of impact melt. These rocks are classified as "fragmental" or "melt" breccias, depending on the proportions of unmelted versus melted material. "Regolith" breccias represent lithified materials that have spent time on the surface, as opposed to breccias from the deeper megaregolith that were buried. Regolith breccias can be distinguished by the presence of implanted solar wind gases and cosmogenic nuclides formed by irradiation.

Chondrites, which comprise most asteroids, are in effect cosmic sedimentary rocks (see Box 4.1). Chondrules and metal grains were sorted by mass in the nebula prior to or during accretion, so that any one chondrite has a restricted range of particle sizes. The aqueous alteration processes that affected carbonaceous chondrites mostly occurred at low temperatures and could be analogous to diagenesis. Lithification of chondrites results from shock or cementation by precipitates from aqueous fluids.

We have previously discussed the erosion, transport, and deposition of sediments by aeolian (Chapter 13) and aqueous (Chapter 14) agents. These processes are only

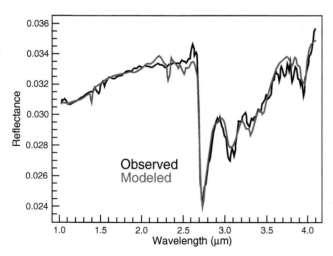

Figure 15.12 Reflectance spectrum of Ceres, compared with a modeled mixture of Mg-serpentine, ammoniated clay, magnetite, and carbonate. Modified from De Sanctis et al. (2015).

manifest on Mars, where current or past conditions have produced abundant sedimentary rocks. However, sedimentary petrology on Mars differs fundamentally from that on Earth (McLennan and Grotzinger, 2008). The planet's basaltic surface composition gives rise to particulate debris that is distinct from that arising from the intermediate to felsic rocks of the Earth's crust. Martian clastic rocks commonly retain the chemistry of their basaltic protoliths, even after chemical weathering. Quartz sands are unknown, and feldspars are more stable in the Mars environment than on Earth. Many clastic rocks are cemented by sulfates, and other evidence of diagenesis (i.e., concretions, casts of dissolved minerals, criss-crossing veins) has been documented by rovers. The evolution of evaporating fluids derived by weathering of basalt is different from terrestrial experience (Tosca et al., 2005), producing Ca-, Mg-, and Fe-sulfates with lesser chlorides and carbonates. Even though clastic and chemical sediments are absent from the martian meteorite collection, rover missions have provided exquisite microscale observations and *in situ* measurements that reveal a great deal about the petrology and stratigraphic context of sedimentary rocks on Mars (Chapter 17).

15.5 Metamorphism

Metamorphism in Earth's crust occurs under varying pressure–temperature conditions, depending on the local geothermal gradient. The various conditions define the metamorphic facies. H_2O is a common participant or catalyst in metamorphic reactions, and CO_2 is important in some metamorphic systems. Fluids and elevated pressure and temperature play similar roles on other planets.

15.5.1 Thermal Metamorphism on the Surface of Venus

The high surface temperature and elevated pressure (~465 °C, 92 bar) on Venus are equivalent to greenschist facies metamorphic conditions. Consequently, rocks on the planet's surface likely have metamorphic mineral assemblages. However, the virtual absence of H_2O would preclude formation of the hydrous minerals that characterize greenschist facies rocks on Earth. Iron in silicate minerals is predicted to become oxidized to form hematite or magnetite.

The mineralogy of venusian surface rocks and regolith may be controlled by reactions with atmospheric gas. It is also possible that such reactions with minerals may even regulate the atmospheric composition (Fegley et al., 1997). Thermodynamic models suggest that the reaction

calcite + silica = wollastonite + CO_2 should occur if the reactants are present, and this reaction could buffer the pressure of carbon dioxide, the dominant atmospheric component. Hydrochloric acid may be controlled by the reaction HCl + nepheline = albite + Cl-sodalite, and hydrofluoric acid by the reaction HF + K-feldspar + enstatite = fluorophlogopite + silica + H_2O. Sulfur dioxide could be affected by the reaction SO_2 + calcite = anhydrite + CO.

15.5.2 Thermal Metamorphism in the Interiors of Asteroids

Most chondrites have experienced early thermal metamorphism, caused by the heat generated from ^{26}Al decay. The degree of metamorphism is incorporated into the chondrite classification scheme as "**petrologic type**" (refer back to Figure 4.8). As noted in Chapter 4, during progressive thermal metamorphism, chondrule textures become blurred by recrystallization, and chemical zoning in minerals is homogenized. Asteroids undergoing metamorphism are thought to adopt the onion shell configuration (see Figure 5.15), in which the intensity of metamorphism – petrologic type – increases toward the center. Heating by ^{26}Al decay occurs faster than the heat can conduct outward, so the thermal profile is preserved. Onion shell bodies are probably not common now, as collisions have disrupted them and they have been gravitationally reassembled as rubble piles.

Thermal metamorphism has occurred in anhydrous asteroids (e.g., ordinary chondrite parent bodies). Without ice melting to inhibit large temperature increases, the temperatures inside dry asteroids can reach 1000 °C or higher. Some chondritic asteroids even experienced partial melting, producing magmas that crystallized in plutons or volcanic flows (achondrites) and leaving behind residues depleted in incompatible elements (primitive achondrites). High temperatures in asteroid-sized bodies lasted for only a few million years because of the limited lifetime of the short-lived radioactive heat source.

15.5.3 Hydrothermal Metamorphism on Mars

The only metamorphosed rocks on Mars so far encountered by rovers are silica-rich rocks at Home Plate in Gusev crater. The Spirit rover analyzed rocks composed of SiO_2 (probably opal) and TiO_2 (probably anatase) (Squyres et al., 2008). Hydrothermal fluids at low pH are hypothesized to have dissolved the basaltic precursor materials, leaving behind these insoluble phases.

A broader perspective on martian metamorphism is provided by hundreds of spectral detections of metamorphic minerals by orbiting spacecraft. These minerals

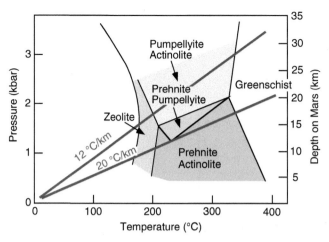

Figure 15.13 Pressure–temperature diagram illustrating geothermal gradients calculated for Mars during Noachian time, along with boundaries for low-grade metamorphic facies. Modified from McSween et al. (2015).

include chlorite and other clays, serpentine, zeolites, prehnite, and epidote (Carter et al., 2013). They occur most commonly on central peaks or in ejecta of craters, implying formation deeper in the crust.

Predicted pressure–temperature conditions in the martian crust are summarized in Figure 15.13 (McSween et al., 2015). The 12 °C/km geothermal gradient is calculated from measurement of radioactive heat-producing elements (potassium and thorium, with an assumed uranium value) in the modern Mars crust by the gamma ray spectrometer on the *Mars Odyssey* orbiter. Heat production was five times greater in the Noachian, before significant decay of these radioisotopes, so this gradient was corrected for the relevant half-lives. The 20 °C/km gradient in Figure 15.13 is the maximum based on heat flow estimated from thickness of the lithosphere and depth to the brittle–ductile transition. These geothermal gradients are less than typical terrestrial values of 25–30 °C/km, and transit several low-grade metamorphic facies (Figure 15.13).

To understand martian metamorphic rocks, we will use the ACF diagram (Figure 15.14), a projection from SiO_2 and H_2O, where $A = Al_2O_3 + Fe_2O_3 - Na_2O - K_2O$, $C = CaO - 3.3 \, P_2O_5$, and $F = MgO + FeO + MnO$, all in molar proportions. This diagram is commonly used in terrestrial metamorphic petrology, especially for mafic rocks. Minerals coexisting at different metamorphic grades lie at the corners of triangles, and the bulk compositions of rocks containing those minerals must lie within the triangles. ACF diagrams depicting the mineralogies of low-grade metamorphic facies are illustrated in Figure 15.14 (McSween et al., 2015). Also plotted are the

bulk compositions of martian meteorites and basaltic rocks analyzed by Mars rovers. Metamorphism of basaltic rocks plotting closer to F should produce chlorite + actinolite + serpentine, or talc, and rocks plotting farther from F should form chlorite + actinolite + laumontite, prehnite, or pumpellyite.

Orbital spectra from the Nili Fossae region indicate the occurrence of prehnite + chlorite (Ehlmann et al., 2011b), although neither actinolite nor any other amphibole has been identified. Laumontite has also not been specifically identified, but unspecified zeolites with similar spectra occur, and modeling suggests pumpellyite could be present. Serpentine and talc have also been noted. The Nili Fossae region has a Noachian age, and its metamorphism appears to have been caused by hydrothermal fluids circulating below abundant large craters.

15.6 Mass Wasting

Regolith and sediments occur on the surfaces of bodies throughout the Solar System. Unconsolidated materials on sloping terrains tend to move under the force of gravity. This process is called **mass wasting** – the movement of masses of regolith or rock debris downslope. Such movements can take a range of forms, including fast motions such as falls, slides, and flows, and slow motions such as creep. The speed and morphology of the mass wasting depends on composition, pre-existing weaknesses, and other factors. The shape of the mass-wasting feature depends on the shape of pre-existing planes of weakness, which may be parallel to the land surface, producing planar slides, or concave up, producing rotational slides or slumps. Materials can also be weakened by weathering and especially by the presence of liquids, which increase pore pressure and reduce friction.

Tectonic structures such as extensional faults have high relief, and so provide an opportunity for slope failure, as do the walls of impact craters. Because all planetary bodies have gravity and at least some topographic relief, if only from impact craters, mass wasting is probably the most common surficial process in the Solar System (after impact cratering, of course!).

Impacts not only often host mass-wasted material, but they commonly trigger mass-wasting events, as do tectonic movements. Rainfall is also a common trigger on Earth. Liquid has also clearly prompted mass wasting on Mars in the past and perhaps on Titan under present-day conditions. However, differentiating dry and wet mass-wasted deposits can be challenging.

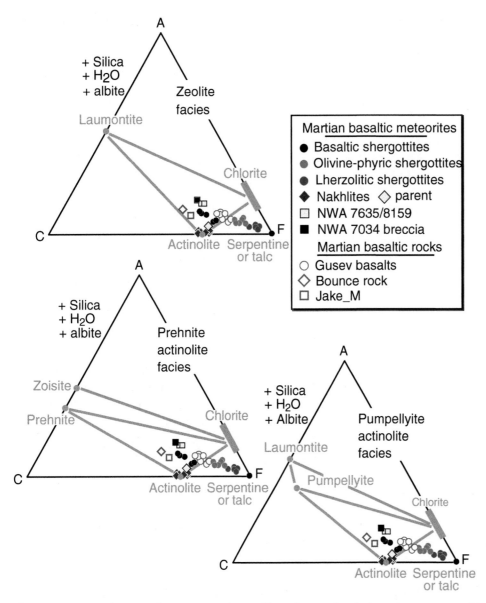

Figure 15.14 ACF diagrams showing mineral assemblages for metabasalts in various metamorphic facies. Bulk chemical compositions of martian meteorites and of basaltic rocks analyzed by Mars rovers are projected onto this diagram. Modified from McSween et al. (2015).

The significant topographic relief of Mars provides a ready context for mass wasting. As a striking example, the ~10 km deep canyon Valles Marineris, formed by global-scale extension related to the uplift of Tharsis, exhibits extensive landslide deposits (Figure 15.15a). These martian landslides have a greater areal extent and longer run-out distance than would be expected from the landslide scar height – a measure of their original potential energy – and are interpreted to have been fluidized by ice or even by acoustic energy from the rocks banging together (Watkins et al., 2015). Although rare on Earth, long-run-out landslides represent a prominent geomorphic process in Valles Marineris, perhaps because of its steep escarpments. Long avalanches have also been detected on Iapetus, a moon of Saturn, where the cause for the exception extent of icy deposits is localized heating during run-out. Because of Titan's low gravity, avalanches might also be extensive.

The flanks of volcanoes are also subject to slope failure, and mass wasting is now recognized as a significant part of volcano evolution. The high scarps around Olympus Mons result from collapse and gravity slide of the outer flanks of

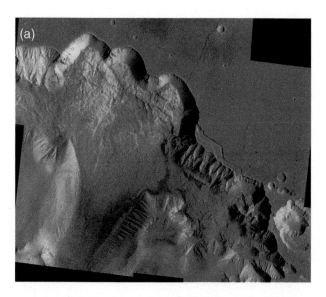

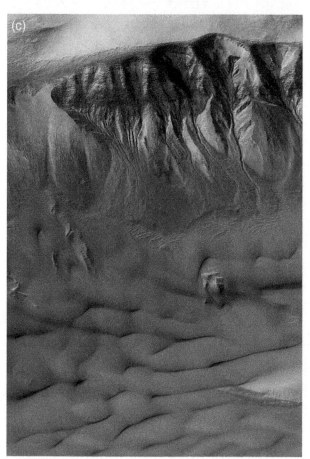

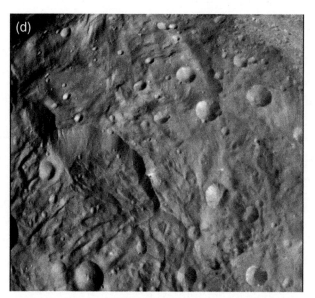

Figure 15.15 Examples of mass wasting. (a) Numerous landslides on graben walls in Valles Marineris, Mars. (b) Collapse of the flanks of Olympus Mons has produced high bounding scarps and rugged debris deposits. (c) Martian "gullies" on a crater wall. (d) Landslides on a fault scarp on asteroid Vesta. NASA images.

the shield, and huge debris deposits surround the volcano (Figure 15.15b). The landslides could have been triggered by marsquakes or stresses induced by the growing volcano.

Martian "gullies" (Figure 15.15c) are also examples of mass wasting, at a much smaller scale (McEwen et al., 2011). These features are characterized by an upper alcove and a depositional apron, linked by a channel. They occur on the walls of craters, generally 30° poleward in each hemisphere. The gullies are very young features, and repeated observations show changes over the course of a martian year. The cause of the gullies is controversial, with advocates for water, ice, dry ice, and dry granular flows. Slope streaks on Mars, which have some relief, are also likely to form by mass wasting.

Mass wasting also occurs on small bodies, such as the asteroid Vesta. Figure 15.15d shows mass wasting along a high-standing scarp, interpreted to be an extension fault. The ridged or stepped morphology of the deposits at the bottom of the scarp suggests their emplacement by rotational sliding, involving rotation backward – toward the cliff – of coherent blocks during their downward movement. In the surrounding impact craters, streaks along the crater walls attest to another type of mass wasting, namely by granular flow. Mass-wasting deposits rest on top of other units, so they are relatively younger, which is typical for units formed by surficial processes that rest on bedrock related to older processes.

Summary

A variety of geologic processes modify rocks from their original forms. Rocks on the surfaces of planets, the Moon, and small bodies are fragmented by impacts (the only common type of physical weathering) to form regoliths. Space weathering, resulting from micrometeorite impacts on airless bodies, modifies reflectance spectra. Because of the absence of liquid water on most planets, only Mars has experienced chemical weathering, producing sedimentary rocks and soils with significant basaltic detritus. Ice-bearing asteroids (carbonaceous chondrites) heated by ^{26}Al decay experienced ice melting that promoted aqueous alteration in their interiors.

Thermal metamorphism occurs by reaction of rocks with atmospheric gas on the surface of Venus, owing to its high temperature and pressure. Thermal metamorphism in the interiors of anhydrous asteroids resulted from the heat produced by ^{26}Al decay. Hydrothermal metamorphism on Mars was likely caused by groundwater heated by large impacts, and has produced a variety of metabasalts and serpentinites.

Downslope movement of sediments occurs by various processes as a result of gravity. Unstable slopes can form by tectonic activity, volcanism, or cratering, and the resultant mass wasting has sculpted the surfaces of large and small planetary bodies. The materials on slopes can be weakened by weathering and, on bodies with past or present volatile cycles, by the presence of fluids.

Aqueous fluids and elevated temperatures have produced changes in rocks on bodies large and small. In the following chapter, we will consider whether those conditions might have allowed life to arise beyond the Earth.

Review Questions

1. What is regolith, and how do its constituents and properties change with maturity?
2. Why is chemical weathering so limited in Solar System bodies?
3. What kinds of metamorphic rocks have been found on Mars? What do we surmise about metamorphism on Venus?
4. What is the heat source for thermal metamorphism and aqueous alteration on asteroids, and what accounts for one process or the other?
5. What is space weathering?
6. What kinds of planetary surface features are subject to mass wasting? What factors facilitate mass wasting?

SUGGESTIONS FOR FURTHER READING

Brearley, A. J. (2006) The action of water. In *Meteorites and the Early Solar System II*, eds. Lauretta, D. S., and McSween, H. Y. Tucson, AZ: University of Arizona Press, pp. 587–624. This is a comprehensive review of aqueous alteration in meteorites.

Ehlmann, B. L., and Edwards, C. S. (2014) Mineralogy of the martian surface. *Annual Reviews of Earth and Planetary Science*, **42**, 291–315. This paper provides an overview and global perspective of the distribution of minerals formed by chemical weathering and hydrothermal metamorphism on Mars.

McKay, D. S., Heiken, G., Basu, A., et al. (1991) The lunar regolith. In *Lunar Source Book: A User's Guide to the Moon*, eds. Heiken, G., Vaniman, D., and French, B. M. Cambridge: Cambridge University Press, pp. 285–386. The nature of the Moon's regolith is described in detail in this chapter.

McLennan, S. M., and Grotzinger, J. P. (2008) The sedimentary rock cycle of Mars. In *The Martian Surface Composition: Composition, Mineralogy, and Physical Properties*, ed. Bell, J. F. Cambridge: Cambridge University Press, pp. 541–577. A thoughtful summary of the differences between martian and terrestrial sedimentary rocks.

McSween, H. Y. (2015) Petrology on Mars. *American Mineralogist*, **100**, 2380–2395. This paper summarizes the petrology of igneous, sedimentary, and metamorphic rocks on Mars.

REFERENCES

Bibring, J.-P., Langevin, Y., Mustard, J. F., et al. (2006) Global mineralogical and aqueous Mars history derived from OMEGA/Mars Express data. *Science*, **312**, 400–404.

Bishop, J. L., Fairen, A. G., Michalski, J. R. et al. (2018) Surface clay formation during short-term warmer and wetter conditions on a largely cold ancient Mars. *Nature Astronomy*. DOI: 10.1038/x41550-017-0377-9.

Brearley, A. J. (2006) The action of water. In *Meteorites and the Early Solar System II*, eds. Lauretta, D. S., and McSween, H. Y. Tucson, AZ: University of Arizona Press, pp. 587–624.

Carter, J., Poulet, F., Bibring, J.-P., et al. (2013) Hydrous minerals on Mars as seen by the CRISM and OMEGA imaging spectrometers: updated global view. *Journal of Geophysical Research*, **118**, 831–858.

De Sanctis, M. C., Ammannito, E., Marchi, S., et al. (2015) Ammoniated phyllosilicates with a likely outer solar system origin on (1) Ceres. *Nature*, **528**, 241–245.

Ehlmann, B. L., and Edwards, C. S. (2014) Mineralogy of the martian surface. *Annual Reviews of Earth and Planetary Science*, **42**, 291–315.

Ehlmann, B. L., Mustard, J. F., Murchie, S. L., et al. (2011a) Subsurface water and clay mineral formation during the early history of Mars. *Nature*, **479**, 53–60.

Ehlmann, B. L., Mustard, J. F., Clark, R. N., et al. (2011b) Evidence for low-grade metamorphism, diagenesis, and hydrothermal alteration on Mars from phyllosilicate mineral assemblages. *Clays & Clay Minerals*, **59**, 359–377.

Fegley, B., Klingelhoefer, G., Lodders, K., et al. (1997) Geochemistry of surface–atmosphere interactions on Venus. In *Venus II*, eds. Bougher, S. W., Hunten, D. M., and Phillips, R. J. Tucson, AZ: University of Arizona Press, pp. 591–636.

Gendrin, A. N., Mangold, N., Bibring, J-P., et al. (2005) Sulfates in martian layered terrains: the OMEGA/Mars Express view. *Science*, **307**, 1587–1590.

Grotzinger, J. P., Arvidson, R. E., Bell, J. F., et al. (2005) Stratigraphy and sedimentology of a dry to wet eolian depositional system, Burns formation, Meridiani Planum, Mars. *Earth and Planetary Science Letters*, **240**, 11–72.

Grotzinger, J. P., Sumner, D. Y., Kah, L. C., et al. (2014) A habitable fluvio-lacustrine environment at Yellowknife Bay, Gale crater, Mars. *Science*, **343**. DOI: 10.1126/science.1242777.

Hahn, T. M., Lunning, N. G., McSween, H. Y., et al. (2018) Mg-rich harzburgites from Vesta: mantle residua or cumulates from planetary differentiation? *Meteoritics & Planetary Science*, **53**, 514–546.

Horowitz, J. A., and McLennan, S. M. (2007) A 3.5 Ga record of water-limited, acidic conditions on Mars. *Earth and Planetary Science Letters*, **26**, 432–443.

Howard, K. T., Benedix, G. K., Bland, P. A., et al. (2011) Modal mineralogy of CM chondrites by X-ray diffraction (PSD-XRD): part 2. Degree, nature and settings of aqueous alteration. *Geochemica et Cosmochimica Acta*, **75**, 2735–2751.

Leshin, L. A., and Vicenzi, E. (2006) Aqueous processes recorded by martian meteorites: analyzing martian water on Earth. *Elements*, **2**, 157–162.

Lucey, P., Korotev, R. L., Gillis, J. J., et al. (2006) Understanding the lunar surface and space–Moon interactions. In *New Views of the Moon*, eds. Jolliff, B. L., Wieczorek, M. A., Shearer, C. K., and Neal, C. R., Chantilly, VA: Mineralogical Society of America, pp. 83–219.

McEwen, A. S., Ojha, L., Dundas, C. M., et al. (2011) Seasonal flows on warm martian slopes. *Science*, **333**, 740–743.

McGlynn, I. O., Fedo, C. M., and McSween, H. Y. (2011) Origin of basaltic soils at Gusev crater, Mars, by aeolian modification of impact-generated sediment. *Journal of Geophysical Research*, **116**, E00F22.

McGlynn, I. O., Fedo, C. M., and McSween, H. Y. (2012) Soil mineralogy at the Mars Exploration Rover landing sites: an assessment of the competing roles of physical sorting and chemical weathering. *Journal of Geophyical Research*, **117**, E01006.

McKay, D. S., Heiken, G., Basu, A., et al. (1991) The lunar regolith. In *Lunar Source Book: A User's Guide to the Moon*, eds. Heiken, G., Vaniman, D., and French, B M. Cambridge: Cambridge University Press, pp. 285–386.

McLennan, S. M., and Grotzinger, J. P. (2008) The sedimentary rock cycle of Mars. In *The Martian Surface Composition: Composition, Mineralogy, and Physical Properties*, ed. Bell, J. F. Cambridge: Cambridge University Press, pp. 541–577.

McSween, H. Y., McGlynn, I. O., and Rogers, A. D. (2010) Determining the modal mineralogy of martian soils. *Journal of Geophysical Research*, **115**, E00F12.

McSween, H. Y., Labotka, T. C., and Viviano-Beck, C. E. (2015) Metamorphism in the martian crust. *Meteoritics & Planetary Science*, **50**, 590–603.

Osterloo, M. M., Hamilton, V. E., Bandfield, J. L., et al. (2008) Chloride-bearing materials in the southern highlands of Mars. *Science*, **319**. DOI: 10.1126/science.1150690.

Squyres, S. W., Arvidson, R. E., Ruff, S., et al. (2008) Detection of silica-rich deposits on Mars. *Science*, **320**, 1063–1067.

Tosca, N. J., McLennan, S. M., Clark, B. C., et al. (2005) Geochemical modeling of evaporation processes on Mars: insight from the sedimentary record at Meridiani Planum. *Earth & Planetary Science Letters*, **240**, 122–148.

Watkins, J. A., Ehlmann, B. L., and Yin, A. (2015) Long run-out landslides and the long-lasting effect of early water activity on Mars. *Geology*, **43**, 107–110.

16

Astrobiology: A Planetary Perspective on Life

At its root, the word "astrobiology" means "biology of the stars." It is the branch of science that concerns the origin and evolution of life on Earth – the only place that, at present, we are certain life exists – and the potential for life to be distributed across the Universe. In this chapter, we explore the evolutionary relationships of life on Earth and review the necessary ingredients and permissible environmental conditions for the origin and evolution of life. We also discuss the characteristics of early life on Earth, and the physical and geochemical evidence for life that might be used to target habitable environments – and potentially to detect evidence of life – elsewhere in the Universe.

16.1 The Diversity of Life

In 2003, the Hubble Space Telescope focused its lens at a single, dark and seemingly featureless spot in the Universe. At the end of 11 days, Hubble had gathered enough visible, near-infrared, and UV light to identify thousands of new galaxies, each containing billions of individual stars (Figure 16.1). This astounding image brings to mind an idea first articulated by Metrodorus in 400 BC, that "it is unnatural in a large field to have only one stalk of wheat, and in the infinite universe, only one living world." This astounding image brought new life to field of **astrobiology**.

Among the many challenges of astrobiology is finding a solution to the seemingly simple question of what is life? Typically, we consider life as a series of observable characteristics: life has an ordered structure (cells), built around a chemical blueprint (genetic material), that responds to its environment; life utilizes energy (metabolism) and shows growth; life is capable of reproduction. Although these characteristics are helpful in considering what to look for when investigating the potential for life in terrestrial and extraterrestrial environments, none of

these characteristics are unique to life. Minerals, for example, have an ordered structure that is built around a chemical blueprint. Incorporation of trace elements into a crystal lattice reflects a mineral's capability to respond to its environment. A forest fire utilizes energy to grow, and can spawn new fires.

In order to define life uniquely, we must also consider the process that is most central to life on Earth – biologic evolution. The evolutionary process (also called "descent with modification") requires that the chemical coding of life (that is, our genetic material) is inherited by successive generations and is susceptible to modification by both random mutation and sexual recombination. Adaptation (also called "natural selection") then occurs when genetic characteristics that are favorable to survival within an environment increase in abundance within a population.

The central importance of evolution to our understanding of life on Earth also provides the opportunity for extraterrestrial life to potentially differ from life on Earth in significant ways, depending on the environments in which evolution occurs. With this in mind, the astrobiology community has coalesced around a popular working definition of life as a "self-sustaining chemical system capable of Darwinian evolution."

16.1.1 Reconstructing the Tree of Life

The acceptance of evolution as a driving mechanism for the diversity of life on Earth provides astrobiologists with a mechanism for deciphering how life might originate and evolve on another planet. To do this, we use our understanding of the process of evolution to look backward through time and to determine how evolution has modified the genetic blueprint of life on Earth. Organisms that have more of their genetic blueprint in common are determined to be more closely related, and by exploring

Figure 16.1 In 2014, NASA released this revised version of the 2003 Hubble Ultra Deep Field (HUDF) image. This image combines visible, near-infrared, and ultraviolet light to produce an image with the full range of colors visible to Hubble, and reveals approximately 10,000 galaxies that extend back to within a few hundred million years of the Big Bang. NASA image.

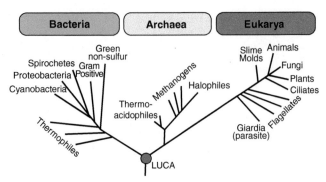

Figure 16.2 Phylogenetic tree of life, after Woese et al. (1990). The three primary domains of life include the Bacteria, the Archaea, and the Eukarya. Detailing the genetic activities of the most deeply branching organisms allows astrobiologists to infer the metabolic capabilities of the earliest life, and environments in which early life may have arisen.

the genetic similarity of all organisms, we can reconstruct what is commonly termed the **tree of life**.

The tree of life is a phylogeny, or a pattern of relationship, that utilizes similarity in genetic coding to define relationships (Gaucher et al., 2010). To construct a tree of life that deciphers relationships across a broad range of biological entities, scientists have focused on **ribosomal RNA** (rRNA). Ribosomal RNA is responsible for the production of ribosomes that catalyze amino acids into the proteins that carry out the most basic functions of biological cells, and therefore occurs in all known organisms. Certain portions of rRNA are also strongly conserved, meaning that rRNA undergoes evolutionary change very slowly. We use these portions to track evolutionary change over long periods of time and to compare diverse species. To build the tree of life, rRNA – specifically the subunits 16S and 18S of the rRNA molecule – is compared between organisms, and mapped to portray the closeness of relationships.

Figure 16.2 shows a basic phylogenetic tree of life (Woese et al., 1990), defined by three primary branches, or domains, of life. These include the Bacteria, the Archeae, and the Eukarya. **Bacteria** are a large and diverse group of single-celled organisms that have cell walls, but lack organelles and an organized nucleus. **Archeae** are similar to bacteria in their basic cell structure, in that they

lack organelles and an organized nucleus, but are distinct in the structure of their cell membranes and their genetic coding. Archeae, like bacteria, also display a broad range of metabolisms (Table 16.1) which permit their dominance in a variety of extreme environments (Table 16.2). The third domain of life, the **Eukarya**, includes all organisms whose cells contain both a discrete nucleus and membrane-bound organelles. Although Eukarya make up only a small part of the diversity of life on Earth, they are commonly the most easily recognized because this group contains all larger, multicellular organisms.

From an astrobiological perspective, one of the most important aspects of the tree of life is the basic pattern of diversity. It is readily apparent that the greatest diversity of life on Earth occurs within the microbial (bacterial and archaeal) branches of the tree, which suggests that extraterrestrial life is most likely to be microbial. Astoundingly, however, we are only now beginning to understand the full range of genetic diversity of microbial life, and the capabilities that this genetic diversity grants to these organisms. In fact, less than 1 percent of microbial life has been successfully cultivated in the laboratory, so biologists have developed a wide range of techniques that focus on detecting the presence of different genes (metagenomics analysis) and the expression of these genes (metatranscriptomic and metaproteomic analysis) to decipher the true diversity represented by these domains (DeLong, 2009).

16.1.2 Complexities in the Tree of Life

In addition to helping us define the full range of diversity of life on Earth, analysis of the tree of life provides astrobiologists with a glimpse into the process of evolution of life on Earth. What we find is both complex and fascinating. We have learned, for example, that although

Table 16.1 Metabolic reactions carried out by bacteria and archaea

Process	Reactants	Products
Photosynthesis – metabolic reactions in which light is used as the source of energy for the synthesis of organic carbon molecules		
Photosynthesis (oxygenic)	Carbon dioxide, water	Organic carbon, oxygen
Photosynthesis (anoxygenic)	Hydrogen sulfide	Organic carbon, sulfate
Chemosynthesis – metabolic reactions in which energy for the synthesis of organic compounds is derived from the oxidation of inorganic compounds		
Sulfur oxidation	Hydrogen sulfide, oxygen	Organic carbon, sulfate
Ammonium oxidation	Ammonium, oxygen	Organic carbon, nitrite
Nitrite oxidation	Nitrite, oxygen	Organic carbon, nitrate
Iron oxidation	Iron (Fe^{2+}), oxygen	Organic carbon, iron (Fe^{3+})
Methane oxidation	Methane, oxygen	Organic carbon, carbon dioxide
Respiration – metabolic reactions in which organic carbon molecules are broken down to release energy for cellular processes		
Aerobic respiration	Organic matter, oxygen	Carbon dioxide, water
Sulfate reduction	Organic matter, sulfate	Hydrogen sulfide
Nitrate reduction	Organic matter, nitrate	Nitrogen
Iron reduction	Organic matter, iron (Fe^{3+})	Carbon dioxide, iron (Fe^{2+})
Methanogenesis	Organic matter, oxygen, hydrogen	Methane
Associated reactions – reactions that do not directly involve organic carbon, but which produce chemical products or energy necessary for metabolic reactions		
Nitrogen fixation	ATP, nitrogen	Ammonium
Anammox	Ammonium, nitrite	Nitrogen

Table 16.2 Extreme environments tolerated by bacteria and archea

Organism type	Environment	Consequences
Thermophiles	50–80 °C	Denaturing of proteins
Hyperthermophiles	>80 °C	Denaturing of pigments, nucleic acids
Psychrophiles	<15 °C	Loss of membrane fluidity, ice damage
Acidophiles	pH < 3	Denaturing of proteins
Alkaliphiles	pH > 9	Insufficient electrochemical energy
Halophiles	3–5 M NaCl	Osmotic stress, low water activity
Piezophiles	To 1000 atm	Loss of membrane fluidity
Radiophiles	To 5000 Gy	Direct damage to proteins, nucleic acids

genes involved in the most basic cellular processes, like protein synthesis and energy generation, produce similar phylogenetic trees, many other genes record distinct evolutionary histories. This provides evidence that the evolutionary process is more complex than a simple Darwinian model in which genetic information is transferred unidirectionally from parent and offspring, and likely contains aspects of either horizontal (or lateral) gene transfer or endosymbiosis.

Horizontal gene transfer is the direct movement of genetic material between extant organisms (McDaniel et al., 2010). Within the bacterial and archaeal domains,

horizontal gene transfer can occur in three distinct ways. The first mechanism is transformation, which is the direct uptake of exogenous genetic material through the cell membrane. Although this process is not well understood, the ability for cell membranes to uptake exogenous material appears to be triggered by adverse environmental conditions. The second mechanism is conjugation, in which cells in direct contact with one another can transfer portions of their genetic material. The third mechanism of horizontal gene transport is conjunction, in which genetic material from a cell is incorporated into a virus, which then transmits this genetic material to a secondary host. Recent studies suggest that the degree of horizontal gene transfer is greater in extreme environments, where microbial populations are under stress, and that gene transfer can result in a large number of common genes between bacterial and archaeal domains that facilitate adaptation to such extreme environments (Fuchsman et al., 2017).

Endosymbiosis provides an additional mechanism for the wholesale horizontal transference of genetic material. Unlike horizontal gene transfer in microbial populations, endosymbiosis occurs in the eukaryotic domain when intact organisms are ingested and incorporated into the metabolic pathway of the cell (see, for example, Archibald and Keeling, 2002). Endosymbiosis is believed to be the primary mechanism by which Eukarya obtained their critical organelles, such as mitochondria used in energy utilization and chloroplasts used in photosynthesis.

Although the driving force behind endosymbiosis is not well understood, both horizontal gene transfer and endosymbiosis provide a means for organisms to rapidly adapt to inhospitable environments.

16.1.3 The Last Universal Common Ancestor

From an astrobiological perspective, the most important aspect of the tree of life may be in understanding the nature of the **Last Universal Common Ancestor (LUCA)**. By definition, LUCA represents a population of organisms from which all modern life is derived. In most modern phylogenies (Figure 16.2), LUCA is placed at the division between the bacterial and archaeal domains. In light of studies indicating widespread transference of genes between Bacteria and Archaea, however, LUCA is unlikely to have been the only organismal population at this time. Rather, we should consider LUCA to represent a distinct population of cells that were well adapted to the environments of the early Earth and from which our current diversity of life originated.

Many of our efforts in understanding LUCA have therefore been focused on defining the environments in which the LUCA population thrived. These environments are derived by exploring the characteristics of organisms that comprise the earliest branches within the bacterial and archaeal domains. The most widely shared trait among these early branching organisms is genetic coding that is necessary to stabilize proteins at high temperatures, suggesting that LUCA may have thrived in thermophilic (50–80 °C) to hyperthermophilic (>80 °C) environments (Stetter, 1996). Similarly, analyses of discrete protein families suggest that LUCA likely inhabited a geochemically active, thermophilic environment that was rich in CO_2, H_2, and a variety of reduced metals (Weiss et al., 2016).

16.2 The Chemistry of Life

Whereas investigation into the character of LUCA has informed our understanding of the environments in which earliest life may have thrived, we must also consider the question of why such environments became the locus for emergence of life. Our first clues come from the chemical makeup of the primary biomolecules of life, which include nucleic acids, proteins, carbohydrates, and lipids.

Nucleic acids and proteins are both composed of **amino acids** (Figure 16.3), which consist of a carbon atom that is bonded with a carboxyl group (COOH), an amine (NH₂), and one of a variety of side chain groups. **Nucleic acids**, specifically deoxyribosenucleic acid (DNA) and ribonucleic acid (RNA), are complex

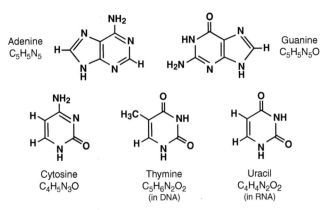

Figure 16.3 Amino acids are considered the building blocks of life, since they are the main component of both genetic material (DNA and RNA) and the proteins that carry out cellular functions. Here we see the structures of the amino acids that comprise genetic material: adenine, guanine, cytosine, thymine (in DNA only), and uracil (in RNA only).

macromolecules that provide the genetic coding to life; they are composed of amino acids bonded with a five-carbon sugar and a phosphate group. Similarly, **proteins** consist of chains of amino acids that coil or fold into three-dimensional shapes and that provide structural support for cells, aid in energy utilization (e.g., adenosine triphosphate, or ATP), and serve as enzymes to catalyze chemical reactions within the cell.

Although simpler in structure, carbohydrates and **lipids** are just as important to biology. **Carbohydrates** consist of simple chains of carbon bonded to a hydrogen (H^+) ion and a hydroxyl ion (OH^-). Short-chain carbohydrates are referred to as sugars, and are broken down in the cell to provide energy. More complex carbohydrates, such as starches, are used as energy storage within cells. By contrast, the basic structure of a lipid is a grouping of simple chains of carbon bonded to two hydrogen (H^+) ions attached to a glycerol ($C_3H_8O_3$) molecule. Lipids serve critical roles in energy storage and in the signaling of cellular functions, and, when augmented by a phosphate group (i.e., phospholipids), in the construction of cellular membranes.

16.2.1 CHNOPS and the Cosmos

The composition of the primary biomolecules of life shows that life on Earth is composed primarily of carbon, hydrogen, nitrogen, oxygen, and phosphorous. These five elements, along with sulfur, which occurs in several amino acids critical to protein formation, are together referred to as **CHNOPS**. In fact, these six elements can make up as much as 97 percent of cellular biomass (Wackett et al., 2004). These elements are also some of the most common elements in the Universe. Recall that hydrogen, the most abundant element in the Universe,

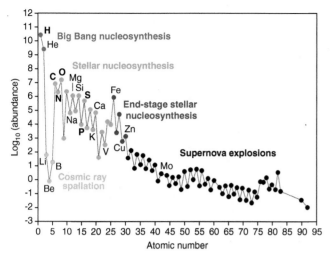

Figure 16.4 The elements that are most critical to life, CHNOPS, are also some of the most abundant elements in the Universe. Hydrogen, formed during the Big Bang, is the most abundant element, followed by oxygen (third), carbon (fourth), nitrogen (sixth), sulfur (ninth), and phosphorous (seventeenth), which are all produced during stellar nucleosynthesis. By contrast, many of the redox-sensitive metals that occur within biological enzymes, such as vanadium, iron, copper, zinc, and molybdenum, are produced only within end-stage nucleosynthesis and within supernova explosions.

was formed by nucleosynthesis during the Big Bang, and the rest of the CHNOPS sequence, as well as essential metals such as Ca, Mg, Na, and K, are the product of fusion reaction in stars (Section 4.2). By contrast, heavier metal ions that are critical components of many biological enzymes, such as V, Fe, Cu, Zn, are formed only in the final moments as stars transition into supernovae, while others, such as Mo, form only in supernova explosions (Figure 16.4, previously shown in Figure 4.9 but with different elements labeled).

Of these elements, carbon is the most critical for life (see Box 16.1). The importance of carbon results, in part, from its chemical versatility. The occurrence of four electrons in its outer electron shell allows carbon to host up to four covalent bonds. Carbon can form stable bonds with both itself and a broad range of other compounds. It can also utilize single, double, and triple bonds with another carbon atom to form a wide variety of stable two- and three-dimensional shapes. Perhaps most critically, the amounts of energy it takes to make (and break) bonds between carbon and CHNOPS are extraordinarily similar, which allows carbon-based compounds to readily exchange ions to drive the metabolic reactions of life.

The importance of carbon in the evolution of life is supported by abundant evidence that organic carbon molecules are common across the Universe. Spectroscopic analysis has detected organic molecules in a wide

BOX 16.1 SILICON AS A BASIS FOR LIFE?

In the original Star Trek series, on Star Date 3196.1, Federation miners of Janus IV began to be repeatedly attacked by a mysterious creature, known as the Horta. The Horta was a silicon-based life form, capable of tunneling through solid rock. The idea of extraterrestrial life based on silicon, rather than on carbon, has intrigued us ever since.

Silicon has many chemical properties similar to carbon. It is the eighth most abundant element in the Universe. Like carbon, silicon is tetravalent, and forms covalent bonds with both itself and other elements, including oxygen and hydrogen (Bains, 2004). Also, like carbon, silicon is capable of forming long-chain polymers and a variety of stable two- and three-dimensional topologies. But although some terrestrial organisms are known to use amorphous silica in making skeletal structures, there are no known instances of silicon–carbon bonding associated with life.

Researchers at Caltech, however, have recently coaxed the evolution of enzymes capable of driving Si–C bonding (Kan et al., 2016), and have suggested that if such evolution can occur in the laboratory, it may very well have occurred in nature. Yet, despite this discovery, there are substantial hurdles to the idea of silicon-based life. The first hurdle is in the behavior of silicon in water. Not only does silicon have a very low solubility in water with a pH less than 12, which restricts its availability, but silicon compounds are also generally unstable in water. Silicones (Si–O polymers) readily break down in water, and silanes (SiH_4 compounds) react explosively with water.

Perhaps the greatest hurdle to silicon-based life is the affinity between silicon and oxygen. Silicon readily oxidizes and forms an insoluble solid, silicon dioxide (SiO_2). The great energy it takes to break the SiO_2 bonds makes the silica tetrahedron a superb base for planetary minerals, such as pyroxenes, amphiboles, and feldspars, but an unlikely participant in life.

variety of planetary atmospheres, stellar envelopes, and in the interstellar medium (Ehrenfreund et al., 2011). These molecules are often termed **prebiotic compounds**, because they are the precursors from which biotic molecules are ultimately formed. While most of these organic carbon and associated molecules are relatively simple in their structure (e.g., CO, HCO, HCN, NH_3, and CH_4),

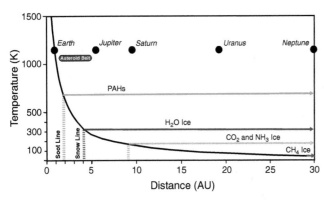

Figure 16.5 The "Soot Line" and the "Ice Line" mark the position in a planetary disk at which temperatures are low enough to retain complex organic matter – specifically polyaromatic hydrocarbons (PAHs) – and water, respectively, in the solid phase. Additional "ice lines" represent stability for other important planetary ices, including carbon dioxide (CO_2), ammonia (NH_3), and methane (CH_4).

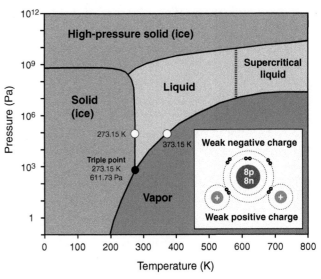

Figure 16.6 The polar structure of water (inset) imparts characteristics that are critical for its role in biological functions, including a broad range of temperature and pressure under which liquid water is stable, and the ability to dissolve a wide range of other chemical nutrients. The density difference between water in its liquid and solid forms may also be critical for the maintenance of habitable terrestrial and extraterrestrial environments. The black circle marks the solid–liquid–vapor triple point for water, at ~273 K and ~612 Pa; white circles mark the freezing point (273 K) and boiling point (373 K) for water at 1 atm.

more complex compounds including amines, long-chain hydrocarbons, and **polycyclic aromatic hydrocarbons (PAHs)** can also form via a series of gas-phase, carbon-insertion, and condensation reactions. In the colder regions of planet-forming accretion disks, these organic molecules condense, along with dust, water, carbon dioxide, ammonia, and methane (Figure 16.5) into icy bodies that are similar to those represented today by Kuiper belt objects and a variety of Main belt asteroids. Experimental evidence suggests that even in these inhospitable and frozen environments, ultraviolet radiation from stars can cause complex and stable organic molecules such as PAHs to transform into more complex organic species (Gudipati and Yang, 2012). Continuation of such chemical interactions may ultimately result in the formation of the most complex prebiotic molecules, such as amino acids, which have been recorded in chondritic meteorites delivered to Earth.

16.2.2 Water, the Elixir of Life

In addition to chemical nutrients (and the energy to drive chemical reactions), life also requires a stable medium in which these chemical reactions can occur. Given the importance of liquid water to all life on Earth (Mottl et al., 2007), most astrobiologists consider evidence for water to be key in our search for extraterrestrial habitability.

In many ways, water is the ideal medium for chemical reactions. The most important characteristic of water is its structure as a polar molecule (Figure 16.6, inset) that produces a slight positive charge on one end of the molecule and a slight negative charge on the other. This polarity produces a number of the key characteristics of

water. First, water molecules have a propensity to bond to each other, which counteracts the relatively low molecular weight of water and results in a broad range of planetary temperatures and pressures under which water remains a stable liquid (Figure 16.6). Second, the polarity of water allows it to dissolve more substances than any other liquid, resulting in water being characterized as the universal solvent. For most polar substances, such as salts and small organic molecules, dissolved ions also distribute uniformly through the water. By contrast, other molecules, such as most lipids, are not readily soluble in water and remain as an immiscible component. However, some lipids, such as phospholipids associated with cell walls, contain an insoluble (hydrophobic) lipid chain attached to a soluble phosphate (hydrophilic) end, which allows them to form bilayer membranes that effectively encapsulate ambient fluids (see Section 16.3).

Another unusual property of water is that it (unlike other potential solvents like ammonia or methane) is less dense in its solid phase than in its liquid phase. This property results from the strength of the hydrogen bonding; in the liquid state, hydrogen bonds readily break and reform, allowing water molecules to freely flow past one another; in the solid state (ice), these hydrogen bonds are restricted in their movement, resulting in greater spacing between adjacent molecules. From an astrobiological perspective, this difference in density is critical for habitability. If ice were the denser of the H_2O phases, it would immediately sink upon its formation. This would result in water bodies freezing from the bottom up, restricting the availability of a liquid component. By contrast, since ice is the less dense phase, it insulates underlying fluids from freezing, thereby aiding in the retention of a liquid component.

16.3 Emergence of Life on Earth

We have seen that the chemistry of life requires nutrients, water, and a source of energy (e.g., light, heat, chemical energy). These requirements would have been satisfied in a number of different environments of the early Earth. Volcanic degassing would have provided a suitable volatile inventory, including nitrogen, carbon dioxide, methane, sulfur dioxide, carbon monoxide, hydrogen, and water (Zhanle et al., 2010), and crustal weathering would have provided phosphorous. Water and organic compounds would also have been delivered from extraterrestrial sources by the continued accretion of asteroidal and cometary material (Chyba and Sagan, 1992).

There is no reason a priori to assume that the origin of life was a unique event, and the prebiotic Earth would have contained a variety of environments in which the chemical reactions of life could have occurred, including within surficial lakes, ponds, or oceans, or in association with surficial or deep-ocean hydrothermal vents. In fact, the tree of life tantalizingly suggests that LUCA may have been only one of a diverse population of early life forms (Delaye et al., 2005). But how did life actually emerge in these environments? Typically, we consider the emergence of life in three stages: the synthesis of complex prebiotic organic compounds, the origination of chemical reaction pathways, and the compartmentalization of these reactions within a cellular membrane.

The first stage in this process – the formation of complex prebiotic organic compounds – convincingly must have occurred in a warm, wet environment. Such a process was first described by Charles Darwin in 1871, when he envisioned "some warm little pond," infused with ions, in which the stuff of life (which we now know as nucleic acids, proteins, carbohydrates, and lipids), could result from simple chemical reactions. In 1953, Stanley Miller and Harold Urey brought this idea to life in an elegant experiment (Figure 16.7): They produced a synthetic "little pond" overlain by a reducing atmosphere containing methane, ammonia, hydrogen, and water vapor (with later experiments also including hydrogen sulfide). An electrical discharge within the atmosphere

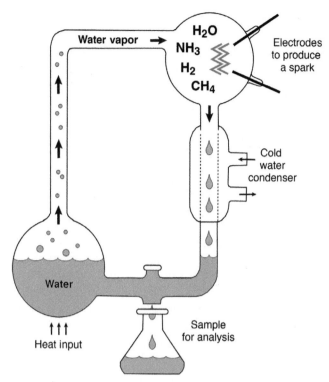

Figure 16.7 Schematic of the 1953 Miller–Urey experiment that showed the potential for formation of complex organic molecules, and specifically the formation of the amino acid building blocks of life, in any environment that contains water, chemical nutrients, and a source of energy.

provided a source of energy. The result was the formation of a wide range of organic compounds, notably a variety of amino acids.

The emergence of chemical reaction pathways is less well understood. Molecular phylogenies tell us that all life forms share basic cellular functions. In modern life, DNA provides the chemical blueprint for metabolism, and RNA acts as the messenger to translate this information to ribosomes and to transport amino acids to the ribosomes for protein synthesis. Proteins then carry out the cellular functions. It has been argued, however, that the complexity of DNA makes it an unlikely participant in the emergence of life. In the **RNA world** hypothesis (Gilbert, 1986), the much simpler nucleic acid RNA acts as both the blueprint for cellular functions and the driver for protein synthesis. Environments rich in prebiotic compounds would provide the necessary building blocks for the production of RNA and subsequent protein synthesis. An alternative hypothesis, based on the similarity between many metabolic reactions and metal-catalyzed chemical reactions (Cody, 2004), suggests that primitive metabolic pathways evolved first and reflected ambient geochemical conditions, and that RNA (and ultimately DNA) then evolved to perpetuate these reactions (Copley et al., 2007).

The final stage in the emergence of life requires the encapsulation of the metabolic and genetic materials. The earliest encapsulation of prebiotic chemistries likely utilized pores and vesicles within geologic substrates. Concentration of prebiotic molecules in such environments could have then facilitated production of more complex molecules, such as the phospholipids that form the basis of modern cellular membranes. Phospholipids are **amphiphilic**, meaning that they contain a polar hydrophilic head and a nonpolar hydrophobic tail. In a polar solvent, like water, such molecules self-organize into both spherical micelles and flexible bilayer sheets, in which molecules align such that the polar heads surround the nonpolar tails, shielding them from fluids (Figure 16.8). Dynamic movement of lipid bilayers can then result in formation of bilayer vesicles. Individual phospholipids readily add to the lipid bilayer, resulting in vesicle enlargement, and even migrate from one layer to another, facilitating the potential for ion transport across this primitive membrane (Schrum et al., 2010).

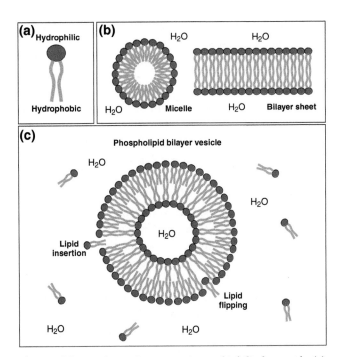

Figure 16.8 Membrane formation via amphiphilic fatty acids. (a) Phospholipids are amphiphilic molecules with a polar, hydrophilic head attached to two, nonpolar, hydrophobic lipid chain tails. (b) When concentrated in fluids, phospholipids self-organize into micelles and bilayer phospholipid sheets. (c) Dynamic movement of bilayer sheets forms vesicles and may represent the earliest cellular membranes. Addition of individual phospholipids can enlarge vesicles, and flipping of phospholipids provides a mechanism for transport of ions through the vesicle wall.

16.4 Earth's Early Biosphere

Relatively little is known about the Earth's biosphere in the Hadean, which spanned the Earth's first 500 million years. It would not have been until the Earth differentiated and cooled, about 100 million years after its formation, that a biosphere could begin to be established. By this point, a combination of volcanic degassing and the chemical weathering of the early crust would have provided the basic ingredients for life. This early biosphere, however, would have continued to experience the effects of extraterrestrial impacts until after the Late Heavy Bombardment (Section 5.6.2), which ended about 3.8 Ga. The largest of these impacts could have imparted enough energy to vaporize portions of the Earth's oceans, to heat the remaining surface fluids, and potentially to drive extinction within Earth's earliest life.

Within this dramatic and volatile environment, however, it is likely that biology gained its first foothold. Our only direct evidence of the Hadean Earth, however, comes from individual igneous mineral grains, zircons ($ZrSiO_4$). These zircons (Figure 16.9) originally crystallized 4.0–4.4 Ga, and were ultimately eroded and redeposited as detrital grains in Archean rocks of the Jack Hills, Western Australia. Analysis of the isotopic composition of these zircons suggests that the Hadean Earth may have retained relatively cool surface waters for extended periods of time between impact events (Wilde et al., 2001). As noted earlier, the tree of life suggests that LUCA was a biological population with the genetic capability of survival in elevated temperatures. Perhaps LUCA represents a population, then, that was capable of surviving heated fluids that derived from these early impact events.

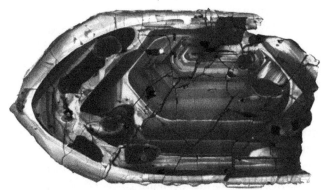

Figure 16.9 False-color image of a 4.4 billion-year-old zircon from Jack Hills, Australia. A small handful of similar zircons represent our only known materials remaining from the Hadean Earth. Image courtesy of J. Valley.

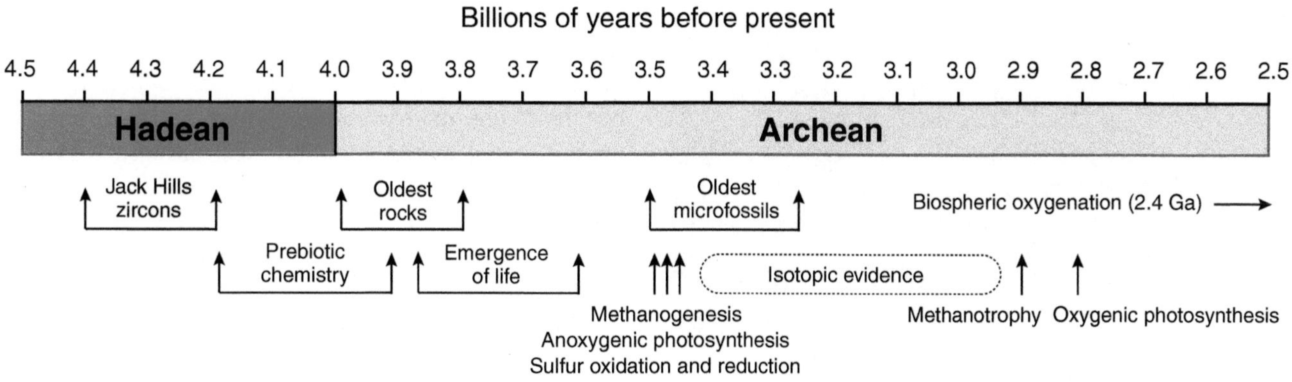

Billions of years before present

| 4.5 | 4.4 | 4.3 | 4.2 | 4.1 | 4.0 | 3.9 | 3.8 | 3.7 | 3.6 | 3.5 | 3.4 | 3.3 | 3.2 | 3.1 | 3.0 | 2.9 | 2.8 | 2.7 | 2.6 | 2.5 |

Hadean   **Archean**

Jack Hills zircons   Oldest rocks   Oldest microfossils   Biospheric oxygenation (2.4 Ga) ⟶

Prebiotic chemistry   Emergence of life   Isotopic evidence   Methanotrophy   Oxygenic photosynthesis

Methanogenesis
Anoxygenic photosynthesis
Sulfur oxidation and reduction

Figure 16.10 Timeline of life on Earth in the Hadean and the Archean. The absence of geologic materials remaining from the Hadean does not allow clear delineation of environments and habitability that could address whether the emergence of life could have occurred prior to the Late Heavy Bombardment. Even with a potentially later emergence of life, the oldest unambiguous physical and chemical evidence for life does not appear for nearly 200 million years, between 3.4 and 3.5 Ga.

Another possibility is that Hadean life gained a foothold in deeper marine environments, such as hydrothermal vents, where disruption by impact events would have been lessened (Martin and Russell, 2007). At mid-ocean ridges, cold seawater circulates through the crust; hydrothermal vents occur where these fluids, now heated, emerge from the seafloor. Modification of these fluids in the subsurface results in strong gradients in both temperature and chemistry between emergent fluids and the surrounding seawater. Of particular interest to astrobiologists is the enrichment of metals in vent fluids, which catalyze the formation of organic molecules either directly or via mineral surfaces (e.g., iron–sulfur minerals).

16.4.1 Recognizing Early Life

A timeline of early Earth (Figure 16.10) highlights the difficulty of reconstructing the history of life on Earth. With the exception of the Jack Hills zircons, plate tectonics has destroyed the Hadean rock record. Even in the Archean (4.0–2.5 Ga), sedimentary rock successions that record environments of the Earth's surface are relatively scarce. Additionally, these sedimentary successions typically represent only shallow marine environments, which limits our potential to explore the breadth of Earth's early biosphere.

Direct evidence for life on the Archean Earth occurs as microbial structures preserved in silica (SiO_2). The precipitation of silica occurs primarily in evaporative, shallow marine environments, where increased silica saturation can overcome kinetic inhibition to precipitation (Manning-Berg and Kah, 2017). In these environments, silica-bearing fluids (or even a silica gel) can permeate the shallow substrate. Where the substrate is colonized by microbial mats, silica can bind to organic

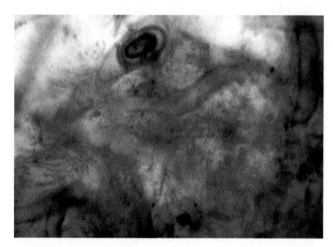

Figure 16.11 An example of a 1.1 billion-year-old microbial mat preserved in silica. Exquisite preservation of microbial morphology – here represented by two distinct sizes of microbial filaments and a coccoidal bacterium – results from binding of silica to organic molecules during mat growth. Image courtesy of L. C. Kah.

matter and effectively entomb the microbial community, resulting in spectacular preservation of microbial features (Figure 16.11). During recrystallization of the silica, however, much of the original morphological detail can be lost, resulting in a greater uncertainty regarding the biological character of preserved organic material (see Box 16.2).

More commonly, our evidence for early life comes from the interaction between microbes and the sedimentary environments in which they occur. Stromatolites, pictured in Figure 16.12, are laminated sedimentary structures that represent the modification of sedimentary substrates by microbial mat communities. Unfortunately, direct evidence for microbial communities is

Figure 16.12 Stromatolites represent the interaction between microbial communities and their sedimentary environment. (a) Irregular, centimeter-scale topographic relief resulting from microbial binding of the sedimentary substrate from the 2.9 Ga Pongola Supergroup, South Africa. (b) Well-laminated, decimeter-scale stromatolites from the 2.52 Ga Transvaal Supergroup, South Africa. Image courtesy of L. C. Kah.

typically lacking from stromatolites. Rather, microbial growth and decomposition are inferred to be reflected in millimeter-scale irregularities of lamina shape that are distinct from laminae formed by the physical transport of sediment. Accretion of successive laminae is similarly inferred to reflect either the trapping and binding of detrital sediment by microbes, or mineral precipitation induced by the microbial activity within mats (Grotzinger and Knoll, 1999). If we accept stromatolites as biological entities, we can confirm that microbial life inhabited shallow marine environments on Earth by at least 3.5 Ga, which agrees with evidence from permineralized microfossils.

16.4.2 The Chemical Record of Life

A more comprehensive understanding of the habitability of the early Earth has been gained from exploring the chemical fingerprints of life. One of the ways in which we can do this is to explore the Earth's stable isotope record. Many of the elements that are commonly utilized by biology have one or more stable isotopes. These include, among others, carbon (^{12}C and ^{13}C), nitrogen (^{14}N and ^{15}N), and sulfur (^{32}S, ^{33}S, ^{34}S, and ^{36}S). The different atomic masses within these isotope families affect both the reaction rate and bond strength within molecules. As a result, the chemical reactions associated with life result in an isotopic fractionation wherein the isotope with the smaller atomic mass becomes concentrated in the reaction products.

As noted earlier, carbon is the most critical element for life, so it should come as no surprise that the most commonly used stable isotope system to explore early life is that of carbon. When examining carbon's stable isotopes, we find that metabolic reactions that produce organic carbon from carbon dioxide in the oceans (Table 16.1) result in rather substantial isotopic fractionations. These fractionations are measured in "delta notation" which describes the ratio of ^{13}C to ^{12}C in a sample and a known standard, and expresses this ratio in parts per thousand (‰, or per mil). The fractionation between reduced carbon (organic matter) and oxidized carbon (carbon dioxide) phases in the oceans is approximately −25‰, meaning that the organic carbon is enriched in ^{12}C (or depleted in ^{13}C) by 25 parts per 1000 relative to the measured standard. Isotopic measurements since 3.5 Ga have shown a consistent fractionation between carbon dioxide in the ocean (as measured in marine carbonate minerals) and coeval organic matter (Figure 16.14), suggesting the continuous microbial occupation of the Earth's surface since this time. Similarly, sulfur isotope fractionation observed between oxidized and reduced phases at this same time suggests biological metabolisms capable of sulfur oxidation and reduction, as well (Figure 16.10).

In addition to measuring the isotopic composition of organic material, we can also explore molecular fossils, or biomarkers. **Biomarkers** are specific organic carbon structures, characterized by rings or chains, that represent the preserved components of biomolecules such as proteins and lipids (Newman et al., 2016). Lipids, in particular, can survive largely intact over billions of years and thus provide a unique glimpse into early life. One particularly important group of lipid biomarkers is the hopanoids – large molecules containing five carbon rings that serve as stabilizing compounds in bacterial cell walls. Hopanoids are widely distributed, but unique to bacteria. Modification of hopanoids, such as the addition of methyl groups, is less common and is hypothesized to be limited to specific bacterial groups, such as cyanobacterial (2β-methyl hopanes) or methanotrophic bacteria

BOX 16.2 ARGUMENTS ABOUT THE EARTH'S EARLIEST ECOSYSTEM

A single question highlights the challenges of astrobiological research: What is the oldest life on Earth? The answer to this question is critical both to our understanding of life's origins, and to determining the best targets for astrobiological investigation.

Schopf (1993) described carbon-rich features preserved in diagenetic silica within the ~3.5 Ga Apex Chert of Western Australia. This discovery was particularly exciting because preserved materials were not just simple spherules that could easily be attributed to a range of biotic and abiotic origins, but complex filaments that showed what appeared to be cellular structures (Figure 16.13). Differences in the size and shape of these cellular structures further suggested preservation of multiple different microbial taxa.

Yet not all in the scientific community agreed on the biogenicity of these microfossils – and even fewer now agree on biogenicity of some even older purported fossils. In 2002, after viewing some of the original material from the Apex Chert, Brasier et al. (2002) published an extensive petrographic and geochemical investigation of new material from the remote Australian site. Using advances in laser Raman spectroscopy (see also Schopf et al., 2002) and energy-dispersive X-ray spectroscopy, the "cellular structures" observed by Schopf were reinterpreted as carbon, synthesized abiotically within hydrothermal fluids, that had been deposited along the margins of mineral grains.

New technology, however, is continuing to advance the debate, and to turn the tables back toward a biogenic origin for structures preserved within the Apex Chert. For example, Schopf et al. (2018) released results of an investigation into the isotopic composition of carbon within the Apex Chert. Earlier work had revealed bulk isotopic compositions that were broadly consistent with production of organic carbon by photosynthetic microbes. The most recent study, which used secondary ion mass spectrometry (SIMS) to investigate the isotopic composition of individual preserved structures, reveals that isotopic compositions are specific to morphologic groups (Schopf et al., 2018). Together, the correlation between differences in isotopic composition and morphology of the structures suggests that, by ~3.5 Ga, the Earth not only had life, but potentially a complex ecosystem consisting of photosynthetic, methanogenic, and methanotrophic bacteria.

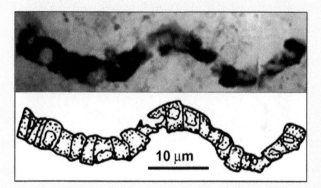

Figure 16.13 A segmented filament preserved within the 3.5 Ga Apex Chert, Western Australia. Purported microfossils from the Apex Chert have been the focus of more than 20 years of debate regarding whether these features represent Earth's earliest preserved life. Image courtesy of J. W. Schopf.

(3β-methyl hopanes). Other hopanoids, such as bis- and tris-norhopanes, have uncertain biologic origins but are strongly correlated to anoxic, sulfur-rich environments.

16.5 Life Beyond Earth

Frigid temperatures, low pressures, and elevated radiation make the survival of biological organisms in space unlikely (Olsson-Francis and Cockell, 2010), but are there habitable worlds beyond the Earth? We can begin to explore the potential for habitable extraterrestrial environments by returning to the key requirements of life: a source of energy, availability of chemical nutrients, and a stable medium in which to carry out biochemical reactions. With this approach, we readily find that sources of energy are abundant in the Universe, and nutrients occur in sufficient abundance to produce prebiotic organic molecules throughout the expanses of space (Ehrenfreund et al., 2011). This leaves a stable medium for chemical reactions as the key component in determining whether an environment may be habitable to life.

16.5.1 Habitable Zones

A **habitable zone** (Figure 16.15) is defined as the region around a star where liquid water is potentially stable on the surface of a planetary body with an atmosphere

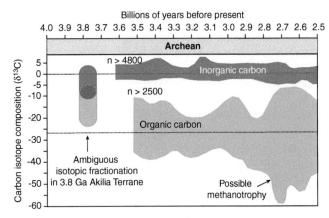

Figure 16.14 Carbon isotope record of life in the Archean. Persistent average fractionation of approximately −27‰ between inorganic carbon and organic carbon supports the biological origin of organic carbon. Isotopic fractionations greater than −50‰ in the late Archean likely result from the biological utilization of biologically formed methane.

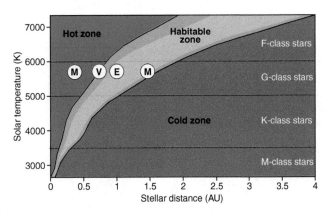

Figure 16.15 Proposed habitable zone around main sequence stars. The habitable zone marks the distance from a star in which liquid water may be present on the surface of a planetary body with an atmosphere. The position of the inner planets (Mercury, Venus, Earth, and Mars) are shown.

(Kasting et al., 1993). The idea of a habitable zone, however, is based primarily on the temperature and heat output of stars, and is therefore only a first-order guide to potential habitability. The stability of liquid water is also strongly dependent on other parameters that affect the surface temperature of a planetary body. These parameters include the composition and density of a planet's atmosphere, as well as the orbital dynamics of a planet, such as whether or not a planet always has the same side facing its parent star.

Our current definition of a habitable zone is also based on the presence of liquid water at the surface of a planetary body. However, even in our own Solar System, potential habitable worlds occur outside the traditional

habitable zone. Liquid water, for example, has been proposed to occur, at least seasonally, in the subsurface of planets on the fringe of the habitable zone, such as Mars (McEwen et al., 2011). Icy moons that orbit gas giants and experience substantial tidal heating, such as Europa and Encedalus, also retain subsurface oceans (Section 12.4.3). The search for life is part of the motivation for the further exploration of these bodies by spacecraft. There is also the special case of Titan (Section 12.4.2). Titan possesses a thick nitrogen–methane atmosphere with a wide range of prebiotic organic molecules (Hörst et al., 2012). Although surface temperatures of Titan are too low to retain liquid water, it has been proposed that a substantial reservoir of liquid water could lie in the subsurface.

In addition to these potentially habitable worlds within our own Solar System, we have also identified thousands of exoplanets. Calculations suggest that there may be as many as 40 billion Earth-sized planets in our galaxy that occur within the habitable zone of their respective stars (Petigura et al., 2013). Exoplanets are described in the Epilogue at the end of this book. What a wonderful opportunity for future astrobiological studies!

16.5.2 Life in a Martian Meteorite?

Have we already found evidence for extraterrestrial life? McKay et al. (1996) stunned the world with their announcement of the discovery of potentially microbial fossils in a meteorite from Mars known as Allan Hills 84001 (ALH 84001). The following day, President Clinton stood outside the White House and commanded the scientific community to put these findings through a thorough peer-review process, and stated his own determination that the USA would put "its full intellectual power and technological prowess behind the search for further evidence of life on Mars."

The ALH 84001 meteorite was discovered in 1984 in Antarctica. Its identification as martian is inferred from the ratio of atmospheric gases trapped within the fusion crust, in conjunction with the oxygen isotope composition of minerals within the meteorite. Radiometric dating determined that ALH 84001 was ejected from Mars approximately 17,000 years ago, but the rock itself is nearly four billion years old, thus dating to a time when most scientists think the martian surface was much more Earth-like.

Preserved evidence of life within ALH 84001 was controversial from the start (McSween, 1997). Although the rock itself is igneous, various materials purported to be of biologic origin are contained in chemically zoned carbonate spherules (Figure 16.16a) precipitated from a

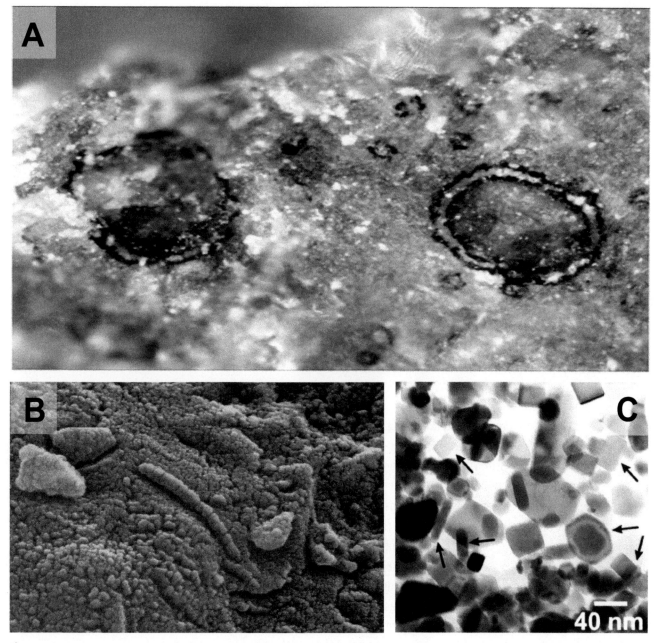

Figure 16.16 Purported evidence for life in Allan Hills 84001 martian meteorite. (a) Zoned carbonate spherules. (b) A segmented filament proposed to be a microfossil. (c) Tiny magnetite crystals proposed to be from magnetotactic bacteria. NASA images.

circulating fluid or evaporating brine. ALH 84001 was also found to contain complex organic molecules, specifically PAHs. PAHs are known from Earth as the combustion products of biological materials, but they also occur across the Universe as abiotic organic molecules; PAHs within ALH 84001 lack any of the side-chain components that would permit us to distinguish their true origin. The bacteria-shaped objects that caught the eye of researchers (Figure 16.16b) are reminiscent of filamentous bacteria, but are ten times smaller than size estimates for the smallest cells (Knoll, 1999). Similarly, tiny magnetite crystals found within ALH 84001 (Figure 16.16c) morphologically resemble the tiny crystals formed by magnetotactic bacteria. At the time, there was no other process understood to form such crystals. In the meteorite, however, these magnetite crystals do not occur in the alignment characteristic of magnetotactic bacteria. Research since the discovery of ALH 84001 suggests that magnetite results from shock decomposition of the iron carbonate, the microfossils may be surface features on shocked carbonate, and the organic matter may be exogenic. Moreover, the meteorite was likely contaminated by terrestrial microorganisms while in the Antarctic environment. The "life in a martian meteorite" controversy thus illustrates how tricky it will be to make a definitive discovery, even once we have samples returned from another planet.

16.5.3 The Ongoing Search for Organic Matter on Mars

Although the controversy over potential biomarkers in ALH 84001 has mostly subsided, the search for potential evidence of life on Mars continues. In 2018, scientists working on the Curiosity rover reported finding a wide variety of organic compounds preserved in lacustrine mudstone of Gale crater (Eigenbrode et al., 2018). Organic compounds were released from powdered rock by **pyrolysis**, where they were converted into gas by stepwise heating.

These compounds included both simple aromatic (ringed) molecules, such as benzene (C_6H_6), methylbenzene (C_7H_7), and alylbenzene (C_8H_9), and a range of sulfur-containing compounds, such as thiophene (C_4H_4S), methylthiophene (C_5H_6S), and dimethylsulfide (C_2H_6S). Critically, many of the organic compounds evolved at high temperatures during pyrolysis, suggesting that these compounds were originally part of larger organic compounds that do not decompose readily. The diversity and composition of these compounds are consistent with the pyrolysis of abiotic organic material that occurs in both carbonaceous chondrites and decomposed organisms in terrestrial sedimentary materials. Although there is no certainty that these organic compounds are biotic in origin, this find is exciting to astrobiologists because these results show that organic molecules that could serve as a carbon source for microbial metabolism can survive the harsh radiation environment at the martian surface.

Summary

Astrobiology is a growing field of science, with opportunities for scientists from disparate scientific disciplines. As biologists and biochemists are working to understand the complex chemistry of life here on Earth, geologists are working to understand both the origins of geologic bodies in the Universe and how to read the geologic record of habitability and life, and astronomers and astrophysicists are working to identify the range of potentially habitable worlds across the Universe.

The present search for extraterrestrial life focuses on identifying the key components to life: a source of energy (e.g., solar, thermal, or chemical), the presence of key chemical nutrients (e.g., CHNOPS), and the availability of a stable liquid medium (e.g., water). Our growing understanding of the Universe suggests that it is the last of these, the presence of liquid water, that is likely to be the main constraint on the habitability of extraterrestrial worlds.

Still, in the search for extraterrestrial life, we cannot forget that the Earth is only a single model for life. LUCA, the Last Common Universal Ancestor of life on Earth, was likely only one population among several early life forms. With potentially more than 40 million habitable planets in the Universe, the geologic evolution of these individual planets could have driven the evolution of life in any number of different directions.

Review Questions

1. What is the Last Universal Common Ancestor (LUCA), and how can we use it to develop targets for astrobiological research?
2. Why is carbon important for life, and how widespread are carbon compounds in the Universe?
3. What difficulties are faced when determining a biologic origin for Earth's early life?
4. Why is the presence of organic matter not always an indication of the presence of life?
5. What lines of evidence can be used to argue for the astrobiological investigation of the moons Titan, Encedalus, and Europa?

SUGGESTIONS FOR FURTHER READING

Arndt, N. T., and Nisbet, E. G. (2008) Processes on the young Earth and the habitats of early life. *Annual Reviews of Earth and Planetary Sciences*, **40**, 521–549. A thorough look at the Earth's earliest environments.

Bosak, T., Knoll, A. H., and Petroff, A. P. (2013) The meaning of stromatolites. *Annual Reviews of Earth and Planetary Sciences*, **41**, 21–44. An exploration of the use of modern analogues in understanding the ancient Earth.

Ehrenfreund, P., Spaans, M., and Holm, N. G. (2011) The evolution of organic matter in space. *Philosophical Transactions of the Royal Society of London*, **369**, 538–554. An excellent primer on the structure and origin of prebiotic organic molecules.

Hazen, R. M. (2012) *The Story of Earth: The First 4.5 Billion Years, from Stardust to Living Planet*. New York: Penguin Books. An exhilarating glimpse at the early evolution of our world and the intimate interweaving of geology and life.

Kasting, J. (2010) *How to Find a Habitable Planet*. Princeton, NJ: Princeton University Press. An eminently readable book outlining the physiochemical conditions necessary for habitability.

Knoll, A. H. (2003) *Life on a Young Planet: The First Three Billion Years of Evolution on Earth*. Princeton, NJ: Princeton University Press. A deftly woven and wonderfully readable tale of life from its origins to the Cambrian explosion of animals.

McKay, C. P. (2014) Requirements and limits for life in the context of exoplanets. *Proceedings of the National Academy of Sciences*, **111**, 12628–12633. A short yet thorough review of the necessary conditions for life.

REFERENCES

Archibald, J. M., and Keeling, P. J. (2002) Recycled plastids: a "green movement" in eukaryotic evolution. *Trends in Genetics*, **18**, 577–584.

Bains, W. (2004) Many chemistries could be used to build living systems. *Astrobiology*, **4**, 137–167.

Brasier, M. D., Green, O. R., Jephcoat, A. P., et al. (2002) Questioning the evidence for Earth's oldest fossils. *Nature*, **416**, 76–81.

Chyba, C., and Sagan, C. (1992) Endogenous production, exogenous delivery and impact-shock synthesis of organic molecules: an inventory for the origins of life. *Nature*, **355**, 125–132.

Cody, G. D. (2004) Transition metal sulfides and the origins of metabolism. *Annual Reviews of Earth and Planetary Science*, **32**, 569–599.

Copley, S. D., Smith, E., and Morowitz, H. (2007) The origin of the RNA world: co-evolution of genes and metabolism. *Bioorganic Chemistry*, **35**, 430–433.

Delaye, L., Becerra, A., and Lazcano, A. (2005) The last common ancestor: what's in a name? *Origins of Life and Evolution of Biospheres*, **35**, 537–554.

DeLong, E. F. (2009) The microbial ocean from genomes to biomes. *Nature*, **459**, 200–206.

Ehrenfreund, P., Spaans, M., and Holm, N. G. (2011) The evolution of organic matter in space. *Philosophical Transactions of the Royal Society of London*, **369**, 538–554.

Eigenbrode, J. L., Summons, R. E., Steele, A., et al. (2018) Organic matter preserved in 3-billion-year-old mudstones at Gale crater, Mars. *Science*, **360**, 1096–1101.

Fuchsman, C. A., Collins, R. E., Rocap, G., et al. (2017) Effect of the environment on horizontal gene transfer between bacteria and archaea. *PeerJ*, **5**, e3865. DOI: 10.7717/peerj.3865.

Gaucher, E. A. J. T., Kratzer, J. T., and Randall, R. N. (2010) Deep phylogeny: how a tree can help characterize early life on Earth. *Cold Spring Harbor Perspectives on Biology*. DOI: 10.1101/cshperspect.a002238.

Gilbert, W. (1986) The RNA world. *Nature*, **319**, 618. DOI: 10.1038/319618a0.

Grotzinger, J. P., and Knoll, A. H. (1999) Stromatolites in Precambrian carbonates: evolutionary mileposts or environmental dipsticks? *Annual Review of Earth and Planetary Science*, **27**, 313–358.

Gudipati, M. S., and Yang, R. (2012) In-situ probing of radiation-induced processing of organics in astrophysical ice analogs: novel laser desorption laser ionization time-of-flight mass spectroscopic studies. *Astrophysical Journal Letters*, **756**, L24. DOI: 10.1088/2041-8205/756/1/L24.

Hörst, S. M., Yelle, R. V., Buch, A., et al. (2012) Formation of amino acids and nucleotide bases in a Titan atmosphere simulation experiment. *Astrobiology*, **12**, 809–817.

Kan, S. B. J., Lewis, R. D., Chen, K., et al. (2016) Directed evolution of cytochrome c for carbon–silicon bond formation: bringing silicon to life. *Science*, **354**, 1048–1051. DOI: 10.1126/science.aah6219.

Kasting, J. F., Whitmire, D. P., and Reynolds, R. T. (1993) Habitable zones around main sequence stars. *Icarus*, **101**, 108–128.

Knoll, A. (1999) *Size Limits of Very Small Microorganisms: Proceedings of a Workshop*. Washington, DC: National Academies Press.

Manning-Berg, A. R., and Kah, L. C. (2017) Proterozoic microbial mats and their constraints on environments of silicification. *Geobiology*, **15**, 469–483.

Martin, W., and Russell, M. J. (2007) On the origin of biochemistry at an alkaline hydrothermal vent.

Philosophical Transactions of the Royal Society of London B, **362**, 1887–1925.

McDaniel, L. D., Young, E., Delaney, J., et al. (2010) High frequency of horizontal gene transfer in the oceans. *Science*, **330**. DOI: 10.1126/science.1192243.

McEwen, A. S., Ojha, L., Dundas, C. M., et al. (2011) Seasonal flows on warm martian slopes. *Science*, **333**, 740–743.

McKay, D. S., Gibson, E. K., Thomas-Keprta, K. L., et al. (1996) Search for past life on Mars: possible relic biogenic activity in martian meteorite ALH84001. *Science*, **273**, 924–930.

McSween, H. Y. (1997) Evidence for life in a martian meteorite? *GSA Today*, **7**, 1–6.

Mottl, M., Glazer, B., Kaiser, R., et al. (2007) Water and astrobiology. *Chemie der Erde*, **67**, 253–282.

Newman, D. K., Neubauer, C., Ricci, J. N., et al. (2016) Cellular and molecular biological approaches to interpreting ancient biomarkers. *Annual Reviews of Earth and Planetary Sciences*, **44**, 493–522.

Olsson-Francis, K., and Cockell, C. S. (2010) Experimental methods for studying microbial survival in extraterrestrial environments. *Journal of Microbiological Methods*, **80**, 1–13.

Petigura, E. A., Howard, A. W., and Marcy, G. W. (2013) Prevalence of Earth-size planets orbiting Sun-like stars. *Proceedings of the National Academy of Sciences*, **110**, 19273–19278.

Schopf, J. W. (1993) Microfossils of the Early Archean Apex chert: new evidence of the antiquity of life. *Science*, **260**, 640–646.

Schopf, J. W., Kudryavtsev, A. B., Agresti, D. G., Wdowiak, T. J., and Czaja, A. D. (2002) Laser-Raman imagery of Earth's earliest fossils. *Nature*, **416**, 73–76.

Schopf, J. W., Kitajimad, K., Spicuzzad, M. J., et al. (2018) SIMS analyses of the oldest known assemblage of microfossils document their taxon-correlated carbon isotope compositions. *Proceedings of the National Academy of Sciences*, **115**, 53–58.

Schrum, J. P., Zhu, T. F., and Szostak, J. W. (2010) The origins of cellular life. *Cold Spring Harbor Perspectives in Biology*. DOI: 10.1101/cshperspect.a002212.

Stetter, K. O. (1996) Hyperthermophilic procaryotes. *FEMS Microbiology Reviews*, **18**, 149–158.

Wackett, L. P., Dodge, A. G., and Ellis, L. B. M. (2004) Microbial genomics and the periodic table. *Applied Environmental Microbiology*, **70**, 647–655.

Weiss, M. C., Sousa, F. L., Mrnjavac, N., et al. (2016) The physiology and habitat of the last universal common ancestor. *Nature Microbiology*, **1**. DOI: 10.1038/nmicrobiol.2016.116.

Wilde, S. A., Valley, J. W., Peck, W. H., et al. (2001) Evidence from detrital zircons for the existence of continental crust and oceans on the Earth 4.4 Gyr ago. *Nature*, **409**, 175–178.

Woese, C., Kandler, O., and Wheelis, M. (1990) Towards a natural system of organisms: proposal for the domains Archaea, Bacteria, and Eucarya. *Proceedings of the National Academy of Sciences*, **87**, 4576–4579.

Zahnle, K., Schaefer, L., and Fegley, B. (2010) Earth's earliest atmospheres. *Cold Spring Harbor Perspectives in Biology*. DOI: 10.1101/cshperspect.a004895.

17

Integrated Planetary Geoscience: A Case Study (Mars)

Planetary exploration typically advances in step with technology. Improvements in spatial and spectral resolution yield discoveries that progress from global to regional scales, and exploration on a planet's surface provides ground truth for remote sensing data and a level of observation and measurement that geologists crave. Samples that can be analyzed in the laboratory provide geochemical and geochronologic information that complements spacecraft data and enhances its interpretation. We illustrate how data at all these scales have been integrated to characterize the complex geology of Mars and to constrain its geologic history.

17.1 Geologic Exploration of a Planet

Planetary geologic exploration normally progresses in stages, from global reconnaissance to regional investigations, and thence to local mapping and analysis. The first stage is begun using ground-based telescopes and flyby or orbiting spacecraft, the second stage utilizes orbiters, and the third focuses on landed spacecraft and rovers. These stages are not always strictly sequential, but changes in the spatial scale of investigation normally happen in step with improvements in technology over time. This progression is opposite to the strategy of terrestrial geologists, where local studies are combined and synthesized to understand the Earth at regional and then global scales. In this chapter we will see examples of how planetary geology is done at each of these scales.

We focus here on Mars, because it is the only planet, besides our own, that has progressed through all three stages of exploration. The Moon has also gone through these stages, with the added benefit of samples returned by astronauts. For the foreseeable future, though, exploration on the surfaces of other worlds will be accomplished robotically, which is the way Mars has been studied.

All the stages of planetary exploration utilize information familiar to terrestrial geologists – maps, topography, stratigraphy, geochronology, geomorphology, geophysics, geochemistry, mineralogy, and petrography. How these data are acquired may differ, but the strategies and principles of geologic exploration are the same.

The laboratory analysis of samples, whether returned or acquired as meteorites, provides a level of petrologic, geochemical, and geochronologic information that is otherwise unobtainable. The value of a sample is magnified if its specific provenance is known or can be determined.

17.2 Planetary Reconnaissance and a Global Geologic Map

Geologic understanding involves the integration of data at all spatial scales. This is usually done continuously. In the following, we consider the insights derived from global reconnaissance, and then illustrate studies of the regional and local geology of a few areas that have been explored by rovers. Then we add the insights gained from martian meteorites. Finally, we utilize all these data to synthesize the geologic history of Mars.

17.2.1 Global Physiography and Structure
The average thickness of the martian crust, as constrained by orbital gravity and topography measurements, is ~50 km, comprising as much as ~6 wt% of the silicate fraction (mantle plus crust) of Mars. The bulk density and moment of inertia factor are consistent with a core comprising ~22 wt% of the planet.

The most notable surface feature in orbital imagery is a global dichotomy that separates terrains that differ in topography, age, and crustal thickness. South of this boundary are the highlands – ancient, heavily cratered units with elevations higher than the global average.

Within the highlands are huge impact basins and magnetic lineations discovered by an orbiting magnetometer. To the north of the dichotomy are the lowlands – younger plains of layered sedimentary and volcanic rocks covering an ancient basement.

Magmatic centers (Tharsis, Elysium, Syrtis Major, Hesperia) host enormous volcanoes and vast outpourings of lavas. Martian tectonics are dominated by the Tharsis uplift, which is surrounded by a moat, radial rifts, concentric ridges, and a system of gigantic canyons.

Superimposed on both highlands and lowlands are layered sedimentary deposits. The accommodation of sediments on Mars is provided by water depth, rather than subsidence as on Earth, with the consequence that the stratigraphic record is discontinuous and is dominated by sediments produced by erosion. A pervasive regolith covers the surface. Polar caps are composed of interlayered ices and dust; the caps exchange icy materials annually and are the most dynamic regions of present-day Mars.

17.2.2 Global Remote Sensing
Orbiters have provided a number of global geoscience datasets. The acronyms are dizzying, and so are summarized below.

- *Mars Global Surveyor* (MGS, 1996–2006) carried a camera (MOC) for imagery, laser altimeter (MOLA) for topographic mapping, and thermal emission spectrometer (TES) for mineral identification; a magnetometer and radio science tracking discovered an ancient magnetic field and allowed the construction of a gravity map.
- *Mars Odyssey* (MO, 2001–present) featured a gamma ray/neutron spectrometer (GRS/NS) for geochemistry and a visible and thermal emission imager (THEMIS) for mineral mapping.
- ESA's *Mars Express* (MEx, 2003–present) operated an imager (HRSC) and a visible/near-infrared spectrometer (OMEGA) for mineral mapping.
- *Mars Reconnaissance Orbiter* (MRO, 2005–present) carried a high-resolution camera (HiRISE), an imaging spectrometer (CRISM) for mineralogy, and ground-penetrating radar (SHARAD).
- *Mars Atmosphere and Volatile Evolution* mission (MAVEN, 2013–present) features instruments designed to analyze the atmosphere to ascertain how Mars lost its water and atmosphere over time.
- *ExoMars Trace Gas Orbiter* (TGO, 2016–present) carries instruments to search for methane and other atmospheric gases that could be signatures of biologic or geologic processes; it also features an ultraviolet/infrared spectrometer (NOMAD) and a neutron detector (FREND) for sensing hydrogen.

The "global" datasets – albedo, topography, thermal inertia, gravity, magnetics, mineralogy (usually one mineral at a time, as shown in Figure 15.8), and geochemistry (only a handful of elements) – were actually obtained at regional scales and then combined to produce global maps. Some examples are illustrated in Figure 17.1.

17.2.3 Global Stratigraphic Timescale and Geologic Map
The observed geologic units on Mars are divided into the Noachian, Hesperian, and Amazonian systems, defined by stratigraphy and crater density measurements. As described in Section 15.3, mineralogical changes accompanying these periods are interpreted to reflect global environmental changes. The first approximately half-billion years of martian history (Pre-Noachian time) are not represented in the currently recognized rock record.

The most recent global geologic map of Mars (Figure 17.2; Tanaka et al., 2014) recognizes 44 geologic units. The map is based primarily on topography and imagery, which allow unit morphologies, stratigraphic relationships, and crater size–frequency data to be defined. Most of the surface of Mars is old, with Noachian units comprising 45 percent, Hesperian units 29 percent, and Amazonian units 26 percent of the planet's total surface.

17.3 Regional Geology from Orbit and Surface Exploration by Rovers

Although many interesting regions on Mars have been studied using multiple datasets as described above, we can only introduce a few. We will illustrate regional and local studies by focusing on three regions that have been studied from orbital spacecraft and subsequently explored from the ground. In these regions, rovers have conducted extensive traverses, providing more representative ground truth than provided by the *Viking* and *Phoenix* stationary landers or the very limited traverse of the *Mars Pathfinder* rover.

17.3.1 Gusev Crater
Orbital Studies
Gusev crater (~165 km diameter) is Noachian in age, but the deposits on its floor formed at ~650 Ma, based on crater counting data (Parker et al., 2010). Gusev was originally thought to be an ancient lakebed, based in part on the drainage of Ma'adim Vallis into it (Figure 17.3). Where the valley enters Gusev are flat-topped hills interpreted to be delta deposits. Using *Viking* orbiter imagery, a number of authors – most recently Kuzmin et al.

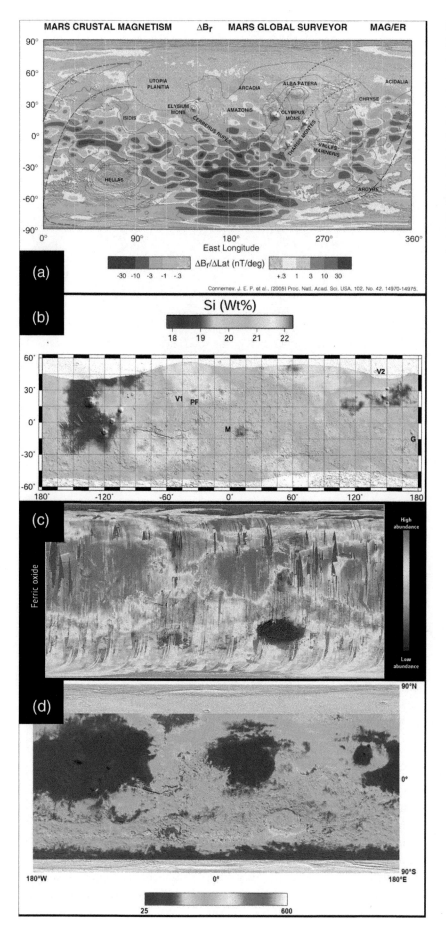

Connerney. J. E. P. et al., (2005) Proc. Natl. Acad. Sci. USA, 102. No. 42. 14970-14975.

Figure 17.1 Examples of global maps of Mars obtained by orbiter instruments. (a) Magnetic field observed by MAG/ER on *Mars Global Surveyor* (Acuna et al., 2008). (b) Silicon abundance by GRS on *Mars Odyssey* (Boynton et al., 2008). (c) Abundance of crystalline ferric oxides by OMEGA on *Mars Express* (Bibring and Langevin, 2008). (d) Thermal inertia by THEMIS on *Mars Odyssey* (Christensen et al., 2008). Reprinted with permission from Cambridge University Press: The Martian Surface: Composition, Mineralogy, and Physical Properties, ed. J.F. Bell III, copyright (2008).

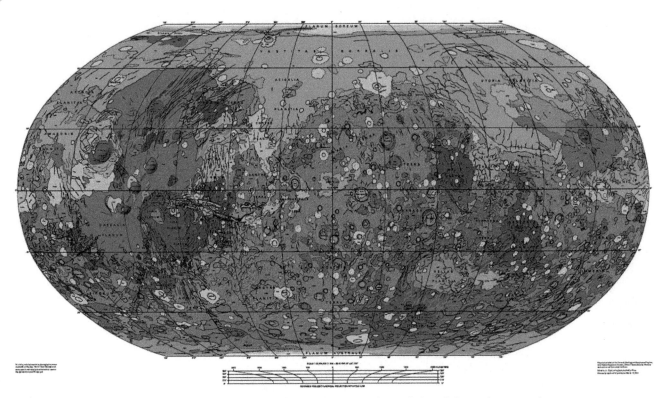

Figure 17.2 Global geologic map of Mars (US Geological Survey). See Tanaka et al. (2014) for explanation of units.

Figure 17.3 Image of Gusev crater and Ma'adim Vallis. NASA image.

(2000) – produced geologic maps of the Gusev region (Figure 17.4). Milam et al. (2003) subsequently utilized MOC imagery, THEMIS thermal mapping, and MOLA elevation data to identify and map seven thermophysical and morphologic units, which were largely correlated.

The existence of so many distinguishable stratigraphic units, as well as observed layering within the units,

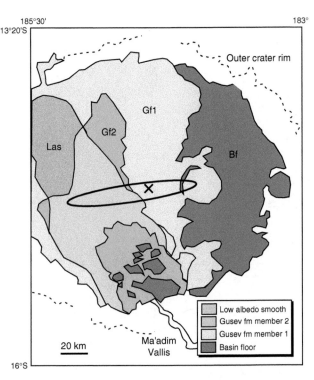

Figure 17.4 Geologic map of Gusev crater, simplified from Kuzmin et al. (2000). Ellipse represents the uncertainty in the landing site for the Spirit rover, prior to arrival, and the actual landing site is marked by X.

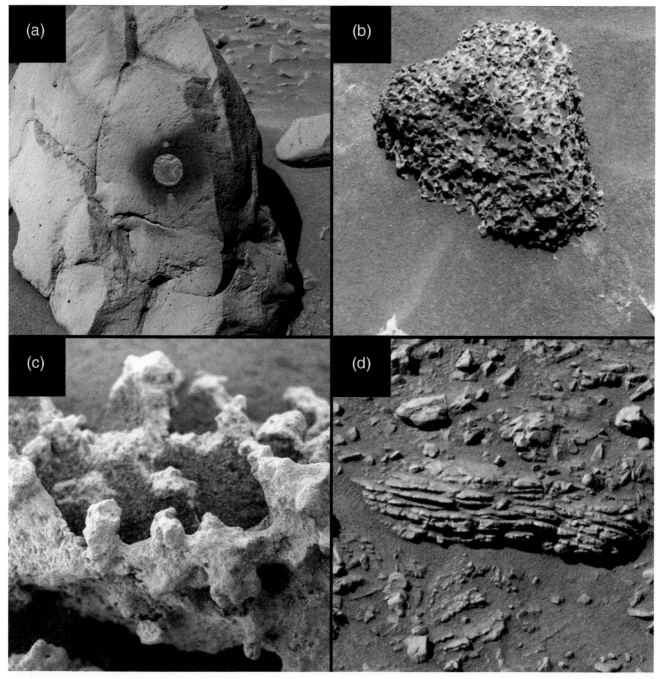

Figure 17.5 Rocks of Gusev crater, imaged by Spirit rover. (a) Plains basalt, with area ground by the rock abrasion tool, (b) scoria from the Inner Basin, (c) highly altered, friable rock from Columbia Hills, (d) layered rock from Columbia Hills. NASA images.

suggest multiple depositional and erosional events. Although previous studies (e.g., Grin and Cabrol, 1997) argued that lacustrine and fluvial processes formed the deposits on the floor of Gusev, Milam et al. (2003) could not rule out volcanic activity, and noted that the nearby Apollinaris volcano could have been the source of pyroclastic materials. Consequently, they favored a combination of volcanoclastic and sedimentary processes. Orbital spectroscopic data could not distinguish between these kinds of materials.

Spirit Rover's Surface Exploration

Spirit landed on ridged plains in Gusev, a surface consisting of abundant volcanic rocks (none of which are outcrops) resting on comminuted basaltic debris (Figure 17.5a). The rocks are fine-grained and porphyritic, sometimes with vesicles and vugs (Figure 17.5b). Compositions measured by APXS and mineralogy identified by spectroscopy indicate the rocks are olivine basalts (McSween et al., 2004), relatively unaltered but with surficial alteration rinds. Gusev contains no obvious

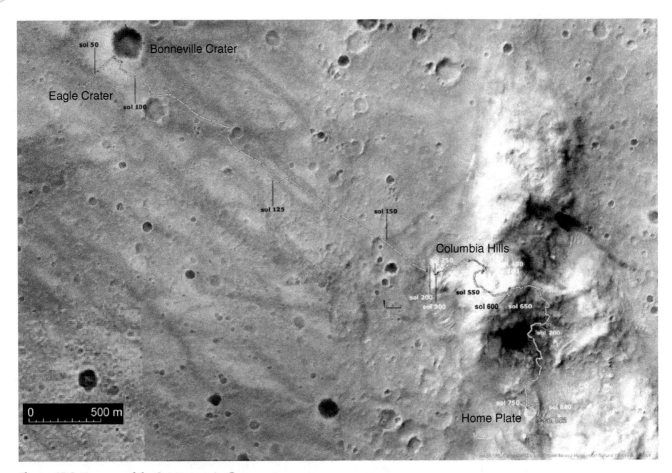

Figure 17.6 Traverse of the Spirit rover in Gusev crater.

eruption centers and flows from Apollinaris cannot be traced into the crater, so the basalts likely result from fissure eruptions within the crater.

After traversing to a nearby crater (Figure 17.6), Spirit trekked to the Columbia Hills, an older (Noachian?) uplifted unit that lies below the plains. Rocks in the Columbia Hills are highly varied in composition and texture (Figure 17.5c,d), and were extensively analyzed by Spirit's instruments. The rocks include lavas and pyroclastic rocks that could have been derived by fractionation of olivine basalt (McSween et al., 2006), as well as clastic sediments or impactites (Squyres et al., 2007).

Following exploration of the Columbia Hills, Spirit traversed through the Inner Basin to Home Plate (Figure 17.7), along the way encountering dark sand dunes and more basaltic rocks. Home Plate is a layered structure formed from pyroclastic materials of basaltic composition (Squyres et al., 2007). The rover wheels stirred up unusual soils rich in silica or sulfate, thought to have formed by hydrothermal processing. Spirit's mission ended when it became mired in soft sand near Home Plate.

The compositions of the alkaline volcanic rocks in Gusev crater were previously illustrated in Figure 10.14,

and their petrogenesis was discussed in Section 10.4. None of the rocks encountered in Gusev are lacustrine sediments, as predicted before landing, indicating that any existing lake deposits must be buried by volcanic and impact materials.

17.3.2 Meridiani Planum
Orbital Studies

Based on orbital imagery, Meridiani Planum was originally interpreted as a layered sedimentary unit deposited in

Figure 17.7 Home Plate, consisting of layered pyroclastic rocks overlain by scoria. NASA image.

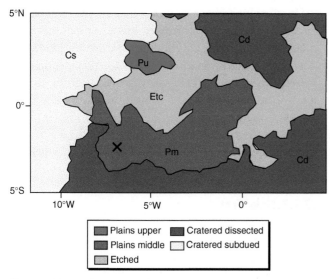

Figure 17.8 Geologic map of Meridiani Planum, simplified from Hynek et al. (2002). Opportunity's landing site is marked by X.

Figure 17.9 Distribution of hematite in Meridiani, The landing ellipse for the Opportunity rover is marked. NASA Mars Odyssey THEMIS map. Reproduced with permission from "Global mineralogy mapped from the Mars Global Surveyor Thermal Emission Spectrometer" by P. R. Christensen et al. within The Martian Surface edited by Jim Bell. © Cambridge University Press 2008.

a closed basin (Edgett and Parker, 1997). A geologic map of the area (Figure 17.8) shows plains units that unconformably overlie etched units of the cratered highlands (Hynek et al., 2002). Arvidson et al. (2003) recognized that the Meridiani surface is dominated by basaltic sands and has been exhumed by erosion. The layered unit does not lie within any clearly defined basin, although a basin boundary has likely been destroyed by erosion. A proposed Late Noachian to Early Hesperian age of the layered unit was poorly defined. Because of erosion and infilling of smaller craters, the size–distribution curve does not follow a crater production curve.

Crystalline hematite covers a huge area ($\sim$150,000 km^2) coinciding with the middle plains unit in Meridiani Planum (Figure 17.9), as identified and mapped by orbital thermal emission spectra (Christensen et al., 2000). This widespread mineral is interpreted to have formed in an aqueous environment. Sulfates were also detected in Meridiani from orbital spectra (Gendrin et al., 2005).

Opportunity Rover's Surface Exploration

Opportunity came to rest in a small, shallow crater on a monotonously flat plain extending to the horizon in all directions. Examination and analysis of layered rocks in the crater wall showed them to be sandstones composed of basaltic detritus cemented by evaporates (mostly sulfates). Scattered throughout the rocks are hematite-bearing spherules (Figure 17.10a), referred to as "blueberries" (they are only blue in false-color images), which weather out and form a concentrated lag deposit on the pavement surface (Figure 17.10b) – this is the hematite detected from orbit. Hematite was identified using the Mössbauer spectrometer.

Traversing to Endurance crater, Opportunity examined a vertical section of the hematite-sulfate unit, named the Burns formation (Figure 17.11). The interpreted stratigraphy is illustrated in the left column of Figure 17.12 (Grotzinger et al., 2005). A lower unit of cross-bedded sandstones formed as aeolian dunes. Finely laminated sandstones in the middle unit truncate cross-bedding of the lower unit. These sediments have been pervasively recrystallized and some dissolution of evaporate minerals has occurred, leaving casts (Figure 17.10a). The middle unit is interpreted as a sand sheet that marks a transition between the lower dune deposits and an upper unit of interdune deposits. The upper unit is characterized by coarse layering with festoon laminations – evidence of formation under subaqueous conditions.

The units of the Burns formation are interpreted to have formed by infiltration of groundwater to the surface to produce playa lakes in which basaltic mud was cemented by evaporates. Multiple groundwater recharge events formed the lower, middle, and upper units. A fourth infiltration produced the hematite concretions

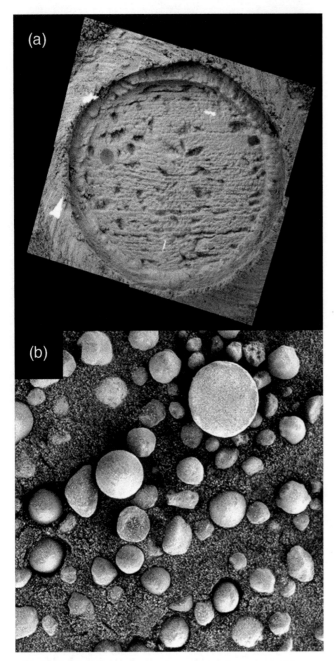

Figure 17.10 Hematite concretions at Meridiani, imaged by Opportunity rover. (a) RAT-ground rock showing layering, casts of dissolved minerals, and a blueberry, and (b) concentration of blueberries forming a lag deposit of hematite on the surface. NASA images.

that occur throughout the sequence (McLennan et al., 2005). The presence of the sulfate mineral jarosite indicates that the waters were highly acidic.

Following the analysis of strata in Endurance crater, Opportunity traversed to Victoria crater (Figure 17.13), an ancient impact scar with a rim scalloped by landslides. The rover circumnavigated a quarter of the way around the crater, investigating Burns formation layering in the walls (Figure 17.14).

Since making the hazardous trek to the even-larger Endeavour crater, Opportunity has explored the Noachian units that underlie the Burns formation (Arvidson et al., 2014). The layered rocks of the Matijevic formation contain smectites, previously discovered from orbital spectra, and are overlain by impact breccias of the Shoemaker formation and clastic sediments of the Grasberg formation (right column of Figure 17.12). As of this writing, Opportunity continues to explore the rim units of Endeavour.

17.3.3 Gale Crater
Orbital Studies

The 150 km diameter Gale impact crater (Figure 17.15), located near the dichotomy boundary, is filled with layered materials. Pelkey et al. (2004) used THEMIS images to map seven units in Gale crater, including crater walls, floor units, sand sheets, and a crescent-shaped central mound. Using visible and near-infrared images and topographic data, Le Deit et al. (2013) prepared the most current geologic map (Figure 17.16), and checked stratigraphic consistency using cross-sections.

The 5 km high central mound, now called Aeolis Mons (informally Mount Sharp), may contain the crater's central uplift at its core. The mound is covered with finely layered strata clearly visible from orbit. Based on geomorphology, almost every conceivable origin for the layered units in the mound has been proposed, including lava flows or ash-flow tuffs, sedimentary deposits of aeolian, fluvial or lacustrine origin, spring deposits, and ancient polar deposits (summarized by Le Deit et al., 2013). Orbital CRISM spectra show interstratified clay- and sulfate-bearing units (Figure 17.17) in the lower mound (bench) member (Thomson et al., 2011), ruling out a volcanic origin. Above an unconformity, the upper (caprock) member lacks signatures of either clays or sulfates (Milliken et al., 2010). Crater counting of the floor at the base of the mound gives an age of 3.6–3.8 Ga (Thomson et al., 2011; Le Deit et al., 2013), corresponding approximately to the Noachian–Hesperian boundary, and younger strata were deposited before ~3.2 Ga.

Curiosity Rover's Surface Exploration

Curiosity landed on a plain of shallowly dipping strata, now called the Bradbury group. The rover traversed these strata (Figure 17.18), eventually reaching the base of Aeolis Mons (Figure 17.19) and passing into the Mount Sharp group. The stratigraphy along the traverse of the Bradbury group is illustrated in Figure 17.20 (right column). During the early part of its traverse, Curiosity also encountered alkaline igneous rocks (Figure 17.21a,

Figure 17.11 Burns formation in Endurance crater. NASA image.

Stolper et al., 2013; Sautter et al., 2014), probably derived from the crater walls.

The Bradbury group consists of fluvial conglomerates (Figure 17.21b) and cross-bedded sandstones, as well as mudstones (Figure 17.21c). The rocks contain high proportions of igneous (basaltic) minerals, and the bulk chemistry of the rocks reflects their basaltic provenance (McLennan et al., 2014). Diagenesis is indicated by concretions and mineralized fractures (Figure 17.21d), and some of the sandstones are iron-cemented (Blaney et al., 2014). Clinoform sandstones in the sequence are interpreted as deltaic deposits formed where braided streams entered standing water (Grotzinger et al., 2015). The multiple clinoform layers at different elevations are thought to have been stacked during stream progradation.

The Murray formation, representing the basal unit of the Mount Sharp group, is a laminated mudstone overlain by cross-bedded or clinoform sandstone, interpreted as interfingering facies of deltaic sediments of the Bradbury group with finer-grained rocks of the Murray formation. Although its mineralogy is mostly basaltic minerals plus clays and amorphous material, an intermediate horizon contains silica (tridymite, cristobalite, quartz, and opal), suggesting a felsic source. In places, the base of the Murray formation is a conglomerate, and a hematite-bearing ridge overlies it. This formation represents a lacustrine facies adjacent to the fluvial–deltaic deposits (Grotzinger et al., 2015).

Above the Murray formation are clay-bearing strata that transition into polyhydrated sulfate-bearing strata (Figure 17.20, left column). Curiosity will continue to explore these units. Above the sulfate units is a caprock unit of unknown composition, with spectral signature similar to martian dust.

The reconstructed history of Gale (Grotzinger et al., 2015) posits that gravels eroded from the crater rim were transported inward by streams and deposited in a moat surrounding the crater's central uplift. These streams transitioned into sandy deltas marking the boundary of an ancient lake within which muds accumulated. Several kilometers of sediments were ultimately deposited. Following that, wind-driven erosion partly exhumed the strata by the Middle Hesperian Period, leaving Aeolis Mons as a surviving record of the sedimentary units that were removed.

17.4 Martian Meteorites: An Added Dimension

We do not know the exact provenance of any martian meteorites, although rayed craters may be likely launch sites (Tornabene et al., 2006). These craters are young, consistent with limited lifetimes of rays and with the short cosmic-ray exposure ages that are thought to date the times of ejection of the meteorites from Mars. The rayed craters are located in volcanic terrains, which accords with the meteorites' petrography. Hamilton et al. (2003) and Ody et al. (2015) identified terrains exhibiting thermal infrared and near-infrared spectra, respectively, that are similar to martian meteorites. The best shergottite analogs they found are Early Hesperian

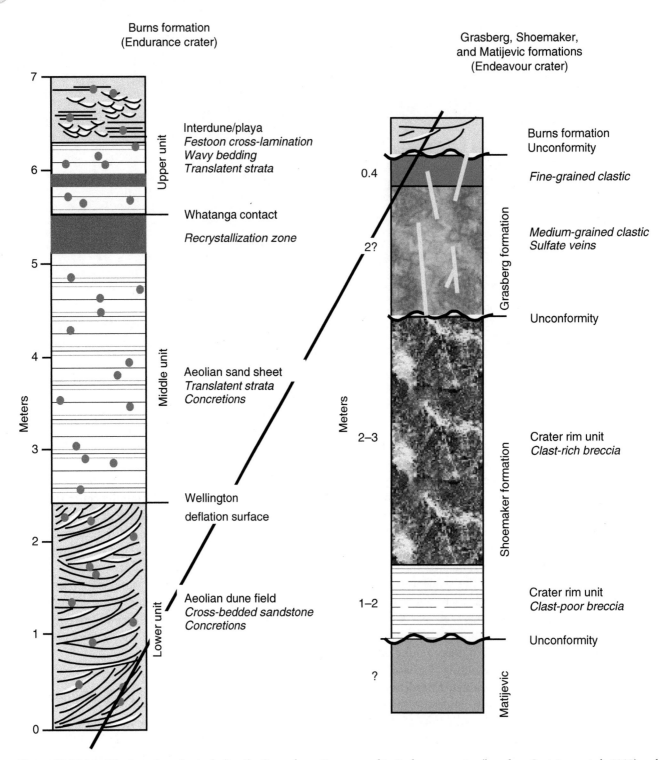

Figure 17.12 Meridiani stratigraphy, including the Burns formation mapped in Endurance crater (based on Grotzinger et al., 2005) and the underlying Noachian strata mapped in Endeavour crater (based on Grotzinger et al., 2015, and other data).

volcanic provinces (Syrtis Major, Hesperia Planum, Thaumasia Planum), but these regions are significantly older than shergottites; no nakhlite analogs were identified.

The petrology of the basaltic meteorites and related ultramafic cumulates was described in Section 10.4. Their predominantly young crystallization ages (<475 Ma for shergottites, 1.3 Ga for nakhlites/chassignites, 2.4 Ga for augite basalts) are Amazonian. Only one Noachian igneous cumulate rock, ALH 84001, is ~4.1 Ga. These ages indicate biased sampling, since most of the martian surface is Noachian. Identical cosmic-ray exposure ages for

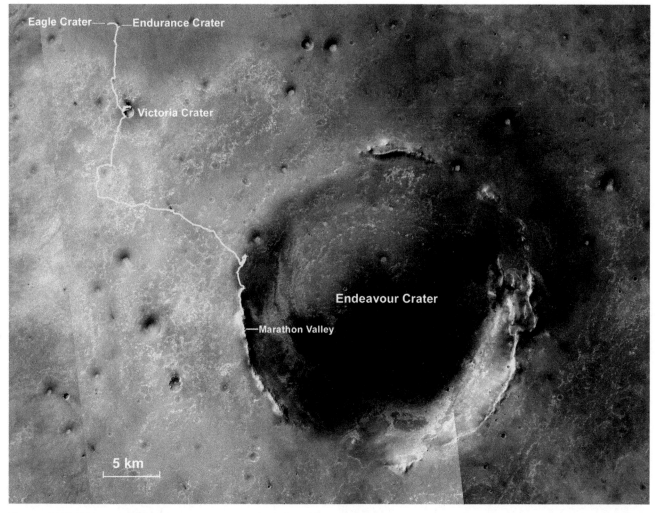

Figure 17.13 Traverse of the Opportunity rover in Meridiani Planum.

some shergottites and augite basalts suggest they sample the same location, and their widely varying crystallization ages imply a volcanic center (Tharsis?) that was intermittently active for billions of years (Lapen et al., 2017). Shergottites with different exposure ages indicate volcanism at a handful of other locations. The nakhlites represent a thick, fractionated flow similar in some respects to terrestrial komatiites, and demonstrate that basaltic shergottites were not the only Amazonian lavas.

Despite the young ages for most martian meteorites, they provide critical information on the origin and differentiation of Mars that is otherwise unobtainable. Radiogenic isotopes define the planet's time of accretion, and geochemical data indicate two distinct source regions that remained isolated for billions of years and were later melted or assimilated to produce the meteorites. It is unclear whether both the enriched and depleted sources reside in the mantle and were produced by solidification of a magma ocean (Debaille et al., 2007) or whether they represent enriched crust and depleted mantle, without the

need for a magma ocean (Humayun et al., 2013). We also noted previously (Figure 10.14) that the meteorites are depleted in alkali elements relative to older rocks analyzed by Mars rovers in Gusev and Gale, implying magmatic evolution over time.

NWA 7034/7533 (Figure 15.6) and a few other paired meteorites are the only sedimentary martian meteorites. These Pre-Noachian regolith breccias are composed of clasts of alkaline igneous rocks (norites, pyroxenites, and monzonites) with crystallization ages of ~4.4 Ga, as well as various impact-melted materials. The ancient age indicates the existence of an igneous crust a little over 100 million years after the planet's accretion (Hewins et al., 2017). Geochemical characteristics of the clasts suggest that precursor rocks were hydrothermally altered, oxidized, and contained phyllosilicates. Abundant fine-grained magnetite and maghemite explain the meteorites' remanent paleomagnetism, and similar rocks may account for magnetism discovered in the ancient southern highlands.

Figure 17.14 Cross-bedded sandstones of the Burns formation in the wall of Victoria crater. NASA image.

Figure 17.15 Image of Gale crater, showing the central mound. The yellow oval represents the landing ellipse for the Curiosity rover prior to arrival. NASA image.

17.5 Integration and Synthesis

The results of decades of exploration by orbiters, landers, and rovers, and of the analyses of martian meteorites, have been integrated to form an understanding of the geologic history of Mars. The summary below is adapted from comprehensive reviews of martian geology (Carr, 2006; Carr and Head, 2010; Tanaka et al., 2014), geochemistry (McSween and McLennan, 2014), and mineralogy/petrology (Bell, 2008; McSween, 2015). It is not

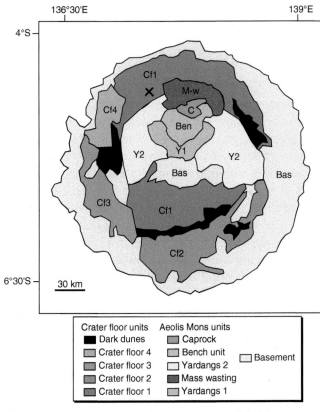

Figure 17.16 Geologic map of Gale crater, simplified from Le Deit et al. (2013). Curiosity's landing site is marked by X.

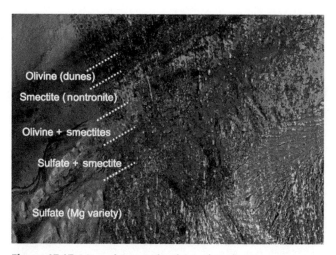

Figure 17.17 Mineralogy on the slope of Aeolis Mons, as mapped by CRISM spectra from the *Mars Reconnaissance Orbiter*. Modified from Milliken et al. (2014).

possible to cite the hundreds of papers that have added meaningfully to this effort, so we won't even try, but those references can be found in the above reviews.

Figure 17.22 presents a simplified view of the geologic evolution of Mars, with each period described below. The crystallization ages of martian meteorites are shown on the right side, and the Earth's geologic eras are compared on the left.

17.5.1 Pre-Noachian Period

No units recognizable at the scale of orbital spacecraft mapping survive from Pre-Noachian time, so little is directly known. Mars accreted and rapidly differentiated into crust, mantle, and core within the first few tens of millions of years of Solar System history, as indicated by the decay products of extinct radionuclides in martian meteorites. It is possible that the planet had an early

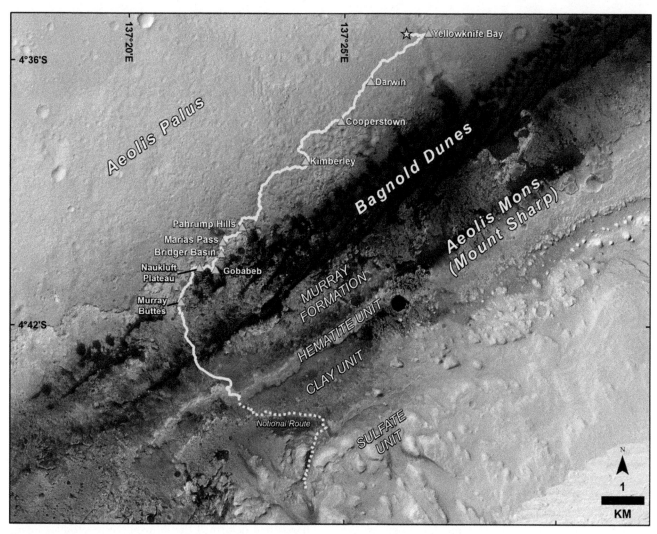

Figure 17.18 Traverse of the Curiosity rover in Gale crater.

Figure 17.19 Approach image of Aeolis Mons, taken by NASA Curiosity rover.

magma ocean that facilitated core segregation and that fractionally crystallized to form the distinct mantle source regions recognized in later-formed basaltic meteorites. The martian core consists of FeNi metal and FeS. An early dynamo resulted in striped magnetic anomalies in the highlands, possibly produced by iron minerals in the deep crust, and then declined by the Noachian. Heat flow peaked and then declined during the first few hundred million years. The ancient crust may have formed by decompression melting of mantle during convective overturn. The compositions of clasts in the only Pre-Noachian martian meteorites are alkaline basaltic rocks. The absence of a martian anorthositic crust, like that on the Moon, likely follows from water that suppressed plagioclase crystallization and from higher pressures that allowed segregation of aluminum into garnet at depth. Similarly, the absence of granitic crust reflects limited magmatic fractionation and no crustal recycling by subduction.

Pre-Noachian time was subject to numerous basin-forming impacts, and the global dichotomy itself may represent a huge impact structure. The end of the Pre-Noachian Period is defined by the formation of the 2400 km diameter Hellas basin variously assigned as ~4.1–3.8 Ga (depending on whether it formed in a Late Heavy Bombardment or in a time of gradually declining impacts).

17.5.2 Noachian Period/System

The Noachian Period was also distinguished by high but rapidly declining rates of cratering. In their global geologic map, Tanaka et al. (2014) distinguished units formed during the Early, Middle, and Late Noachian Epochs in the highlands, respectively containing 65, 15, and 2 impact basins >150 km in size. These impacts would have distributed ejecta over the globe, brecciated the upper crust, and promoted hydrothermal metamorphism beneath and around craters.

Magmatism during this period was likely concentrated in Tharsis, resulting in accumulation of a volcanic pile 9 km high and 5000 km across, equivalent to a global layer of lavas 2 km thick. If the magmas contained amounts of volatiles similar to Hawaiian basalts, they could have outgassed water equivalent to a global layer 120 m deep. By the end of the Noachian, the formation of Tharsis had deformed the lithosphere to produce a surrounding trough and a pronounced gravity anomaly, as

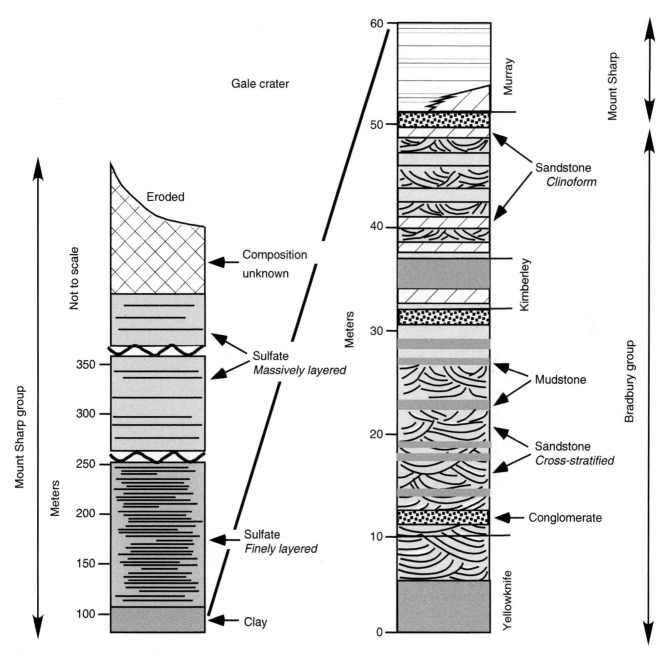

Figure 17.20 Stratigraphy of Gale crater. The right column shows the already-explored Bradbury group and lower portion of the Mount Sharp group in Gale crater, modified from Grotzinger et al. (2015). The left column (note change in vertical scale) shows the as-yet unexplored stratigraphy of Aeolis Mons, modified from Milliken et al. (2014).

predicted by loading of a spherical elastic shell. Extensional tectonics produced grabens, including rifting that initiated the formation of Valles Marineris. Most of the rocks in the southern highlands are probably Noachian basalts and pyroclastics reworked by impacts and sedimentation. Orbital spectra indicate the basalts are dominated by low-calcium pyroxene and sometimes olivine. The NWA 7034/7533 meteorite breccia (4.4 Ga) contains clasts of alkaline mafic igneous rocks, and the ALH 84001 martian meteorite (4.1 Ga) is an ultramafic cumulate from a basaltic magma, but the Noachian is otherwise

so far unsampled as meteorites, likely because of the difficulty of ejecting brecciated rocks.

Noachian surface conditions enabled rapid erosion and surface degradation that produced large amounts of debris. Craters of this age generally have highly eroded rims and are filled with sediments. Estimated erosion rates were 2–5 orders of magnitude higher than in subsequent periods, but even so, they were lower than erosion rates on Earth. Noachian terrains are dissected by valley networks, which have been attributed to both groundwater sapping and precipitation. The networks drain into

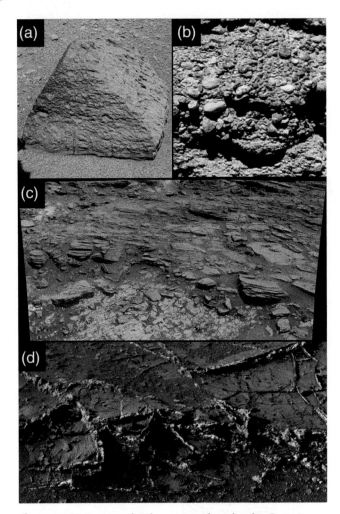

Figure 17.21 Images of Gale crater rocks, taken by Curiosity rover. (a) Jake_M, an alkaline basaltic rock, (b) conglomerate, (c) mudstone, and (d) sulfate veins. NASA images.

topographic lows (mostly craters) that have inlet and outlet valleys, suggesting the formation of lakes.

The formation of smectite clays was widespread during Noachian time. The Mg,Fe^{2+}-rich clays occur in rocks excavated from depth, suggesting they may have formed by diagenesis or hydrothermal metamorphism of mafic rocks, whereas Al,Fe^{3+}-rich clays likely formed by surficial chemical weathering (Ehlmann et al., 2011; Bishop et al., 2018). Serpentine and talc also occur. Although much speculation about Noachian climate is based on the assumption of weathering, the evidence for groundwater and lakes is most consistent with only episodically warm conditions. Even with an atmosphere enhanced with greenhouse gases from volcanic outgassing, climate models cannot produce the global average temperatures that would sustain liquid water at the surface. The lack of significant carbonate deposits does not support ideas of a dense CO_2 atmosphere that would support greenhouse warming. Perhaps large impacts or volcanic eruptions released H_2O that could account for precipitation to form

valley networks and intermittent greenhouse heating. Like the rates of impact, erosion rates dropped precipitously toward the end of the Noachian, and the formation of clays essentially ceased.

17.5.3 Hesperian Period/System

The Hesperian Period began at ~3.7 Ga, and has been divided into Early and Late Hesperian Epochs. The Hesperian System preserves a greater diversity of processes than the Noachian, because of reduced impacts and lower rates of erosion.

Widespread volcanism formed extensive lava fields around major volcanic centers, including Syrtis Major and Hesperia Planum. Orbital spectra indicate that these rocks are basaltic, with abundant low-calcium pyroxene. Hesperian volcanic rocks are also interbedded with sediments in highlands basins such as Gale (~3.7 Ga) and Gusev (~3.65 Ga). The ancient igneous rocks in Gale crater analyzed by Curiosity and in Gusev crater by Spirit indicate alkaline basalts and their fractionation products. Although most Hesperian volcanoes were constructed by effusive lavas, some show evidence of pyroclastic eruptions, and crater counts on volcanoes suggest a transition from explosive to effusive eruption style occurred at ~3.5 Ga (Robbins et al., 2011).

Hesperian tectonics was generally compressional, perhaps related to planet cooling, and wrinkle ridges occur in most units of this age. However, limited extension continued to form Valles Marineris. At least four basin-forming impacts occurred, but cratering was more limited during this period.

Catastrophic outflow channels developed during Hesperian time, producing massive floods that extensively scoured the regions through which they passed and formed sedimentary deposits. Large bodies of water must have been left at the termini of outflow channels. The presence of a Hesperian ocean in the northern lowlands has also been advocated, although compelling evidence is as yet lacking. The formation of valley networks declined significantly during the Hesperian, indicating a progressive change in environmental conditions. Sulfate- and chloride-rich deposits, as found in Meridiani and Gale, suggest evaporation of lakes, possibly due to oscillations in the groundwater table. The orbital detection of widespread olivine in Hesperian units indicates that weathering was very limited.

The oldest ice-bearing units in the martian polar deposits are of Hesperian age. These may have formed by cryovolcanic eruptions or from accumulation of a glacial ice sheet. The end of the Hesperian is defined as the cessation of deposition of sedimentary plains in the northern lowlands.

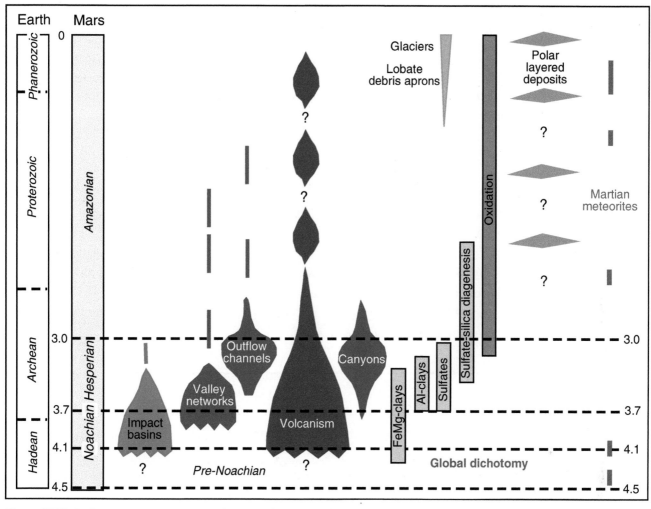

Figure 17.22 Geologic activity on Mars, as a function of time. Modified from Carr and Head (2010) and Ehlmann and Edwards (2014).

17.5.4 Amazonian Period/System

The Amazonian Period is the longest, comprising two-thirds of martian geologic time. Its beginning at ~3 Ga coincides approximately with the Earth's Archean. The Amazonian is divided into Early, Middle, and Late Epochs. Despite the period's long duration, the geomorphic effects of cratering, volcanism, and tectonism are modest. The effects of ice and wind are more prominent in Amazonian units, owing to the low rates of erosion and landscape construction by other processes. The environment has been cold, dry, and oxidizing, leading to widespread formation of hematite that gives Mars its red color.

Amazonian volcanism was mostly confined to Tharsis and Elysium, where large shields continued to grow episodically. Lava flows are prominent and suggest more limited outpourings relative to the flood basalt style of the Hesperian volcanic plains. In contrast to Noachian and Hesperian basalts dominated by low-calcium pyroxene, the Amazonian basalts contain mostly high-calcium

pyroxene. The ages of martian meteorites span much of Amazonian time. The youngest shergottite ages (175 Ma), along with crater-density ages of tens of millions of years for the youngest flows in Tharsis and Elysium, suggest that volcanism has persisted nearly to the present. All of the basaltic meteorites and associated cumulates formed from tholeiitic rather than alkaline magmas. Most Amazonian flows are not deformed by wrinkle ridges, but some limited development of narrow grabens and wrinkle ridges may have resulted from local crustal loading by lavas.

The effects of ice include ice-dust veneers covering the planet's surface at mid- to high latitudes, ground ice, possible glacial effects on some volcanoes, ice-cored debris aprons at mid-latitudes, and polar deposits. The presence of present-day near-surface ice has been detected by orbital gamma ray/neutron measurements of hydrogen, and the *Phoenix* lander confirmed the occurrence of a shallow ice table. Ice stability is sensitive to the obliquity cycle, which varies considerably on Mars. Thinly layered polar deposits are accumulations of ice and dust moderated by recent orbital and obliquity changes. Such deposits have probably been accumulated and removed repeatedly over geologic time.

Gullies are the most common type of fluvial features in Amazonian time. They occur on steep, poleward-facing slopes at high to mid-latitudes. These recent features most likely form by melting of snow or ice.

Aeolian activity has been the dominant sedimentary agent during the Amazonian and forms dunes almost everywhere. Thick deposits with fluted surfaces such as the Medusae Fossae formation may have formed from wind-blown materials, and rover images of etched rocks and ventifacts demonstrate the wind's erosional power.

Summary

The geologic history of Mars has been unraveled systematically, as data have been acquired at ever-improved spatial and spectral scales. Global reconnaissance, progressing to regional geologic analysis, and finally to the exploration of local areas by landed spacecraft, have been integrated to reveal the planet's geologic evolution. Martian meteorites analyzed in the laboratory also provide information on processes that occurred prior to the preserved rock record, as well as geochemical and geochronologic constraints on the unseen interior and on magmatism.

The focus of Mars exploration, at least in the modern era, has arguably been astrobiology, but so far the discoveries have been geologic. What new surprises await is for future explorers to discover.

Review Questions

1. How has the interpretation of Gusev crater as an ancient lakebed from orbital data been changed by the Spirit rover's exploration of the crater floor?
2. How has the exploration of the surface of Meridiani Planum by the Opportunity rover augmented the interpretation of hematite detected from orbit?
3. How has the Curiosity rover's exploration complemented orbital remote sensing of Gale crater?
4. Give two examples of how the analysis of martian meteorites has improved our understanding of the unseen interior of Mars.
5. What are the first-order changes in geologic processes that characterized the Noachian, Hesperian, and Amazonian Periods?

SUGGESTIONS FOR FURTHER READING

Bell, J., ed. (2008) *The Martian Surface: Composition, Mineralogy, and Physical Properties*. Cambridge: Cambridge University Press. These 27 chapters are a goldmine of information about the geologic exploration of the surface of Mars.

Carr, H. Y. (2006) *The Surface of Mars*. Cambridge: Cambridge University Press. An authoritative book on the geology of Mars.

Carr, M. H., and Head, J. W. (2010) Geologic history of Mars. *Earth and Planetary Science Letters*, **294**, 185–203. A clear and concise summary of the geologic evolution of Mars.

McSween, H. Y., and McLennan, S. M. (2014) Mars. In *Treatise in Geochemistry*, Vol. 2, 2nd edition, eds.

Holland, H. D., and Turekian, K. K. Oxford: Elsevier, pp. 251–300. The geochemistry of Mars, from A to Z.

Tanaka, K. L., Skinner, J. A., Dohm, J. M., et al. (2014) Geologic map of Mars. US Geological Survey Scientific Investigations Map 3292. The most recent iteration of a global geologic map for Mars. An accompanying pamphlet provides details on how the map was crafted.

REFERENCES

Acuna, M. H., Kletetschka, G., and Connerney, J. E. P. (2008) Mars' crustal magnetism: a window into the past. In *The Martian Surface: Composition, Mineralogy, and Physical Properties*, ed. Bell, J. Cambridge: Cambridge University Press, pp. 242–262.

Arvidson, R. E., Seelos, F. P., Deal, K. S., et al. (2003) Mantled and exhumed terrains in Terra Meridiani, Mars. *Journal of Geophysical Research*, **108**(E12), 8073.

Arvidson, R. E., Squyres, S. W., Bell, J. F., et al. (2014) Ancient aqueous environments at Endeavour crater, Mars. *Science*, **343**. DOI: 10.1126/science.1248097.

Bell, J., ed. (2008) *The Martian Surface: Composition, Mineralogy, and Physical Properties.* Cambridge: Cambridge University Press.

Bibring, J-P., and Langevin, Y. (2008) Mineralogy of the martian surface from Mars Express OMEGA observations. In *The Martian Surface: Composition, Mineralogy, and Physical Properties*, ed. Bell, J. Cambridge: Cambridge University Press, pp. 153–168.

Bishop, J. L., Fairen, A. G., Michalski, J. R., et al. (2018) Surface clay formation during short-term warmer and wetter conditions on a largely cold ancient Mars. *Nature Astronomy*. DOI: 10.1038/s41550-017-0377-9.

Blaney, D. L., Blake, D. F., Vaniman, D. T., et al. (2014) Chemistry and texture of the rocks at Rocknest, Gale crater: evidence for sedimentary origin and diagenetic alteration. *Journal of Geophysical Research*, **119**, 2109–2131.

Boynton, W. V., Taylor, G. J., Karunatillake, S., et al. (2008) Elemental abundances determined via the Mars Odyssey GRS. In *The Martian Surface: Composition, Mineralogy, and Physical Properties*, ed. Bell, J. Cambridge: Cambridge University Press, pp. 105–124.

Carr, H. Y. (2006) *The Surface of Mars.* Cambridge: Cambridge University Press.

Carr, M. H., and Head, J. W. (2010) Geologic history of Mars. *Earth and Planetary Science Letters*, **294**, 185–203.

Christensen, P. R., Clark, R. N., Kieffer, H. H., et al. (2000) Detection of crystalline hematite mineralization on Mars by the Thermal Emission Spectrometer: evidence for near-surface water. *Journal of Geophysical Research*, **105**, 9623–9642.

Christensen, P. R., Bandfield, J. L., Fergason, R. L., et al. (2008) The compositional diversity and physical properties mapped from the Mars Odyssey Thermal Emission Imaging System. In *The Martian Surface: Composition, Mineralogy, and Physical Properties*, ed. Bell, J. Cambridge: Cambridge University Press, pp. 221–241.

Debaille, V., Brandon, A. D., Yin, Q. Z., et al. (2007) Coupled ^{142}Nd-^{143}Nd evidence for a protracted magma ocean on Mars. *Nature*, **450**, 525–528.

Edgett, K. S., and Parker, T. J. (1997) Water on early Mars: possible subaqueous sedimentary deposits covering ancient crater terrain in western Arabia and Sinus Meridiani. *Geophysical Research Letters*, **24**, 2897–2900.

Ehlmann, B. L., and Edwards, C. S. (2014) Mineralogy of the martian surface. *Annual Reviews of Earth and Planetary Science*, **42**, 291–315.

Ehlmann, B. L., Mustard, J. F., Murchie, S. L., et al. (2011) Subsurface water and clay mineral formation during the early history of Mars. *Nature*, **479**, 53–60

Gendrin, A. N., Mangold, N., Bibring, J-P., et al. (2005) Sulfates in martian layered terrains: the OMEGA/Mars Express view. *Science*, **307**, 1587–1590.

Grin, E. A., and Cabrol, N. A. (1997) Limnologic analysis of Gusev crater paleolake, Mars. *Icarus*, **130**, 461–474.

Grotzinger, J. P., Arvidson, R. E., Bell, J. F., et al. (2005) Stratigraphy, sedimentology and depositional environment of the Burns formation, Meridiani Planum, Mars. *Earth and Planetary Science Letters*, **240**, 11–72.

Grotzinger, J. P., Gupta, S., Malin, M. C., et al. (2015) Deposition, exhumation, and plaeoclimate of an ancient lake deposit, Gale crater, Mars. *Science*, **350**. DOI: 10.1126/science.aac7575.

Hamilton, V. E., Christensen, P. R., McSween, H. Y., et al. (2003) Searching for the source regions of martian meteorites using MGS TES: integrating martian meteorites into the global distribution of igneous materials on Mars. *Meteoritics & Planetary Science*, **38**, 871–885.

Hewins, R. H., Zanda, B., Humayun, M., et al. (2017) Regolith breccia Northwest Africa 7533: mineralogy and petrology with implications for early Mars. *Meteoritics & Planetary Science*, **52**, 89–124.

Humayun, M., Nemchin, A., Zanda, B., et al. (2013) Origin and age of the earliest martian crust from meteorite NWA 7533. *Nature*, **503**, 513–517.

Hynek, B. M., Arvidson, R. E., and Phillips, R. J. (2002) Geologic setting and origin of Terra Meridiani hematite deposits on Mars. *Journal of Geophysical Research*, **107**(E10), 5088.

Kuzmin, R. O., Greeley, R., Landheim, R., et al. (2000) Geologic map of the MTM-15182 and MTM-15187 quadrangles, Gusev crater–Ma'adim Vallis region, Mars. USGS Geologic Investigations Series.

Lapen, T. J., Righter, M., Andreasen, R., et al. (2017) Two billion years of magmatism recorded from a single Mars meteorite ejection site. *Science Advances*, **3**. DOI: 10.1126/sciadv.1600922.

Le Deit, L., Hauber, E., Fueten, F., et al. (2013) Sequence of infilling events in Gale crater, Mars: results from morphology, stratigraphy, and mineralogy. *Journal of Geophysical Research*, **118**, 2439–2473.

McLennan, S. M., Bell, J. F., Calvin, W. M., et al. (2005) Evidence for groundwater involvement in the provenance and diagenesis of the evaporaite-bearing Burns formation. *Earth and Planetary Science Letters*, **240**, 95–121.

McLennan, S. M., Anderson, R. B., Bell, J. F., et al. (2014) Elemental geochemistry of sedimentary rocks at Yellowknife Bay, Gale crater, Mars. *Science*, **343**. DOI: 10.1126/science.1243480.

McSween, H. Y. (2015) Petrology on Mars. *American Mineralogist*, **100**, 2380–2395.

McSween, H. Y., and McLennan, S. M. (2014) Mars. In *Treatise in Geochemistry*, Vol. 2, 2nd edition, eds. Holland, H. D., and Turekian, K. K. Oxford: Elsevier, pp. 251–300.

McSween, H. Y., Arvidson, R. E., Bell, J. F., et al. (2004) Basaltic rocks analyzed by the Spirit rover in Gusev crater. *Science*, **305**, 842–845.

McSween, H. Y., Ruff, S. W., Morris, R. V., et al. (2006) Alkaline volcanic rocks from the Columbia Hills, Gusev crater, Mars. *Journal of Geophysical Research*, **111**, E09S91.

Milam, K. A., Stockstill, K. R., Moersch, J. E., et al. (2003) THEMIS characterization of the MER Gusev crater landing site. *Journal of Geophysical Research*, **108**(E12), 8078.

Milliken, R. E., Grotzinger, J. P., and Thomson, B. J. (2010) Paleoclimate of Mars as captured by the stratigraphic record in Gale crater. *Geophysical Research Letters*, **37**. DOI: 10.1029/2009GL041870.

Milliken, R. E., and the MSL Science Team (2014) Mineral mapping of Gale crater using orbital data: results from visible-near infrared reflectance spectroscopy. Available at: www.searchanddiscovery.com/documents/2014/51013milliken/ndx_milliken.pdf.

Ody, A., Poulet, F., Quantin, C., et al. (2015) Candidate source regions of martian meteorites as identified by OMEGA/MEx. *Icarus*, **258**, 366–383.

Parker, M. K., Zegers, T., Kneissi, T., et al. (2010) 3D structure of the Gusev crater region. *Earth and Planetary Science Letters*, **294**, 411–423.

Pelkey, S. M., Jakosky, B. M., and Christensen, P. R. (2004) Surficial properties in Gale crater, Mars, from Mars Odyssey THEMIS data. *Icarus*, **167**, 244–270.

Robbins, S. J., Di Achille, G., and Hynek, B. M. (2011) The volcanic history of Mars: high-resolution crater-based studies of the calderas of 20 volcanoes. *Icarus*, **211**, 1179–1203.

Sautter, V., Fabre, C., Forni, O., et al. (2014) Igneous mineralogy at Bradbury Rise, the first ChemCam campaign at Gale crater. *Journal of Geophysical Research*, **119**, 30–46.

Squyres, S. W., Arvidson, R. E., Blaney, D. W., et al. (2006) The rocks of the Columbia Hills. *Journal of Geophysical Research*, **111**, E02S11.

Squyres, S. W., Aharonson, O., Clark, B. C., et al. (2007) Pyroclastic activity at Home Plate in Gusev crater, Mars. *Science*, **316**, 738–742.

Stolper, E. M., Baker, M. B., Newcombe, M. E., et al. (2013) The petrochemistry of Jake_M: a martian mugearite. *Science*, **341**. DOI : 10.1126/science.1239463.

Tanaka, K. L., Skinner, J. A., Dohm, J. M., et al. (2014) Geologic map of Mars. US Geological Survey Scientific Investigations Map 3292.

Thomson, B. J., Bridges, N. T., Milliken, R., et al. (2011) Constraints on the origin and evolution of the layered mound in Gale crater, Mars using Mars Reconnaissance Orbiter data. *Icarus*, **214**, 413–432.

Tornabene, L. L., Moersch, J. E., McSween, H. Y., et al. (2006) Identification of large (2–10 km) rayed craters on Mars in THEMIS thermal infrared images: implications for possible Martian meteorite source regions. *Journal of Geophysical Research*, **111**, E10006. DOI: 10.1029/2005JE002600.

Epilogue:
Geologic Processes in Other Solar Systems?

Only a quarter of a century ago, the only planets known to exist were those orbiting our Sun. Now we know that other stars host extrasolar planets, or "**exoplanets**." The earliest discoveries were made using Earth-based telescopes that detected stellar wobbles caused by nearby orbiting planets. To be detectable, these exoplanets must be large and orbit very close to their stars. Using a different method, NASA's Kepler orbiting space telescope (2009–2013) observed partial eclipses as planets transited in front of stars (Batalha, 2014). Kepler's high-precision measurements – only possible in space where stars don't twinkle – have enabled the discovery of ~4000 candidate planets; of these candidates, perhaps 90 percent are thought to be real planets (Lissauer et al., 2014). The list includes nearly 700 multiple planet systems (Figure E.1), and the large numbers imply that these must be very common. Planetary orbits in multi-planet systems are usually coplanar, consistent with their formation within accretion disks (like our own Solar System's ecliptic plane).

As a planet flits across the face of a star, Kepler measures its size from the dip in brightness. Exoplanet radii are compared with the radii of Jupiter, Neptune, Earth, and Mars in Figure E.1. These data reveal that planets smaller than Neptune are far more common than giant planets like Jupiter. This is surprising, since planets 1–4 times the size of Earth are uncommon in our Solar System.

Geologists are especially interested in what exoplanets are made of, and that requires knowing their densities, which in turn necessitates determination of their masses. Ground-based telescopes use the Doppler technique, which measures how starlight shifts to redder and bluer wavelengths as the star rocks back and forth in response to the gravitational tugs of the planet. Kepler cannot make Doppler measurements, except in a few instances of massive planets in close-in orbits, but it can indirectly determine the masses of planets in multi-planet systems by analyzing how the planets tug on each other and thus disturb their otherwise elliptical orbits. Mass measurements for ~200 exoplanets (Howard, 2013: Marcy et al., 2014) are compared with models for planets consisting of pure hydrogen, water, rock (assumed to be olivine), and iron metal in Figure E.2. These data are biased toward large planets because they are easier to measure.

What can we infer about the geology of exoplanets from these limited data? The smallest planets, up to 1.5 times Earth's radius, have densities ($>5\,g/cm^3$) expected for rocky bodies with metal cores (Earth's bulk density is $5.5\,g/cm^3$), and their densities increase with increasing radius, explainable by gravitational compression (Marcy et al., 2014). Larger planets, "mini-Neptunes" with radii 1.5–4 times Earth, have densities similar to or greater than Uranus and Neptune (1.27 and $1.63\,g/cm^3$, respectively) that decline with increasing radius, indicating increased amounts of low-density gas or ices in envelopes surrounding rocky cores. Even larger planets are gas giants with densities near $1.0\,g/cm^3$, more like Jupiter and Saturn. Some of these are "puffier" than expected, and their inflated radii are not yet well explained. Attempts to add further details about the internal structures of exoplanets rely mostly on what is known about planets in our own Solar System (Spiegel et al., 2014) and as yet have offered more speculation than insight.

In Chapter 4 we learned that stars vary in composition (we used the term "metallicity," defined as elements heavier than hydrogen). Metal-rich stars are more likely to host giant planets. This is taken as evidence for the core accretion model of planet formation, wherein planetary cores grow from solids in the disk, which then undergo runaway accretion of nebular gases. Rocky planets occur around stars with a range of metallicities (Marcy et al., 2014).

Although attention has focused on main sequence stars, exoplanets have also been discovered around red

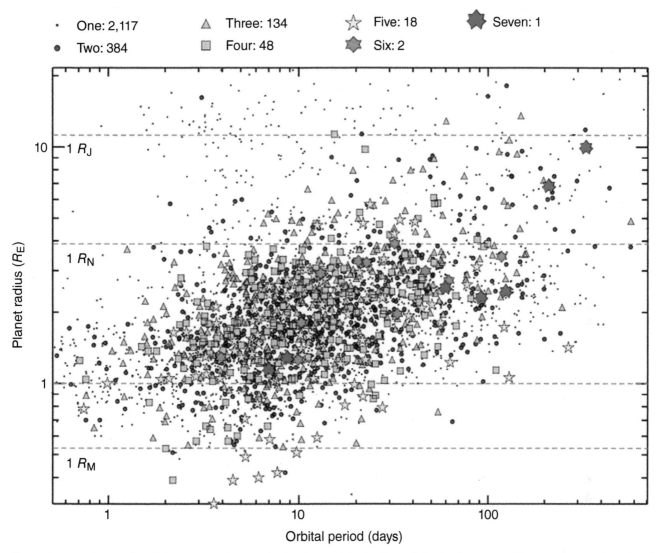

One: 2,117 Two: 384 Three: 134 Four: 48 Five: 18 Six: 2 Seven: 1

Figure E.1 Exoplanet radius (relative to Earth) versus orbital period (in Earth days) for the thousands of known objects. The radii of Jupiter, Neptune, Earth, and Mars are shown by dashed horizontal lines, from top to bottom. Tiny dots are systems with only one detected planet, and the other symbols indicate multi-planet systems. Reprinted by permission from Springer Nature: Nature, Advances in exoplanet science from Kepler, Jack J. Lissauer, Rebekah I. Dawson, Scott Tremaine, copyright (2014).

giants, neutron stars, and binary star systems (Tasker, 2017). Because accretion disks have compositions similar to their host stars, the oxygen in disks around carbon-rich stars would react to form CO, leaving little or no oxygen to make H_2O. Such disks would have no snow line, and planets formed within such disks are speculated to have crusts of graphite and interiors of diamond.

Of particular interest is the occurrence of Earth-like planets, especially those that could harbor life. The zone of habitability is a region around a star where a planet can have surface temperatures consistent with the presence of liquid water (Seager, 2013). The energy input from the host star (stellar effective temperature) mostly controls the climates of planets with thin atmospheres, and the surface temperatures decrease with orbital distance. Small stars have habitable zones closer to them than do Sun-like stars. However, the habitable zone can be expanded or shrunk based on greenhouse gases in the atmosphere. Two different definitions of the habitable zone are illustrated in Figure E.3 (Kopparapu et al., 2014). Kepler has detected dozens of nearly Earth-size planets (less than two times Earth's radius) that are interior to their stars' habitable zones, and a few within habitable zones (Figure E.3). Most of the planets within habitable zones detected so far are somewhat larger than Earth. However, a system of seven Earth-size bodies orbits TRAPPIST-1 (Gillon

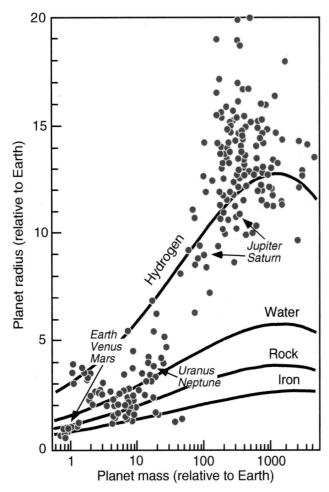

Figure E.2 Exoplanet radius versus mass, compared with models for planets composed of hydrogen, water, rock, and iron. The planets of our Solar System are shown by red dots. Modified from Howard (2013), with additional data on small planets from Marcy et al. (2014) and Gillon et al. (2017).

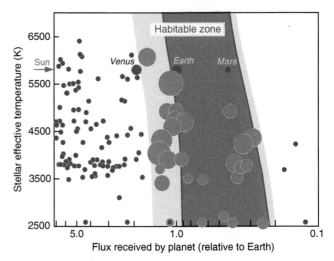

Figure E.3. Stellar effective temperature (a measure of a star's energy) versus the flux (insolation at the planet's semi-major axis) received by nearly Earth-size extrasolar planets (<2× Earth's radius) detected by Kepler. Most of these planets lie inside the habitable zone. Symbols for exoplanets within the habitable zone are sized relative to Earth, Venus, and Mars. The seven planets at the bottom of the figure all revolve around the low-mass star TRAPPIST-1. Modified from Batalha (2014), with additional data from Gillon et al. (2017).

the study of the Earth and its neighboring planets will provide the greatest insights into geologic processes on other worlds.

SUGGESTION FOR FURTHER READING

Tasker, E. (2017) *The Planet Factory: Exoplanets and the Search for a Second Earth*. London: Bloomsbury Sigma. This engaging account is written at an accessible level for astrophysically challenged geoscientists. It describes the rapidly advancing field of exoplanets, focusing on their discoveries and varieties, with speculation on whether they could be abodes for life.

REFERENCES

Batalha, N. M. (2014) Exploring exoplanet populations with NASA's Kepler mission. *Proceedings of the National Academy of Sciences USA*, **111**(12) 647–654.

Gillon, M., Triaud, A. H., Demory, B.-O., et al. (2017) Seven temperate terrestrial planets around the nearby ultracool dwarf star TRAPPSIT-1. *Nature*, **542**, 456–460.

Howard, A. W. (2013) Observed properties of extrasolar planets. *Science*, **340**, 572–576.

Kopparapu, R. K., Ramirez, R. M., Schottelkotte, J., et al. (2014) Habitable zones around main-sequence stars:

et al., 2017), a dwarf star about the size of Jupiter. These planets are plotted at the bottom of Figure E.3, and four of them are within the habitable zone. A future goal is to investigate the spectra of Earth-size planets in the habitable zone to identify gases that could be biosignatures (Seager, 2013). On Earth, a global biosignature in the atmosphere is O_2 produced by photosynthesis, although CH_4 produced by microorganisms might have filled that role on the early Earth. A caution is that such gases can also be produced abiotically and can be affected by UV radiation. Comparison of possible biogenic gases with their environmental contexts and replenishment rates will be required to argue for life.

Exoplanets demonstrate that our own Solar System does not consist of planetary archetypes, and that the outcomes of planetary formation are much more varied than we could have surmised. What we can hope to learn about the geology of exoplanets is limited, however, and dependence on planetary mass. *Astrophysical Journal Letters*, **787**(2), L29.

Lissauer, J. J., Dawson, R. I., and Tremaine, S. (2014) Advances in exoplanet science from Kepler. *Nature*, **513**, 336–344.

Marcy, G. W., Weiss, L. M., Petigura, E. K., et al. (2014) Occurrence and core–envelope structure of 1–4× Earth-size planets around Sun-like stars. *Proceedings of the National Academy of Sciences USA*, **111**(12),655–660.

Seager, S. (2013) Exoplanet habitability. *Science*, **340**, 577–581.

Spiegel, D. S., Fortney, J. J., and Sotin, C. (2014) Structure of exoplanets. *Proceedings of the National Academy of Sciences USA*, **111**(12), 622–627.

Tasker, E. (2017) *The Planet Factory: Exoplanets and the Search for a Second Earth*. London: Bloomsbury Sigma.

Glossary

Because this book was written for undergraduate geoscience students who are already familiar with many of its subdisciplines, this glossary does not include many common geologic terms that would be familiar to any geoscience major.

abrasion: grinding by small particles

accretion disk: rotating, flattened disk of gas and dust that settles onto the nebular midplane and eventually forms planetesimals and planets

achondrite: rocky meteorite that either crystallized from a magma or represents the solid residue from partial melting; the latter is called a "primitive achondrite"; Achondrites may be samples of asteroids, Mars, or the Moon

active lid: lithosphere is decoupled from the body interior and consists of segments that move relative to each other, creating abundant surface tectonic structures

active remote sensing: remote sensing technique for which the source of illumination or irradiation is provided by the experiment

adiabatic lapse rate: rate at which a vertically displaced parcel of atmosphere will neither gain nor lose heat with its surroundings

admittance: ratio of gravity to topography

aeolian: process or landform formed by wind

agglutinates: glass-cemented clusters of clastic grains in a regolith

albedo: amount of light reflected from the surface

alluvial channel: landform produced by channelized fluid flow over loose sediment

alluvial fan: sedimentary landform developed at the mouth of a canyon that contains significant transportable sediment

amino acids: carbon atoms bonded with a carboxyl group (COOH), and amine (NH_2), and one of a variety of side chain groups

amphiphilic: molecule having a polar hydrophilic head and a nonpolar hydrophobic tail

angle of repose: the steepest angle at which a sloping surface formed of loose material is stable

Archeae: single-celled organisms that lack organelles and a nucleus, but have distinctive membrane structure and genetic coding from bacteria

asteroid: rocky or metallic object orbiting the Sun that is smaller than a planet or dwarf planet

astrobiology: biology on other worlds

astronomical unit (AU): the mean distance between the Earth and the Sun

atmophile: elements that remain in gaseous form, except at extremely low temperatures

Bacteria: single-celled organisms that have cell walls, but lack organelles and an organized nucleus

bajada: coalescence of multiple alluvial fans

bands (on Europa): troughs with hummocks parallel to the trough boundary that may be 30 km wide and hundreds of kilometers long, accommodating lithospheric dilation where new surface ice formed

biomarker: organic carbon structures that represent the preserved components of biomolecules

blackbody: a substance that emits radiant energy at the maximum possible rate per unit area for any given temperature

blackbody spectrum: the radiance of a perfect emitter at a given temperature as a function of wavelength

bolide: projectile that impacts a target

boundary layer: region of decreasing wind speed close to the surface

breccia: rock composed of broken, angular fragments; can be composed of one rock type (monomict) or multiple rock types (polymict)

calcium–aluminum inclusion (CAI): refractory grains that may represent nebular condensates or residues from evaporation; they occur in chondrites and are the first-formed solids in the Solar System

carbohydrates: simple chains of carbon bonded to a hydrogen ion and a hydroxyl ion, used to provide and/or store energy

central uplift: peak formed by elastic rebound in the interior of a complex crater

chalcophile: elements that are preferentially incorporated into sulfides

chaos terrain: area of jumbled blocks formed during catastrophic breakout of confined subsurface fluid

charge transfer features: absorptions produced by transfer of a portion of an electronic charge between two molecules or parts of a molecule

chasma or chamata: complex population of normal faults that trend sinuously across the surface, having large cumulative displacements

CHNOPS: shorthand notation for the Earth's biogenic elements carbon, hydrogen, nitrogen, oxygen, phosphorus, and sulfur

chondrite: the most common type of rocky meteorite; its mineralogical composition is ultramafic, and its chemical composition is approximately that of the Sun for non-volatile elements; a sample of an asteroid

chronostratigraphic units: bodies of rock formed during a specific interval of geologic time

comet: object orbiting the Sun that exhibits "cometary behavior" caused by outburst of volatiles from ice sublimation

cometary particles: dust shed off comets and collected by spacecraft or other means

complex crater: a large crater with a flat floor, terraced rim, and central uplift

condensation: the formation of solids (minerals) from hot nebular gas as it cooled in the early Solar System; the sequence of condensed phases can be calculated from thermodynamics and observed in CAIs

condensation flows: movement of atmospheric volatiles caused by sublimation followed by recondensation

continuum: wavelength regions in a spectrum that have reflectance or emissivity values at or near 1

continuum mechanics: deals with the analysis of kinematics and mechanical behavior of materials modeled as a continuous mass rather than as discrete particles

core segregation: sinking of dense, molten metal to the center of a body

Coriolis force: effect whereby a mass moving in a rotating system experiences a force acting perpendicular to the direction of motion and to the axis of rotation

coronae: large circular patterns of ridges and troughs on Venus

cosmic abundances: the abundances of elements in the Solar System, approximately equivalent to solar abundances

cosmochemistry: study of the chemical composition of the Universe and the Solar System and the processes that produced that composition

cosmogenic nuclide: isotope formed by interaction with cosmic rays

crater basin lake: fluid-filled impact crater

crater production function: rate of impact crater formation through time

crater size–frequency distribution: numbers of craters of certain size ranges in a given area; sometimes called "crater density"

creep: movement of particles by rolling across the surface

cryovolcanism: volcanoes that erupt liquids or slurries of water, ammonia, methane, or other volatiles at frigid temperatures

crystal field absorptions: absorptions arising from placing a transition metal ion in the crystal field generated by a set of bonds, i.e., an octahedral site in a mineral stabilizes certain electronic orbitals

cycloidal trace: fracture trace in ice that consists of a series of convex segments connected by apices formed in response to stress from diurnal tides

delta: sedimentary deposit at the mouth of a river

dielectric constant: a material property that controls how radio waves propagate into or off of the target

differentiation: separation of an initially homogeneous body into crust, mantle, and core

digital terrain model (DTM): planetary topographic map in digital format

dispersing element: a device that disperses light into its spectrum; either a prism or diffraction grating

drainage basin: an enclosed area circumscribed by the highest topographic elevations, into which fluids converge

ductile: pliable, not brittle

dwarf planets: objects orbiting the Sun that are massive enough to have been pulled into spherical shapes but have not cleared their orbits of other objects, e.g., Pluto and Ceres

ecliptic: the plane in which the planets orbit

eddies: stationary atmospheric turbulences due to pressure gradients

ejecta: materials excavated from impact craters and redeposited

elastic (behavior): a solid that regains its original shape and size when deforming forces are removed

elastic collisions: collisions between neutrons and a nucleus that do not convert kinetic energy to any other form

electromagnetic radiation: radiation including visible light, ultraviolet and infrared waves, radio waves, gamma rays and x-rays, in which electric and magnetic fields vary simultaneously

electromagnetic spectrum: the range of wavelengths over which electromagnetic radiation extends

end member spectra: spectra of pure substances that can be combined into a mixed spectrum

endosymbiosis: ingestion of intact organisms that are incorporated into the metabolic pathway of the cell

entrainment: putting stationary grains into motion

equilibrium temperature: temperature just above the solid surface, reflecting a balance between solar heating and heat reflected back to space, and assuming no internal heating

escape velocity: the velocity required for an object (either a rock or a spacecraft) to escape a planet's gravitational grasp

Eukarya: organisms whose cells contain a discrete nucleus and membrane-bound organelles

exoplanets: planets orbiting stars other than the Sun

exosphere: tenuous or transitory gaseous envelope

extinction (of light): amount of light at any given wavelength that is removed from a beam passing through the atmosphere

fluvial: processes or materials deposited by flowing liquid (river)

flyby: reconnaissance spacecraft mission that passes near a target body without going into orbit

framing camera: framing is the presentation of visual elements in an image, especially in relation to other objects; a framing camera employs a set of filters placed in front of the imaging array, each of which passes light of a certain wavelength range

Fraunhofer lines: absorption bands in the solar spectrum, indicating the presence of certain elements

free oscillations: stationary standing waves that vibrate in different modes

freestream wind speed: wind speed that can be experienced or measured by instruments

friction wind speed: wind speed that represents the strength of gas flow at the interface with the surface

galactic cosmic rays: highly energic particles that come from outside the Solar System

Galilean moons: the four largest moons of Jupiter (Io, Europa, Ganymede, Callisto) that were discovered by Galileo

gas giant: Jupiter and Saturn, which are composed mostly of hydrogen and helium gas and their ultradense forms

geochemical cycle: description of the behavior of a specific volatile component in a planetary system

geoid: the surface of a body over which the gravitational potential is constant

geothermal gradient: rate of increase of temperature with depth; often abbreviated as geotherm

giant planet: massive outer planets: Jupiter, Saturn, Uranus, and Neptune

gravity anomaly: local variation in gravity due to mass variations in the subsurface

greenhouse effect: warming of a planetary surface by atmospheric absorption of heat energy

ground penetrating radar (GPR): a technique that employs long wavelength radio waves to map subsurface compositional heterogeneities

habitable zone: region around a star where liquid water is potentially stable on the surface of a planetary body with an atmosphere

Hadley cell circulation: planetary atmospheric circulation pattern in which hot air at the equator flows to the poles, where it cools and descends, and returns along the surface to the equator

heat flux: thermal energy flowing out of a planetary body

HED meteorites: howardites (H), eucrites (E), and diogenites (D), meteorites from asteroid Vesta

Herzsprung–Russell (H-R) diagram: a plot of luminosity (a proxy for mass) versus surface temperature for stars; the diagram illustrates the stages of stellar evolution

Hjulström diagram: plot of flow velocity versus grain size, with curves illustrating the fluid velocities required to erode and transport grains

horizontal gene transfer: direct movement of genetic material between extant organisms

Hugoniot elastic limit (HEL): maximum stress that a material can experience during shock without permanent deformation

hydrocarbons: organic molecules formed of carbon and hydrogen, often with oxygen, nitrogen, and sulfur

hyperspectral: data product containing enough spectral channels to oversample the width of a typical absorption feature

hypervelocity impact: impact of an object large enough to not have been decelerated during passage through the atmosphere

ice giant: Uranus and Neptune, which are composed mostly of ices of water, ammonia, and methane

ices: condensed (solid) forms of gaseous molecules, such as water, methane, and ammonia

image cube: a stack of images, all spatially registered to each other, taken at multiple wavelengths

impact erosion: blowing off of atmosphere by very large impacts

incompatible elements: elements with large ionic size and/or high charge that do not fit easily into the crystal structures of common rock-forming minerals; they are concentrated in partial melts and fractionated liquids

insolation: illumination by the Sun

interplanetary dust particle (IDP): tiny grains that once orbited in interplanetary space; particles of an asteroid or comet

iron meteorite: a meteorite composed primarily of FeNi metal; generally represents a sample of a solidified core of an asteroid

isochron: line on a diagram showing how radioactive isotopes, relative to a stable isotope, change with time

isostacy: a statement of buoyant equilibrium of lithospheric blocks

Jeans escape: atmospheric loss mechanism in which lighter molecules are not gravitationally bound to the planet

Kirchoff's Law: emissivity is inversely related to reflectance at any given wavelength

KREEP: acronym for lunar rocks rich in the incompatible elements potassium (K), rare earth elements (REE), and phosphorus (P)

Kuiper belt: region in the ecliptic plane beyond the orbit of Neptune containing thousands of cometary bodies; Kuiper belt objects are abbreviated "KBOs"

lacustrine: processes and materials deposited in standing bodies of liquid (lakes)

Lagrange points: position in the orbital configuration of two large bodies, at which a small body can maintain a stable position; the Trojan asteroids occupy Lagrange points in Jupiter's orbit

lander: spacecraft that remains stationary after landing on the surface of another object

lapse rate: variation of the temperature of atmospheric gas with altitude

Last Universal Common Ancestor (LUCA): population of organisms from which all modern life is derived

Late Heavy Bombardment: period of intense planet bombardment by large planetesimals that produced impact basins; may have been caused by migrations of the giant planets and appears to have ended at ~3.9 Ga

lipids: grouping of simple chains of carbon bonded to hydrogen ions attached to a glycerol molecule; these have roles in energy storage and in the signaling of cellular functions

lithophile: elements that are preferentially incorporated into silicate or oxide minerals

lithostratigraphic unit: a geologic unit defined by its lithology (rock type)

lobate scarp: contractional structure (fault-related fold) that forms an asymmetric topographic ridge with a typical amplitude of a few kilometers and length of up to thousands of kilometers

long-lived radioisotopes: radioactive isotopes with long half-lives

long-lived radionuclide: a radioactive isotope with a long half-life

long-wavelength undulations: folds with amplitudes of 1–3 km and wavelengths of 800–1300 km

magma ocean: a huge volume of magma, formed by total or near-complete melting of a planetesimal or planet

magnetic anomaly: local variation in the strength of the magnetic field, which can provide information on compositions and structures

Main asteroid belt: the region between the orbits of Mars and Jupiter that is the locus of most asteroids

maria: large outpourings of magma (mare basalt) that fill impact basins on the Moon

marine: processes or materials deposited in an ocean or sea

martian meteorites: meteorites from Mars, including shergottites (basalts), nakhlites (clinopyroxenites), chassignites (dunites), and a few unique samples

mass wasting: geomorphic process by which soil, sand, regolith, and rock move downslope typically as a solid, continuous or discontinuous mass, largely under the force of gravity

mean density: average density (mass/volume) for an entire planetary body

meteorite: extraterrestrial object of rock or metal that has impacted on Earth; prior to impact, it is a meteor

molecular cloud: concentration of gas and dust in interstellar space

moment of inertia factor: dimensionless quantity that characterizes the distribution of mass within a body, calculated from the largest principal moment of inertia divided by the mass and mean radius squared

multi-ring basin: gigantic impact basins with two or more concentric mountain rings and intervening ring grabens

multispectral: operating in or involving several regions of the electromagnetic spectrum

near-Earth objects (NEOs): asteroids with orbits that approach or cross the Earth's orbit

neutron capture: absorption of a neutron by an atomic nucleus

non-elastic scatter: transfer of a portion of a neutron's energy to an atomic nucleus

nonsynchronous rotation: outer portion of a body rotates at a different rate than the interior, typically parallel to latitude, caused by tidally related torques

nucleic acids: complex macromolecules, like DNA and RNA, that provide the genetic coding of life; composed of amino acids bonded with sugar and phosphate groups

Oort cloud: theoretical spherical cloud of cometary bodies located beyond the Kuiper belt

orbital elements: parameters that describe how bodies in space move relative to each other

orbiter: spacecraft that orbits the target object, to conduct scientific studies or relay data from landed spacecraft

organic molecules: hydrocarbons composed of carbon, hydrogen, sometimes with oxygen, nitrogen, and sulfur, that occur commonly in chondrites and are found on some planets

outflow channel: erosional channel produced by sudden and massive deluge of liquid

outgassing: release of volatiles from the interior, normally accompanying melting, and producing a secondary atmosphere

overland flow: runoff of liquid over a surface

paleomagnetism: rock magnetism that can provide information on ancient magnetic fields

passive remote sensing: remote sensing technique that relies on naturally occurring photons or particles

paterae: dish-shaped volcanoes, often eroded and inferred to be pyroclastic in origin

pedestal craters: craters surrounded by lobate aprons of ejecta, formed by impacts into targets containing groundwater or ice

permafrost: subsurface ice that does not melt

permittivity: a material property that governs how much of a radio beam is transmitted into the target versus being scattered at its surface

petrologic type: designation of degree of thermal metamorphism or aqueous alteration in chondritic meteorites

photons: the smallest discrete packets of light

Planck's Law: a description of the spectral density of electromagnetic radiation emitted by a blackbody at a given temperature

planetary embryos: large planetesimals ranging up to 1000 km in diameter, which eventually grow to become planets; synonymous with protoplanet

planetesimals: accreted objects ranging in size from meters to a few hundred kilometers

polyaromatic aromatic hydrocarbons (PAH): organic compounds composed of multiple carbon rings

polymorphs: solid phases with the same composition but different crystal structures, commonly resulting from changes in temperature and pressure in the deep interiors of planets or in response to impact shocks

prebiotic compounds: chemical precursors from which biotic molecules formed

presolar grain: tiny particles formed as dust around another star and incorporated into chondritic meteorites

primary atmosphere: planetary atmosphere composed of captured nebular gas

primary crust: primordial crust formed through planetary differentiation

proteins: chains of amino acids that coil or fold into three-dimensional shapes and provide structural support for cells

protoplanetary disk: the flattened disk from which the planets accreted

push-broom imager: imager that uses a linear array of detectors, and the spacecraft motion sweeps this array along the target

pyrolysis: breakdown of organic matter into smaller molecules by heating

radiogenic isotope: isotope formed by radioactive decay

radiometric age: absolute age determined from measurement of radioactive isotopes

refractory: elements that condense from a gas of solar composition at high temperatures

regolith: unconsolidated surface layer formed by impacts over geologic time

relative humidity: ratio of the partial pressure of vapor to that in saturated air

Reynolds number: dimensionless number that describes the degree of laminar or turbulent flow

ribosomal RNA: nucleic acids responsible for the production of ribosomes that catalyze amino acids into proteins

rilles: sinuous channels that feed lava flows

RNA world: RNA, rather than DNA, acts as the blueprint for cellular functions and the driver for protein synthesis

rover: landed spacecraft that is mobile and can traverse beyond the landing site

saltation: movement of grains by bouncing or hopping across the surface

sample return mission: spacecraft that acquires and returns samples to Earth

sapping: erosional process that forms amphitheater-headed valleys by headward migration resulting from fluid discharge

scale height: vertical distance over which atmospheric pressure falls by a factor of $1/e$

scanning system: a system that captures an image into digital or electronic form

secondary atmosphere: planetary atmosphere formed from volatiles outgassed from the interior

secondary craters: lines or arcs of small craters emanating from a large crater, formed by dislodged blocks ejected by the impacting bolide

secondary crust: crust produced from magmas formed by melting of the mantle

seismic tomography: three-dimensional imaging of velocity changes compiled from travel time measurements of seismic waves

seismic waves: energy in the form of vibrations that travel through a planet

shape model: three-dimensional model illustrating topography

shatter cone: conical-shaped rock with fractures that radiate from the top; diagnostic of impact

Shields Curve: an equation for shear stress, made nondimensional using other boundary conditions for flow

shock metamorphism: mineralogical and physical changes to target rocks imposed by extreme pressure and temperature excursions caused by impact

shock wave: energy that radiates outward from the point of impact at high velocity; faster than a seismic wave, it is marked by an abrupt increase in pressure, temperature, and density

short-lived radioisotopes: radioactive isotopes that decay rapidly and become extinct nuclides

short-lived radionuclide: a radioactive isotope with a short half-life; also called extinct radionuclide; if sufficiently abundant, its decay can be a potent heat source

side-looking radar (SLR): imaging radar pointing perpendicular to direction of flight, used to map surface topography and scattering properties

siderophile: elements that are preferentially incorporated into iron metal alloys

signal-to-noise ratio: the ratio of the strength of a signal carrying information to that of interference

simple crater: bowl-shaped depressions formed by small to modest sized impacts

snow line: the radial distance from the Sun at which ices begin to condense from the nebular gas; the giant planets formed beyond the snow line, but its position likely varied with time

solar insolation: exposure to the Sun's rays

solar nebula: the cocoon of gas and dust surrounding the infant Sun that eventually became our Solar System

solar spectrum: the spectrum of light emitted from the Sun, from which its elemental composition is determined

solar wind: particles, mostly hydrogen, that stream outward from the Sun and can be deflected or entrained by magnetic fields or implanted into surface regoliths on airless bodies

space weathering: changes in the spectral properties of the surface of an asteroid or planetary body, caused by mineralogical modifications during small impacts

spallation: ejection of high-energy neutrons resulting from irradiation

spatial resolution: scale of the smallest feature that is resolvable

spectral absorption features: lines or bands in an electromagnetic spectrum due to absorption at specific wavelengths

spectral index: parameter created by manipulating the reflectance values for each pixel to highlight the presence of a mineral

spectral radiance: radiance of a surface per unit frequency or wavelength

spectral resolution: ability of a sensor to define fine wavelength intervals

sputtering: acceleration of atoms when hit by another fast atom or ion; a common phenomenon in cosmic ray irradiation

stagnant lid: lithosphere is coupled to the body interior so that it is relatively tectonically inactive and typically decorated by abundant craters

stellar nucleosynthesis: the formation of new, heavier elements in the interiors of stars by fusion of neutrons with other nuclei

stereo imaging: two images of the same area taken under similar illumination but with different viewing geometry; provides for 3D viewing

Stokes' Law: an expression giving the terminal velocity at which grains fall

stony-iron meteorite: meteorite composed of roughly equal parts silicate rock and FeNi metal

strain: deformation caused by stress

strain rate: continuous rate of deformation

stratigraphy: geologic layers whose relative ages can be distinguished by superposition; on planetary bodies, strata can be ejecta from craters, as well as sedimentary and volcanic units

stress: force divided by the area over which the force acts

stress gradients: changes in stress levels that affect deformation

subsumption: process for the return of the brittle ice lithosphere to the underlying ductile lithosphere, accommodating horizontal shortening and volume loss

subsurface flow: movement of liquid beneath the surface

supernova: explosion of a giant star at the end of its evolution

tektites: glass beads, often with aerodynamic shapes, formed by melting of distal ejecta from large impacts on the Earth

terrestrial planet: rocky, inner planets: Mercury, Venus, Earth and Mars

tesserae: oldest parts of the Venus lithosphere that contain sinuous ridges that are antiformal folds

thermal inertia: slowness with which the temperature of a body approaches that of its surroundings

tholi: small volcanoes on a planet

tidal force: gravitational force exerted by one body on another

tiger stripes: parallel linear troughs in the South Polar Terrain of Enceladus that are relatively warm and erupt plumes of vapor

topography: surface shape based on elevation

transient cavity: the originally excavated impact crater, before partial filling with brecciated material that falls back into the cavity

tree of life: depiction of evolutionary connections among terrestrial life forms

true polar wander: rotation of the outer portion of a body at a different rate than the interior due to torques induced by thickness and/or density changes in the outer layer; motions can travel in any direction across the body

uncompressed mean density: mean density corrected for self-compression in large planets

ventifact: outcrop-scale rock abraded and shaped by wind-driven particles

vibrational features: spectral absorptions resulting from vibrations in mineral lattices or molecules

vibrational transitions: molecular vibrations that emit or absorb photons

visco-elastic: linear elastic response initially, followed by ductile flow with increasing stress

visible spectrum: portion of the electromagnetic spectrum that is visible to the human eye; approximately 400–670 nm

volatile: elements that only condense from a cooling gas of solar composition at low temperatures

wavelength: the distance between successive crests in a waveform

wrinkle ridges: contractional structure (fault-related fold) that forms an asymmetric topographic ridge with a typical amplitude of hundreds of meters and length that tend to a few hundred kilometers

yardang: erosional landform in which rock has been abraded by wind-driven particles into a streamlined form

zonal winds: atmospheric flow along lines of constant pressure, producing latitude-confined cloud bands on Jupiter; on Earth these are called "jet streams"

Index